AF587659

Scientific Revolutions Series

PHILOSOPHICAL INSIGHTS ABOUT MODERN SCIENCE

SCIENTIFIC REVOLUTIONS SERIES

Unfolded Proteins: From Denatured to Intrinsically Disordered
Trevor P. Creamer (*Editor*)
2008. ISBN 978-1-60456-107-4

Philosophical Insights about Modern Science
Eva Žerovnik, Olga Markič, and Andrej Ule (*Editors*)
2009. ISBN 978-1-60741-373-8

Scientific Revolutions Series

PHILOSOPHICAL INSIGHTS ABOUT MODERN SCIENCE

EVA ŽEROVNIK
OLGA MARKIČ
AND
ANDREJ ULE
EDITORS

Nova Science Publishers, Inc.
New York

For permission to use material from this book please contact us:
Telephone 631-231-7269; Fax 631-231-8175
Web Site: http://www.novapublishers.com

LIBRARY OF CONGRESS CATALOGING-IN-PUBLICATION DATA

Philosophical insights about modern science / [edited by] Eva Zerovnik, Olga Markic and Andrej Ule.
p. cm.
Includes index.
ISBN 978-1-60741-373-8 (hardcover)
1. Science--Philosophy. I. Zerovnik, Eva. II. Markic, Olga. III. Ule, A. (Andrej)
Q175.P5116 2009
501--dc22
2009009166>

Published by Nova Science Publishers, Inc. ✢ New York

CONTENTS

PREFACE

Among the various human endeavors, science and the development of technology are two domains which clearly distinguish humans from the animal world. Perhaps science arose from the eternal quest of humans to know more, to pose bold questions about the universe they live in. And not just questions about how things happen, but also the why they do. Most distinguished scientists, regardless of their religion, were convinced that science will ultimately be able to answer all the questions that humanity wants to know.

Modern science has become so specialized that staying abreast of all the disciplines would seem to be a utopian fantasy. This book is aimed at bridging the gap between specialists views and a common understanding of "modern science". A desirable goal would be for all educated people to know something about the humanities, literature, and the arts as well as the latest developments and discoveries in the natural sciences. One fears what one does not understand. Polymaths such as Michelangelo Buonarroti, Isaac Newton, Janez Vajkard Valvazor, Albrecht Dürer, Rudjer Bošković and Nikola Tesla, among others, mastered several fields of science in their time; they often were as creative in technical innovations as in the arts.

One aim of this book is to point out the main messages of a given scientific field, to identify that which is really new and beyond the average educational level, in order to broaden our horizons. For this reason, at the end of each chapter there is a discussion of possible future contributions and ethical concerns from this field.

Many fields of modern science, especially the natural sciences, are covered in this book: brain plasticity (Chapter 1) by M. Bresjanac and G. Repovš, gene regulation (Chapter 2) by J. Ule et al., the molecular background of Alzheimer's disease (Chapter 3) by E. Žerovnik, artificial intelligence (Chapter 6 and Chapter 13) by M. Gams and I. Kononenko, respectively, new sustainable energy sources (Chapter 9) by A. Detela, and the concept of the biofield (Chapter 10) by I. Jerman et al.

There even are two chapters in part A on stem cell research, which offers promising prospects for tissue and organ replacement therapies (Chapters 4 and 5) by A. Atala and F.J. Müller et al., respectively. This is an interesting subject as one can imagine that nearly all body parts and tissues could be "replaced" without destroying the core of personality and awareness of oneself. Will that really be the case? This is a question relevant both for physicians and philosophers. And a similar question: if robots are once made "human-like", i.e. made from living tissue and cells, what would they feel, if anything—with the brain made of "wires"—as this organ is not foreseen to be easily created in all of its complexity. But what if it one day became possible even to make the brain, with all of the neurons and numerous and correct neural connections? Would this then give rise to the mysterious "consciousness"?

Several chapters deal with consciousness and free will: Chapter 1 by M. Bresjanac and G. Repovš discusses the miraculous plasticity of the brain, while the chapters dealing with artificial intelligence (Chapter 6 by M. Gams and Chapter 13 by I. Kononenko) both doubt that simulating human intelligence will ever be possible. Can one predict which scientific questions can be answered and which ones are beyond the capabilities of the human mind? This is the question asked by I. Kononenko in Chapter 13. M. Perus (Chapter 12) talks about "visual consciousness" as the basis of any conscious experience. U. Kordeš (Chapter 11) describes an attempt of a person researching firsthand his own experience. Z. Pirtošek (Chapter 7) asks how to see "free will" from the point of view of the neurologist dealing with patients who cannot control movements and behaviors.

Admittedly, some important fields were left out. One is psychiatry, obviously an important field of medicine and a developing science, which is expected to bring deeper understanding of the mind-body-spirit connections (perhaps it will one day integrate old traditions and new treatments while at the same time respect personal integrity and show enough compassion for sufferers). Much progress is expected from understanding of the nervous system in molecular, cellular (Chapters 1 to 5 of part A) and even quantum-holistic terms (partly covered in Chapter 10 and Chapter 13 of part B). Another important field which was left out and which no doubt will contribute to a new picture of the world is cosmology and particle physics.

In addition to scientists, three philosophers were asked to extract "philosophical insights" from some subjects of the modern science as covered here, or simply express their point of view on some other aspects of modern science.

An introductory and at the same time the summary chapter "Science on the Path towards New Horizons and Beyond" was written by Andrej Ule (specialist in analytical philosophy and theory of science). Part A concludes with the chapter "On Neuroscience and the Image of the Mind" written by Olga Markič (specialist in philosophy of mind and philosophy of cognitive science). Part B, which presents authors' own opinions on subjects and concepts which do not yet have a clear answer in contemporary "mainstream" science, ends with the chapter "Philosophical Reflections on the History and Future of Science and Spirituality" written by Thomas Clough Daffern (specialist in the history of philosophy and religions).

Overall, our aim as the editors remains modest: to overcome the prejudice that philosophical views on contemporary science are not possible due to overspecialization, and to start a respectful discussion among scientists of different attitudes and beliefs.

Thanks are due to all the authors who responded to the invitation to write such a book and took the time to write their contributions.

In: Philosophical Insights about Modern Science
Eds: Eva Žerovnik, Olga Markič and Andrej Ule
ISBN: 978-1-60741-373-8

Introduction

SCIENCE ON THE PATH TOWARDS NEW HORIZONS AND BEYOND

Andrej Ule
Department of Philosophy, Faculty of Arts, University of Ljubljana, Aškerčeva 2, 1000 Ljubljana

SYNOPSIS

This book presents contributions from scientists working in a range of different disciplines (neuroscience, bio-medicine, artificial intelligence, cognitive science, energy technology, philosophy) exploring the ethical relevance of past discoveries and their implications for the future of human society. The contributions in this book provoke many questions regarding the future of modern sciences, as well as their effects on the paths of human society as a whole.

I concentrate on two questions: first, are modern scientific discoveries leading to a stronger integration of ethical sensitivity within the scientific mind? Second, do new emerging lines of research and scientific theorizing change or even dissolve the lines of division between established science and its alternatives?

In answering the first question I will follow up on the notion of a "cooperation of reason and heart", propounded by Bertrand Russell in the 1920s as the solution to crisis in science and human society. Scientific progress has led to the rapid development of modern technology and to a more rational approach to many problems. However, this development often requires a concomitant development of ethical awareness, which does not necessarily arise from scientific progress. Ethical awareness is necessary for the responsible application of science and technology. I discuss some obstacles to the growth of ethical sensitivity of scientists: the idea of value-free science, the gap between pure and applied sciences, belief in the unbridgeable gap between facts and values and the (possibly too) heavy burden of scientists' responsibility for the consequences of their discoveries.

A deeper ethical awareness and sense of responsibility should be assumed by scientists as well as those who direct scientific research or use scientific discoveries on a large scale (such as new technologies, industry, agriculture, medicine, social policy and defense). However, by accepting a certain extent of ethical responsibility for the consequences of their discoveries in the human domain and in our environment, scientists should not bias scientific explanation and the formulation of problems and theories by these concerns.

In approaching the second question, I start with a critical discussion of the criteria for demarcation between science and non-science. Such criteria depend on the models of science in the historical or epistemic contexts. For example, in order to apply falsification to a particular hypothesis, criteria of falsification are required. There are no a priori given criteria of falsification because a hypothesis can be defended in principle ad infinitum, in spite of any seeming counter example. We cannot find cases of science that could serve as a model for all kinds of modern sciences. However, we can determine the necessary conditions of elementary scientific correctness and rationality, e.g., openness to the review process, differentiation between the actual observations and their interpretation, public accessibility of data, correct argumentation, and reproducibility of results. Scientific disciplines differ in some aspects, and resemble each other in others. It is thus better to present sciences as an inter-related network of disciplines and research fields, rather than as one single discipline.

What is the essential quality of a scientist? In my opinion, knowledge is less essential than the willingness and ability to present hypotheses and their supporting evidence to intersubjective, critical analysis. Thus, new scientific research and paradigms may change the lines of division between established science and some not established kinds of investigation in cases where they are correctly presented, interpreted and can be reproduced. On the other hand, "normal" science (in Kuhn's sense) needs stability in its research methods and delineation of research problems. Only in some rare cases of scientific revolutions may it seem that "anything goes", that the lines of division between established science and its alternatives dissolve. However, absence of an established scientific framework over a longer period of time inevitably leads to an accumulation of non-testable speculations or simple, pre-tested "recipes".

Keywords: scientific progress, ethics, responsibility of scientists, criteria of demarcation, falsifiability, normal science, alternative science.

1. Russell's Skepsis on the Future of Science

In the 1920s, Bertrand Russell published a radical and a worrisome outlook on the future of science and modern civilization in the booklet "Icarus, or The Future of Science". In it, he saw the necessity of globalization, understood as the organization of the world as an economic unit and provided some very astonishing ideas on the prospects for "cultivating" humanity (Russell, 1925, p. 16).

He warned that the highly rationalized scientific and technical enterprises of modern civilization may lead us astray. These enterprises may even promote interests and practices that are not themselves rational, such as artificial curbing of human instincts of power and rivalry (Ibid., p. 13), cruelty of wars (p. 20), mass propaganda for extreme nationalist and party interests (p. 26-7), and high dependency of masses on strong leaders (p. 34). Reason alone can not solve all problems. Russell thus pleaded for what he calls a 'cooperation of reason and the heart'. According to Russell, modern science is no substitute for virtue; the heart is as necessary for a good life as the head (p. 58). By the heart he means the sum-total of kindly impulses. Russell emphasized that where kindly impulses exist, "science helps them to be effective; where they are absent, science only makes men more cleverly diabolic." (p. 59). It is a pity that Russell did not write more about the cooperation of reason and the heart, and that he emphasized some highly suspicious ideas like massive use of eugenics and a despotic world government armed with scientifically and technically perfected weapons as a means of solving global world problems. He knew these means are dangerous and can be easily misused but he believed they are necessary in order to prevent the dangers of overpopulation and nationalism.[1]

In essence Russell was skeptical about the prospects of improving humans through science. At the end of the booklet, Russell summed up the discussion in a few words: "Science has not given men more self-control, more kindliness, or more power of discounting their passions in deciding upon a course of action. It has given communities more power to indulge their collective passions, but, by making society more organic, it has diminished the part played by private passions. Men's collective passions are mainly evil; far the strongest of them are hatred and rivalry directed towards other groups. Therefore at present all that gives men power to indulge their collective passions is bad." (Russell, 1925:62-3) After World War II he became a bit more optimistic about the future of science and humanity but in general he remained skeptical regarding any easy prospects of a scientifically supported politics or economy.

I am far from the conclusion that Russell's thoughts are simply obsolete or present a case of exaggerated negative utopias (similar to the messages of famous negative utopias of Huxley or Orwell). Some of them seem even more relevant today than they were at the time of their first publication. I think first of all of Russell's idea of the cooperation of "the heart and reason" (ethical sensitivity and critical scientific intellect) and his warning that without this cooperation, science could enhance individual and collective egoistic impulses. It seems that the development of new scientific and technical findings and technologies (e.g., biomedicine, genetic engineering, artificial intelligence, nanotechnology) is usually faster than the ethical awareness that adjusts to the new possibilities of use and misuse and associated challenges.

When I write on philosophical insights from modern science, especially those based on the new emerging sciences, I think of the insights that may help to support or develop ethical awareness. With this purpose I will thus here emphasize some ideas presented in the essays collected in this book.

2. On The Contributions to this Book

The first part of the book consists of articles on some hot topics in contemporary science: brain and consciousness (Bresjanac and Repovš), genes and molecular biology (Barborič et al; Žerovnik), stem cell research (Atala; Müller et al), robot intelligence (Gams), brain and free will (Pirtošek), and on the neuroscience and the image of the mind (Markič).

Maja Bresjanac and *Grega Repovš* present a paper on the neuroplasticity of the human brain. They discuss the marvelous ability of the human brain for learning and reshaping its processing: mechanisms of plasticity. This ability enables the brain to regenerate even after suffering serious lesions. They describe two connected and interdependent levels of plasticity:

[1] Russell e. g., defended eugenics even later although he obviously became much more critical but he was ever defending a strong unitary world government as a necessary means to prevent evils like nationalist or inter-religious conflicts, genocides, overpopulation, etc. (Russell, 1952).

synaptic and cognitive plasticity. The first enables learning and the formation of novel neuronal processing patterns and behaviors, and the second leads to the rise of cognitive control and ability of voluntary processing which enables on-line planning and reprogramming of our behavior. The authors are aware of the great difficulty to obtain a clear picture of plasticity mechanisms because of the subtle biochemical and neurological processes and a massively parallel processing in brain. They present highly interesting mechanisms of neurogenesis which spontaneously occur in the infant brain and even in certain regions of the normal adult brain, and the mechanism of maladaptive neuroplasticity. A better knowledge of neuroplasticity opens up some exciting possibilities for effective brain-machine interfaces which may replace lost sensory and motor abilities as well as extend existing ones.

The authors are justified in their belief in advances in neuroscience and especially in the better insight into possibilities to "mold the brain in order to shape the mind" but they warn against the danger of unequal distribution of the new means for cognitive enhancement in society. Better access to effective cognitive enhancement in certain socio-economic groups can increase inequalities and raise social tensions. At the end of the article the authors refer briefly to new moral and ethical concerns associated with manipulations of our brains in order to modify our cognitive abilities, personality traits and subjective experiences.

Matjaž Barborič, Tina Lenasi, Nika Lovšin and *Jernej Ule* present the new discoveries on the importance of non-coding regions of our genome, which used to be referred to as the so-called 'junk'. For instance, transposable elements (TE) used to be regarded as intracellular genetic parasites that exploited cellular machineries for their own survival. However, newer studies show that they may play important roles in gene regulation. The authors discuss the important roles of pre-mRNA alternative splicing, microRNAs and other types of non-coding RNAs. They present evidence for the thesis that TEs and our genome have co-evolved into a state of mutual dependence, and that therefore the repetitive elements are after all not all "junk".

Because junk parts of the genome seem like "genomic parasites" similar to many viruses, the authors discuss the mechanisms used by the HIV virus to invade human cells. Current therapies can quite successfully inhibit viral replication and spreading, but cannot completely eradicate the virus from the human body. The authors report on newly discovered natural healing substances (known even long ago by healers in some traditional communities) that may directly eliminate even latent viruses. It seems that nature and human traditions might still be our best teachers.

The overwhelming presence of non-coding transcripts, and their unusual genomic organization are challenging the notion of the gene as a distinct region of the genome. We cannot divide the genome into independent portions of genes like pearls on a string. Instead, it appears that the genome functions more like the control deck of an airplane, where the instruments are crammed up all around the pilot's seat. In an analogy to a computer, the authors compare the cell's structural proteins to the hardware, and the regulatory factors to the software. Our genome seems to have built within itself the potentiality for the unforeseen future.

Eva Žerovnik analyzes in some detail different hypotheses regarding the possible causes of and possible therapies for multigene diseases like Alzheimer's, Parkinson's and prion diseases. The central hypothesis here is that they might have some common causes in misfolded proteins. It is thus essential to find mechanisms which block the generation of misfolded proteins in the brain cells. Success would enable the discovery of effective therapies and preventive measures of neuro-degenerative diseases which might lead to significant improvements in the health of elderly people and prolong human life. The author does not see anything principally wrong in the human wish for longevity and even immortality (assuming people live in such a way as to enjoy life – healthy and bright-minded). Some multigene diseases raise heavy ethical concerns (e.g., when and what to inform the patient about her disease). Through early genetic testing (e. g. ApoE testing) we may at least significantly relieve some diseases but genetic testing raises ethical, legal, and social questions for which there are few answers. The author sees a problem in the fact that despite some confidentiality laws which prevent the dissemination of the information obtained from ApoE testing, it may not remain confidential if it becomes part of a person's medical records. Thereafter, employers, insurance companies, and other health care organizations could find out this information, and discrimination could result.

Anthony Atala and *Franz Mueller* together with *Jeane Loring* and *Paul Baier* presented two papers on stem cell research, a very relevant and controversial topic of research. Atala presents some major advances in stem cell research and Mueller et al. reflect on the use of stem cells for repairing neurodegenerative damage. Atala provides an overview of methods that may generate stem cells from sources other than human embryos. Thus we might avoid ethical and religious concerns which try to limit or even prohibit stem cell research and medical treatments. After an analysis of different methods of production and the therapeutic use of stem cells, Atala stresses that many obstacles remain (e. g. the tendency to form teratomas) before stem cells will become a viable and widely used form of clinical treatment. Mueller et al. present the possibilities of treating Parkinson's disease by stem cell grafts. They point out that the key requirement for any stem cell therapy, proposed or actual, is that the outcomes have to be predictable: predictability in the case of hematopoietic stem cells is due to their bioequivalence. However, this requirement in today's stem cell research is far from realization. There are many other difficulties which hinder the use of stem cells in treating neurodegenerative illnesses: for one thing, the stem cell state is not the stage in which a cell type has been considered to be useful for mitigating impairment by neurodegenerative disorders. We need mature, differentiated progeny derived from neural stem cells and not "pure" stem cells, and regeneration by stem cells does not occur at functionally relevant levels in the adult brain or after pathological insults. The authors conclude the paper with a cautious conclusion: stem cell research most certainly will lead to scientific insights and eventually to novel therapeutic strategies. Yet today we probably have no clue what these therapies might be.

Matjaž Gams in his paper on robot intelligence presents the idea of the robotic mind, which works on the principle of multiple knowledge acquisition which may enable a really creative intelligence. The major hypothesis of his paper is that the human mind, the brain and the body are different from those of computers and robots; that the human mind is a supermind compared to digital computing powers. He believes the human mind works according to multiple-world theory in many worlds/dimensions. He hopes the robots may reach the level of human intelligence. In future they may gain free will, individuality, unpredictability, and feeling. They would enter into "personal" relationships and even know some kind of sexuality, be developing over time and they might even somehow "die". Gams states that even if the human mind is a "supermind", robots (and computers too) could in principle acquire computing power which might be stronger than the universal Turing machine, for example, if it works according to the principle of multiple knowledge as formulated by the author in some of his works. This principle states that on average it is reasonable to expect improvements over the best single model when single models are sensibly combined. The "creative" version of the principle of multiple knowledge states that multiple models are an integral and necessary part of any creative process.

Zvezdan Pirtošek reviews the concept of free will in the areas of neurological sciences. He observes that the concept of 'freedom of the will' was for a long time in the domains of philosophy, theology and law. It was only recently that cognitive scientists and neurologists entered the debate. The author discusses neurological and psychiatric disorders affecting decision making and free choice. He ascertains that in clinicopathological settings, neurologists and psychiatrists have for a long time observed and described syndromes of a 'sick will', characterized by inactivity, poverty of movement and thought (various movement disorders, schizophrenia, depression, autism, ADHD syndrome, dementia, parkinsonism, hysteria, apraxia). The main methods that neuroscientists use in the study of volition and free will are functional imaging and electrophysiological methods. Electrophysiological techniques provide characteristics of brain activity in temporal terms. Pirtošek briefly discusses their usage in Libet's experiment. The study implies that conscious awareness of willing the action actually comes after the initiating the action. Libet's suggestion is that voluntary acts are unconsciously initiated but are subject to conscious control which can either promote or veto the unconsciously initiated process. Pirtošek continues with a discussion of the functional anatomy of volition. He argues that clinical case reports, electrophysiological studies and new brain imaging techniques confirm the assumption that free will is 'localizable' to a certain extent. The author concludes the paper with the observation that the results of neuroscientific studies support a certain skepticism about the existence of 'free will' and do, at least, plead for a different concept and terminology.

Olga Markič concludes the first part with a philosophical reflection on neuroscience and the mind (*Neuroscience and the Image of the Mind*). She presents Flanagan's two competing images of who we are: the humanistic image and the scientific image. The humanistic image is the heir of theology and dualistic philosophy but it is also much in accordance with everyday thinking about the mind. The scientific image says that we are animals that evolve according to the principles of natural selection and can not circumvent the laws of nature. It takes consciousness, cognition and volition as natural capacities of embodied creatures that live in natural and social environments. It seems that these two approaches are incompatible. The author argues that although there is much fear that evolutionary biology, cognitive science and especially cognitive neuroscience lead to an image of the mind that will not support human beings able to deliberate and live a moral life, such a pessimistic attitude is not justified. She discusses Wegner's proposal that conscious will is an illusion and suggests that Wegner is influenced by the humanistic image and sets standard for free will so high that only a supernatural being can reach it. In the final part of the paper she examines neurophilosophy as an approach that links neuroscience and mind.

Cautious statements are a common trait of the authors in Part A of the book. In the second part (Part B) of the book the authors discuss their own points of view and make somewhat more adventurous statements. They present a new synergetic view on sustainable energy and technology development (Detela), or discuss new hypotheses and concepts which at present lack broader scientific consensus. Areas covered deal mainly with holistic or non-reductivist approaches to life (Leskovar et al.), phenomenological research of experience

(Kordeš), mind and consciousness (Peruš; Kononenko), and philosophical reflections on science and spirituality (Daffern).

The views in these approaches (Leskovar et al., Kononenko) open the doors to some alternative sciences (e. g. homeopathy, theory of bio-energy, parapsychology) and metaphysics. For example, Kononenko introduces a kind of dualism between spiritual and material reality. But this position is not dominant because a new kind of 'physicalism', let us say 'spiritual physicalism', which is grounded in some holistic interpretations of quantum physics (e.g., on Bohmian views of implicate order) is also present (Peruš, Leskovar et al.). We know consciousness is still beyond the full grasp of the current science.

Andrej Detela introduces a holistic approach to the modern dilemma of energy balance in human society which opens some important consequences for the future of modern science and technology. After an analysis of the main proposals for the solution of the energy crisis and of alternative sources of energy, he concludes that we need a holistic approach to understand the working of complex technological systems or human society. Detela states that, in our approach to these complex systems, human values like ethics and sincerity are of paramount importance. If we lose support in these values, we also lose every criterion of truth. He argues that it is a total illusion to believe that observation of merely technical facts will give a complete answer to our question. Our human instrumental mind that is so much adored in modern times is not enough: Here we must also adore genuine human sensitivity that functions only through an intuitive, holistic approach to what we dare to call reality. This demands a new ethical and even spiritual consciousness in order to rationally combine and use new sustainable energy sources.

Robert T. Leskovar, *Rok Krašovec* and *Igor Jerman* ask for a central theoretical concept which would integrate different biological energy processes in an inclusive fashion. They argue for the concept of an emergent and potentially all-encompassing biofield. The authors strongly distinguish the concept of a biofield from seemingly similar vitalistic views. They present some hypotheses on the nature of a biofield and give some evidence for them. Some of those hypotheses transcend standard physics and biology (e.g., a vacuum field consisting of hypothetical magnetic monopoles, electric dipoles, axions, unknown dark matter particles, zero field energy). The authors then connect the idea of a biofield with the idea of holistic morphogenetic fields which (according to different defenders of this idea) might shape the developing organism and work in the orientation of cellular division and integrative function, providing the wholeness of the organism. They concede that the biofield is at the most indirectly empirically accessible, e.g., by computer controlled electrophotography of human body, and by conductivity distribution functions of the skin. I agree with the very important conclusion of the authors: if the hypothesis of a biofield could be empirically verified it would call for admitting some kind of "formal causes" in science.

Urban Kordeš pleads for the broadening of the concept of scientific research with the honest and systematic first person research of the researcher's own experience with no a priori set goals of knowledge. Kordeš exposes the interesting epistemological and methodological problem of explaining consciousness. He is quite justified in the statement that in the case of observing one's own experience, situating it into a theoretical framework (i.e. into a general, statistical context) does not have the same effect as in the case of phenomena in which I do not existentially take part. I pass from observing the experiential here-and-now to thinking about concepts. He concludes that the "explanation" of experience in turn makes part of experience. Thus it has not (merely) explained the experience but changed (or even replaced) it! In researching consciousness we thus need some method which does not "replace" experience but lets our experience "present itself precisely as it presents itself". This is his main reason for the introduction of phenomenological first person research as a serious research approach even if it could not be reckoned among sciences.

Mitja Peruš propounds the holistic and non-local traits of quantum phenomena as a proto-model in explaining mental phenomena (visual experience, intelligent learning). Peruš develops the hypothesis to expand Pribram's hologram idea of the brain to the quantum domain. He believes that purely quantum processes are necessarily involved in conscious processes. The main reasons for this hypothesis are neuronal coherence, which is probably, at least indirectly, bi-directionally related to coherence-phenomena at subcellular (e.g., dendritic) and quantum levels, a holistic nature of awareness which seems similar to the phenomenon of quantum wholeness (or quantum entanglement) and some mathematical-physical observations and considerations which show interesting formal similarities between the processing of visual experiences in the brain and quantum net processing. Although Peruš presents the hypothesis that quantum processes are essentially involved in conscious experience, he emphasizes that consciousness is a complex multi-level and multi-aspect phenomenon where physical, biochemical, dendritic, neuronal, network-dynamical, informational, psychical processes are all essential and irreducible.

The paper by *Igor Kononenko* has two parts. In the first he presents major findings on artificial learning and its similarity to human learning processes. Although computer models of learning in AI successfully simulate some significant "procedural" traits of human learning and memorization, they do not and probably cannot explain its "experienced" traits that mean conscious experience. Kononenko speculates that a system can in principle be more or less intelligent but without consciousness (e. g., a robot), or on the other hand can be conscious but much less intelligent (e.g., less intelligent people, animals etc.). He relates (phenomenal) consciousness (as a state of being) to life, intelligence, and free will. This leads him to the second part of his article, which is more speculative and much more "ambitious", namely to the relation of scientific objectivity and spiritual subjectivity. He is quite skeptical regarding any possibility of scientific testing or detecting of purely subjective experiences such as spiritual experiences. We can approach them by a kind of intuition which remains the domain of gifted individuals. The situation of today is, according to Kononenko, a strong opposition between "extreme objective science" and "extreme subjective spirituality", which are mutually exclusive even though they are somehow complementary. Using the analogy between rational numbers as a tiny subset of real numbers and the rational world as a tiny part of the real world, he suggests the speculative idea that objectivity is a tiny part (a special case) of experiential subjectivity and that intellect is a tiny part of consciousness. He thus indirectly indicates an interesting possibility of less extremely scientific objectivity and/or less extremely subjective spirituality.

Along with Kononeneko, *Thomas Daffern* also pleads for a reconsideration of the standard antagonism between modern science and spirituality. Daffern comes to this conclusion on the basis of his comprehensive historical analysis of relationships between scientific achievement and spiritual insights in major human civilizations. Humans have a deep spiritual hunger which can be satisfied in different ways at different times and its satisfaction follows some universal laws. Daffern refers to maps and a cartography of altered states of consciousness (even if some of them might seem quite exotic from the viewpoint of standard science) that different thinkers have succeeded in producing in recent years, in which it is becoming possible to map the terrain of the transpersonal dimensions which surround and ensoul our human mundane realm.

His study concludes with the important statement that "the notion of science has a deep and complex intercultural history, and that the mono-syllabic pursuit of science as a 'Western' invention is a false idea which needs revision." I must add here that this idea has no support among modern historians of science because they regularly recognize and appreciate the essential contributions of Chinese, Indian and Arabic thinkers and researchers to the emergence of modern science. The author considers the possibility of a "science of peace", which would be both a social and a natural science, and whose implications and reverberations would reach into all other domains of knowing. It may offer the reconciliation of the natural and social sciences with the spiritual sciences under the condition that (as Bacon proposed) our "mind be enlarged, according to its capacity, to the greatness of the mysteries, and not the mysteries contracted to the narrowness of the mind."

Even if we cannot give satisfying answers to all the questions raised in this part of the book, they do serve as important points for discussion. For me the most important point here is the thesis which is more or less explicitly stated in many articles: we need some completely new theoretical and methodological concepts in order to overcome current impasses in scientific understanding of the cosmos, life and consciousness, and to link this understanding to the ethical consciousness needed to solve global problems of modern human civilization.

The broad spectrum of ideas, statements, evaluations of the actual state in different fields of science, assessments of the prospects for the future and concerns about possible (mis)uses of new discoveries provided in the articles is impossible to reduce to a few points, and I will not attempt to provide this kind of 'synthesis'. However, I would like to emphasize two crucial points: the contribution of modern scientific research and discoveries to the realization of Russell's idea on the cooperation of reason and the heart, and the relationship of established sciences to new emerging sciences and to different 'alternatives' which appear at and outside the limits of science. More precisely, I ask two questions: do modern scientific research and discoveries significantly contribute to an internal connection of scientific mind and ethical sensitivity, and do new emerging lines of research and scientific theorizing change or even dissolve the lines of division between established science and its alternatives?

3. Growth of Scientific Knowledge, Development of Human Potential and Ethical Sensitivity

Let us first discuss the first point a bit. The scientific results which have been presented in the contributions to this book and the promises of important new discoveries to come, especially in bio-medicine (more efficient gene analysis and gene therapies, methods for prolonging life, potentiality for new evolutionary jumps) and artificial intelligence (development of intelligent robots with a kind of autonomy and ever greater dependency of humans on artificial intelligence systems without consciousness) have a clear ethical implication.

Science has helped very much in the development of the human ability to control our surroundings, to improve our living conditions and especially in developing new, non-mythological and secular views of the world, the origin and evolution of life, human societies and history. It helps us (at least in principle) to look on our life situations, fears and prospects from a more objective perspective and consider more possibilities and alternatives, both hopeful and dangerous, than when we look on the world from dogmatic worldviews taken from religion, mythology, metaphysics and political ideology. This aspect of the scientific approach has surely contributed decisively to the rapid development of modern technology and to a more rational approach to many problems, but it is questionable as to whether it sufficiently supports the growth of ethical sensitivity and ethical awareness (Smart, 1981, 449), which is necessary in order to establish the responsible use of science, and especially to prevent the danger of misuses of science and technology.[2]

A very serious obstacle to the internal connection of scientific research and ethical consciousness is the thesis that value-bounded reasoning in science necessarily leads to biases in the scientific explanation of facts and the testing of hypotheses and theories, and to the ideologization of science. It follows that science has to be based on value-free research and reasoning. These theses support the idea of a gap between "pure" and "applied" science, or between basic research and technology. According to this idea, pure science or basic research has to be value-free, or better, it may be bounded at most to internal epistemic values and norms of "good" science, but applied science and technology may be linked with "external" values and interests of potential and actual users of scientific results.[3] Such ideas are usually supported with arguments on the impassability from the area of scientific facts to the area of ethical norms and values (and vice versa). Values represent subjective phenomena, preferences or utilities. They are considered to be only articulations of personal preferences not open to rational appraisal (Lacey, 1999:7). According to Carl Hempel, value judgments lack truth value; they do not express assertions (Hempel, 1965:86). As Hugh Lacey presents this view, a person's making value judgments "is open to scientific investigation and explanation, but not fundamentally to critical evaluation. In this view, they cannot be among a theory's logical implications, not just on the grounds that theories lack value categories, but because (lacking truth value) no proposition at all can have them among its entailments. Similarly, a value judgment, in principle, cannot cognitively affect either empirical data or scientific inferences." (Lacey, 1999:7).

If these views are objectively justified than we have two negative consequences:

- any claim on the internal ethical responsibility of scientists for their research and for possible misuses of their scientific discoveries leads to the "unacceptable" value-boundedness of science, and
- "leaving" science free of internal ethical demands transfers all ethical responsibility for the possible negative consequences of scientific research onto its users.

---

[2] Hans Jonas, a German philosopher of bio-ethics, saw very clearly that the global instrumentalization of nature leads to the global instrumentalization of men. In order to prevent this danger he formulated a new version of a categorical imperative for scientific research: "Act so that the effects of your actions are not destructive of the future possibilities of genuine human life" (Jonas, 1984, p. 4).

[3] The dispute on value-free science formally began with Max Weber's classical defense of value freedom and the neutrality of social science (Weber, 1917) Weber tried to insure the objectivity of social science against any impact of religion and politics. He opposed the view that one can derive all values from science - that science can answer all value questions. According to him, scientists should not say categorically what one ought to do. But even Weber did not endorse the extreme variant of the idea of a value-free science according to which it does not matter which research problem you choose, since all are something positive, and that a scientist is not responsible for the possible applications of his or her research, but that the responsibility lies with those who apply it in practice.

It seems that the first consequence leads to the danger of ideologization of science, that is, to a kind of research which may be ethically correct but scientifically partial or biased. It may impede scientific progress. And similarly, it seems that the second consequence leads to the economic or political instrumentalization of science, that is, to a technically and economically highly efficient kind of research which depends on huge financial resources, the privatization and secrecy of research, the commodification of research results, and the bureaucratization of scientific institutions, subjecting science to extra-scientific interests. Such research may relieve scientists of any responsibility even if their scientific results contribute significantly to the use of science in deeply anti-human practices (such as wars of aggression, ethnic cleansing, manipulation of the human genome, and performing total control over people).

The ideologization and instrumentalization of science can be connected, since subordination of scientific research under some explicit political aims and demands may at the same time include a high degree of ideologization and instrumentalization. We know how often scientific research was subordinated to some very bad ideology and politics (e. g. science in the former USSR under Stalin and in Germany under Nazi ideology and politics). Such cases point to the alternative, but value-free or value-bounded science is not a real alternative at all. Seemingly value-free research may be underpinned by explicit ideological premises and used for explicitly anti-human aims. On the other hand, ethical reflection of the premises, aims and possible consequences of research may prevent the "unconscious" mix of political and ideological interests and scientific work and thus it defends scientific objectivity.[4]

The argumentative answer to the opposition of value-free vs. value-bounded science needs an exhaustive study of the subject, e.g., fact-value relationships, external and internal values in sciences, objectivity of ethical judgments and different aspects of the seeming value freedom of science.[5] I cannot enter into this complex discussion. My short answer to the

---

[4] The division of science into "pure" and "applied" science is questionable because it assumes the purity of some research in advance, e.g. of theoretical astrophysics or theoretical mathematics. But such divisions forget the frequent mix of epistemic and practical motivations of actual researchers in all disciplines. Philip Kitcher strongly objected to these assumptions: "Very frequently, the complex intertwining of the epistemic and the practical and the mixed motivations of actual researchers will make the application of any simple distinction (or set of distinctions) impossible, but, even when we separate out these complications, the links to past projects and to future possibilities have to be assessed *before* we can count the inquiry as a piece of pure science. Flourishing the badge of purity isn't automatic. The label has to be earned." (Kitcher, 2001: 90).

[5] Lacey's book "Is Science Value Free" (1999) and Kincaid et al.'s book "***Value-Free Science***: Ideal or Illusion?" (2007) present a very good review and critical analysis of all of these topics. They present how positing the alternatives of value-free vs. value-bounded science is much too simple in order to conceive the real relationship between ethical presuppositions and the "internal" normative demands of scientific research. Lacey proposes a weakened variant of value- free science which consists of three aspects of value freedom: impartiality, neutrality and autonomy. *Impartiality* means "that a theory is accepted (of a domain) if, and only if, in relation to the appropriate empirical data, it manifests the cognitive values to a high degree according to the most rigorous available standards (Lacey, 1999:255)." *Neutrality* means "that, for any viable value complex, there are theories (accepted in accordance with impartiality) which may be applied so as to further significantly the manifestation of the values that constitute it (ibid.)". *Autonomy* means "that scientific practices proceed, and scientific communities and their institutions that support them are constituted, for the sake of furthering manifestations of impartiality and neutrality …" (ibid.). These three aspects are values, and thus the very idea of value freedom is value-bounded. Lacey believes that real science cannot sustain autonomy because values pervade, and must pervade, scientific practices and scientific inquiry. Modern science can only partially sustain neutrality. At least it is not neutral regarding "materialist strategies" which constrain theories to those that may represent phenomena in terms of being generated from underlying material structure, process and law, and select empirical data that may bear on such theories. However, impartiality

given opposition would be: science allows a kind of weak ethical boundedness which implies ethical responsibility of working scientists for the negative consequences of the incautious use of scientific discoveries in the human domain and our environment, and for the intertwining of science in anti-human practices, but it does not allow value statements in advance which might bias scientific explanation, formulation of problems and theories. It is impossible to obtain a general description of such a model of scientific work. We are necessarily bound to concrete analyses of scientific procedures and to a clear ethical awareness where we cannot be relieved of the responsibility for our decisions and deeds. I do not see any help in the introduction of more "holistic" or even "spiritual" views into science as some of our authors assume because we cannot say how ethical (or spiritual) values which different holistic (or spiritual) views propound can evade biases in scientific explanations, formulation of problems and theories. We still need a subtle balance between the empirical content of our claims and explanations and value judgments in order to evade pseudo scientific apriorism and (ethical and scientific) conventionalism. Certainly on the programmatic level we may believe that the introduction of holistic or spiritual concepts in science may be a very positive achievement for science and it promises the desired ethical and intellectual revolution, but this is still far from a working model of science with a creative balance of value-free and value-bounded theoretical and methodological approaches.[6]

We need a deeper ethical awareness and sense of responsibility by scientists as well as by all who try to direct scientific research or use scientific discoveries on a large scale (e.g., in new technologies, industry, agriculture, medicine, social policy and defense matters). One may believe scientists need some well considered spectrum of ethical codes which might constrain scientific research and the use of science to ethically admissible objects, methods, and aims. However, ethical codes and other institutions for the development of ethics in science are not enough for the implementation of this idea.[7] This also goes for the

seems a defensible and obligatory value of scientific practices. It is rooted "in the very objective of gaining understanding, a requirement of being able to separate the genuinely possible from the merely conceived or desired to be possible." (ibid.:258).

[6] Let me quote a vivid example of such demands of "holistic" and "compassionate" science from Mark Bekoff's article on "redecorating nature": "We need science with a heart — a compassionate science. Solid science can be driven by one's heartstrings - solid science can be done even if one goes to the beat of a different drummer. Saturating science with spirit and compassion will help bring science, nature, and society together into a unified whole. Questioning science will help insure that we will not repeat past mistakes, that we will move towards a world in which humans and other animals share peaceably the beneficence of nature. Magnificent nature — the cacophony of her deep and rich sensuality — will be respected, cherished, and loved." (Bekoff, 2000). The question remains: how to achieve such science? Aren't we going rather to a kind of metaphysics and not science? Science may help in the evolution of new, deeper and more interactional attitudes and relationships of humans with nature but it cannot substitute for it.

[7] I give some examples of such codes for science in general. In the 1980s, a group of leading Swedish scientists formulated "The Uppsala Code of Ethics for Scientists" (1984), which stressed the responsibility of individual scientists for the possible consequences of his/her research.

In 1994, the international group of scientists who attended the conference "Constraints on the Freedom of Scholarship and Science" in Toronto proclaimed the resolution "Ethics in Science and Scholarship" (1994), which contains the key elements of different codes of ethics in science. They stressed that all scientists should make a determined individual and collective effort to foresee the implications and possible consequences of their scholarly and scientific work, and avoid studies that are likely to harm the quality of life.

In 2006, the Science Council of Japan proclaimed the Code of conduct for sciences, which in its first principle stresses the responsibilities of scientists for assuring the quality of specialized knowledge and skills, and for using their expert knowledge, skills and experience for the health and welfare of humanity, the safety and security of society and the sustainability of the global environment.

In 2007, the British government edited the universal ethical code for scientists, which stresses responsibility, rigor and respect.

implementation of Russell's idea on the cooperation of reason and the heart, not only for the ethical regulation of research. Research in the frontiers of science and its promising use in technology, medicine and the economy demand a high level of responsible research and use in order to prevent catastrophic effects or misuses. It seems that humans are too weak to bear and clearly reflect this responsibility. Thus we sometimes oscillate between the demands for preventing or even blocking further research in "dangerous" areas and the blind hopes that science itself will resolve all problems if only we leave it "free" and without internal and societal control.

I would like to compare this situation with the situation of a good surgeon who may have used his best knowledge and experience in order to save the life of his patient but the patient nevertheless died. The surgeon was not guilty of any mistake but still feels responsible for the patient's death. This experience may help him in being more deeply aware of possible complications in similar situations in the future and it may prevent complications in operations. In contrast, if he forgot his past experience and the need for a deeper awareness of the situation he would be really guilty of the complication or the patient's death. Similarly, a scientist may accept her responsibility for bad or questionable consequences or misuses of her discoveries even if she was doing her best and without any malicious intent. The extension of her responsibility depends on the role of the discovery in the practices which have led to the bad consequences. Even if she holds herself responsible, she is not guilty of the consequences. But if in the next similar situation she forgets this experience, and is not aware of the situation and possible complications and does not inform all the relevant persons and institutions of possible bad or questionable consequences of the incautious use of the discovery, she will be taken as guilty of, and not only responsible for, the bad consequences of her work.

4. Criteria for Science

Questions regarding the ethical or even spiritual reshaping of science and technology open up further and deeper questions on the nature of science. Thus we arrive at the second main question which I put: the relationship of established sciences to new emerging sciences and to different 'alternatives' which appear at and outside the limits of sciences. The question of what is science and what is not, or better, what is within the domain of science and what lies outside it, or at least on its borders, is first of all the question of the criteria of science and secondly the question of which social and cultural conditions support or deny the acknowledgement of a theoretical or research activity as a scientific activity. It is obvious that if the given social and cultural conditions support the recognition of a given activity as a scientific one, then its results are given a highly prestigious status.

The mere labeling of some activity as scientific research, or in contrast as non-scientific, or pseudo-scientific research, is already highly value- and even emotion-laden. Moreover, the differences are vague and far from clear. No wonder that a broad area of activity called

---

UNESCO is also preparing an ethical code for science which may present a kind of synthesis of earlier ethical codes from the Hippocratic Oath onwards (see the report of the special expert group of UNESCO, 2005).

In many countries there exist a number of ethical codes for individual sciences and kinds of research, especially for medicine, physics, biotechnology, information sciences and social sciences (see the UNESCO report on existing ethical codes in science, 2006).

alternative science(s) is trying to establish its place in the realm of science but near to the "borders" of 'definite' sciences reaching from parts of modern ecology, alternative medicine, bio-agriculture, non-standard theories in physics, cosmology or evolution to many variants of holistic science and (serious) research of paranormal phenomena. Some of our authors, e. g. Kononenko, Jerman et al., Daffern, Peruš and Detela quite directly and openly propound such views or allow them at least as rational and scientifically relevant. Are they simply non-scientific? Or are they perhaps visionary regarding the future of science?

Demarcation of science and non-science, often use some ready-made model of definite science and then classify different kinds of research and theorization into categories such as science, alternative-, para-, pseudo-, or even counter-science. Generally speaking, a theory presents a scientific empirical theory if it describes empirical regularities or natural laws and if it can be justified or falsified by observed facts (Detel, 2007: 95). This seems sound but depends on the concepts of empirical regularities or natural laws, justification and falsification. These are not neutral concepts but stem from scientific models of the world and scientific practices. We have to consider some models of "good" science but they depend on historic or epistemic contexts of reasoning and evaluation.

The idea of physics as the paradigm of a good empirical science was dominant in the last century. We forget that this idea is far from self-evident. It is the product of a long historical process of the development of science where physics was in the forefront of scientific efforts. Before the birth of modern physics, astronomy and medicine were the models for good science, and perhaps now the emerging new synthesis of biology, physics, computer sciences and psychology will become the model for good science.

If one takes a model science with a physical type of natural regularities and laws and with it a physical type of experimentation and measuring as the paradigm for justification or falsification of hypotheses, then she will have a very strict concept of science (i.e. physicalism) and a lot of actual scientific research would "become" non-scientific. Conversely, if one take as a model science some more "liberal" kind of research, then she will obtain a larger set of "scientific" theories and research.

For Popper the ability of hypotheses and theories to be falsified was the necessary condition of empirical science. A statement or system of statements, in order to be scientific, must be capable of conflicting with possible, or conceivable, observations (Popper, 1963: 51). But the question is, how to apply this criterion: when shall we take a statement or a set of statements as falsifiable or not? In order to apply falsification to a particular hypothesis, criteria of falsification are required. There are no a priori given criteria of falsification because a hypothesis can be defended in principle ad infinitum, in spite of any seeming counterexample.[8] We cannot find cases of science that could serve as a model for all kinds of modern sciences. We can take the models of falsification from the history of science (e. g. Lavoisier's famous falsification of the phlogiston hypothesis) or from actual scientific practices. Fine, but then we are in a circle because we refer to some model(s) of (established) science in order to demarcate science(s) from non- or pseudoscience(s).

[8] This follows from the so-called Duhem-Quine thesis of the impossibility of testing a scientific hypothesis in isolation. It says it is impossible to test a scientific hypothesis in isolation, because an empirical test of the hypothesis requires one or more background assumptions. The consequences of the hypothesis typically rest on background assumptions from which to derive predictions. We cannot falsify any hypothesis or theory through empirical means if the background assumptions are not proven (Grünbaum, 1963).

For Lakatos, who extended Popper's theory of falsification into a complex theory of research programs, the criterion of demarcation between science and non-science lies in the ability for the creation of "progressive" research programs. The continuity of "degenerative" research programs, that is, programs which know only post-hoc explanations either of chance discoveries or of facts anticipated by, and discovered in, a rival program, indicate a pseudo-scientific research (Lakatos, 1978: 5, 88, 112). Pseudosciences do not allow the refutation of a degenerative program with a better, concurrent one, but try to support the established program with ad hoc or circular arguments. But how do we know when a program becomes or is degenerative? How do we know it "uses" pseudo-arguments and no real arguments in its defense? Again, we need the comparison with some established cases of "good" science. There are no absolute criteria of progressivity or regressivity of research programs.

Even the most "balanced" criteria of demarcation experience a similar difficulty. Let us take Paul Thagard's criterion of demarcation, which he formulated in his analysis of astrology as pseudoscience:

"A theory or discipline which purports to be scientific is pseudoscientific if and only if:

1 It has been less progressive than alternative theories over a long period of time, and faces many unsolved problems; but
2 The community of practitioners makes little attempt to develop the theory towards solutions of the problems, shows no concern for attempts to evaluate the theory in relation to others, and is selective in considering confirmations and disconfirmations." (Thagard, 1998: 70f)

The basic problem of this definition of pseudoscience is that modern scientific research is very heterogeneous. It is impossible to find some common necessary and sufficient condition of progressivity over a long period of time and successful solutions to the problems. We again need some "model sciences" in order to evaluate a theory or discipline as scientific.

We cannot find cases of definite science that could be the models of science for all kinds of modern sciences. In the best cases we can find some necessary conditions of elementary scientific correctness and rationality, e. g. openness in the review of possible weak points in observation and experiments, differentiation between actual observations and reporting of them, accessibility and publicity of data, correct argumentation, and repeatability of results in similar conditions. Sciences are different in many aspects, and similar in others. It is thus better to present sciences as a network of disciplines and research fields, ordered in a reasonable way. We choose the criteria of ordering according to the goal of comparison. If one asks whether some "alternative" research program or theory may evolve into science or whether it represents a pseudoscience, then the realm of comparison is very large because almost any science has its "alternative" (or even several of them).

I agree with Newton-Smith, who ends the discussion on the criteria of demarcation with the following harsh statements: "The enthusiastic manipulation of the rhetoric of science and pseudoscience reveals a failure to appreciate that science is not the only form of activity governable by reason. Scientific inquiry is a particular form of rational inquiry and there is simply no reason to think that it is the only form of inquiry that so qualifies. Consequently there is no reason to condemn some investigation just because it fails to meet some criterion of demarcation. It is trite but true to say that all forms of investigation should be examined on their merits to see what insights they embody and what understanding they provide." (Newton-Smith, 1981: 90).[9]

I stress that it is more appropriate to speak of many "sciences" than of one "science", and similarly of many "alternative sciences" than of one "alternative science". The unification of many sciences into one science, and similarly, of many alternatives into one alternative, is an ideological and rhetorical move which disguises the struggles for power and influence over people, privileges, money etc., and it has nothing to do with the real conditions and differences between different human practical and epistemic activities. Not only is there is no one single science, there is also no one single basic science which could be the basis, norm, goal or ideal of all other sciences. It is true that in the past some theoreticians tried to establish "unified science", or some "basic" science, such that all other sciences might be reducible to it. They tried to simplify the discussion on science and non-science. According to that idea, all scientific knowledge, scientific facts and scientific laws have to be reduced to physical knowledge, physical facts and physical laws. It is far from evident that all sciences can be evaluated and ordered by strict physicalist criteria of scientific knowledge.[10]

The development of the sociological and humanistic sciences as well as the growing differences between natural sciences make this comprehension of science questionable, as does the thesis of limiting all reality to physical reality and consequently the principle of transfer of all sciences to physics because the criteria of scientific character known in physics cannot be expanded to all the sciences. In the same way the criteria for some research or theorizing that exists at the edge or beyond some established science cannot be summarized according to the criteria of physics. Seen from the physicalist or naturalist standpoints some non-physical sciences would not be considered as real sciences because they do not enable clear comparison with the methods and results of natural sciences. So for example, a great part of psychology would be eliminated from the field of science, since it deals with experimental and observational, inaccessible processes and conditions that neither recognizes statistical nor strictly universal laws and no completely invariable and accurate explanations or forecasts can be made for them.[11] A great part of sociology would also be eliminated, since

[9] Barry Barnes, David Bloor and John Henry put a similar claim but with an important sociological point. They stated that the demarcation of science from pseudoscience can be fully understood only in sociological terms. By any future attempt of fixing demarcation criteria we must ask: why has this attempt been made? We have to consider the social and historical context because "scientific boundaries are defined and maintained by social groups concerned to protect and promote their cognitive authority, intellectual hegemony, professional integrity, and whatever political and economic power they might be able to command by attaining these things." (Barnes et al, 1996: 168). In the remark attached to this quote they emphasized "there is little hope of demarcating science from non-science by looking at procedures and arguments used by scientists themselves to draw up and maintain cognitive boundaries. Their rhetoric varies from one context to another … In some cases genuine scientific knowledge claims are said to be obvious, in others they are said to be understandable and confirmable only by an élite few." (Ibid: note 213).

[10] Especially some leading logical positivists (e. g. Carnap, Feigl, Neurath, Frank, Hempel) strived some time for the unity of science based on physics as basic science and logical reduction of other sciences to physics. The "Journal of Unified Science," which began in 1940, was the public "organ" of this movement (and was later reshaped in the program of "The International Encyclopaedia of Unified Science". It lasted until the end of the 1960s). Proponents of actual "reductive physicalism" speak of the reducibility in principle of all non-physical phenomena and non-physical regularities in physical entities, processes and laws. There are also non-reductive physicalists who allow ontologically distinct properties as being explicable and predictable in principle from their physical basis (Silberstein, 2002: 103).

[11] The proponents of eliminative materialism declare that the main part of modern psychology presents "folk psychology" which will in the course of time be replaced by a strictly scientific, naturalized psychology, similar to the replacement of the old qualitative physics with modern quantitative physics (Churchland, 1981).

it does not recognize the scientific (or natural) laws of its field nor accurate measurements or tests. The same applies to the humanities (history, linguistics, art and literature, philosophical sciences), which even disagree on the facts, since they depend on the interpretation and understanding of the experts.

It has to be said that many science theorists of the older school have also come to the same conclusion and have defined limits between the real "sciences" and those "being developed" or "being at the very beginning". The Anglo-Saxon division of the recognizable disciplines into "sciences", "humanities" and "technology" rather accurately reflects such comprehension. This division also hides the evaluative differences that are reflected in the social reputation of the researchers, the awards they are given, money and funds for research work, etc.

5. Conflicts on the Status of Science

One has to be first acquainted with the methods of observation, tests, scientific explanations, forming of hypotheses, and laws in the given discipline or science, and only then can an assessment be made of the scientific or non-scientific working methods, hypotheses and theories of examples of a research study which occurs on the "border" of sciences. This is not a simple matter, as the supposed or actual phenomena of "bordering research" frequently overlap with numerous areas of science (e. g. some phenomena of "alternative" medicine or biology may reach into the domain of medicine, psychology, biology and physics). Some critics consider even ecology an alternative science, or even as a pseudoscience because it "irrationally" opposes already established technologies such as nuclear energy, the oil industry and modern intensive agriculture. The boundaries between the real sciences and related "border" research (sometimes imprudently called "alternative", "complementary" or "para"-sciences) are not only unclear, but also changeable, depending on the circumstances, ratio of power between the defenders of one or the other discipline, realistic and imaginary social needs.

More important than the question of what (still) does or does not belong to science is the query as to why all these questions and conflicts are important for determining the scientific status. These questions are undoubtedly not only epistemological, but also deeply practical. This statement applies both to the defenders of "established" sciences as well as to the defenders of "alternative" sciences. The reasons for this can be found in the social position and power given by the status of a socially recognized science (and technology) to their supporters and defenders. A discipline that has reached this status can apply for a share of government support and other funds for research of its activity, it can disseminate its scientific findings in schools, assert its own system of attracting and instructing new experts and it has an impact on important social decisions, etc. Sometimes it is a matter of "business", i.e. who will take over the problems which are solved by one or the other discipline. This point of view is undoubtedly strongly present in the conflict between modern medicine and alternative methods of healing. This is not only a health matter or who is in the right, but also a question of where patients will go. It is no coincidence that throughout the world as well as in our country the request to "settle accounts" with alternative medicine becomes greater at a time when "going to see a doctor" becomes too expensive for the average patient and they start searching for less expensive and less institutionalized forms of healing.

6. The Capacity of Being Falsified

The basic question presented in the discussion of "border" research is which kinds of non-agreement with established theories, methods and concepts of modern science are still acceptable and do not reject the scientific value of some not yet established kinds of research, e. g. border research. In my opinion, the question is not whether the scope of science should be expanded (or even reduced) to include some alternative research programs among the sciences or to push some of the existing disciplines beyond the scientific edge, but to what degree the search for truth and scientific recognition can be principally limited and socially codified with some well determined set of epistemological, technical and historical reasons and norms. In my opinion this cannot be done, at least not in advance, for future research work, as it would put a halt to our own development. The search for truth and recognition does not promise identifiable achievement and progress; it has to be constantly established in "theory and practice".

It is not essential how much knowledge a person has gained but whether she is ready to be repeatedly exposed to critical testing of her sources of evidence, hypotheses, theoretical assumptions, and how much of this activity is based on intersubjectively valid experiences. If she allows and undertakes this, then her efforts are a serious cognitive activity irrespective of whether they belong to established sciences or not. Sooner or later she will have to be confronted with complex scientific findings, for the greatest part of empirical science is research that requires ongoing testing, discussion and discovery of new facts. Popper's well-known refutation criterion is still worthy of being considered at least as preliminary advice which makes a distinction between claims which are worthy of closer scientific examination and claims which may be interesting but have yet to be evolved into scientifically testable hypotheses. My claim differs from Popper's requirement in that it is more a reasonable suggestion than an all-embracing norm of science. It suggests that we first try to make (empirical) scientific hypotheses and theories to be refutable in principle and then to make testing conditions as precise as possible. It is good for the defender of a given hypothesis to strive towards such forms of claims and testing procedures that enable the formulation of accurate and rigorous conditions for possible falsification of the hypothesis. The "goodness" of falsifiable hypotheses is that we preserve through them the link with experience and in case of finding relevant counter-examples we can search for new, better hypotheses. They do not differ, however, principally from non-falsifiable hypothesis regarding their "power" of systemizing our knowledge in theoretical wholes or in their potential role of premises in scientific explanations.

I do not claim that adherence to the falsifiability principle is either a necessary or sufficient condition of being a scientific hypothesis but it keeps alive the practice of sound scientific skepticism.

Certainly this practice works well under conditions of "normal" science in the Kuhnian sense, that is, where a clearly defined scientific paradigm without serious challenges obtains. It is more like "puzzle solving" than the production of radically new views or theories (Kuhn,

1970: 36-38). It does not work well in times of scientific revolutions where a more "ideological" kind of battle between the proponents of the old paradigm and the proponents of the possible alternatives appear. Kuhn thus stated that the way of science as a continued line of research depends on the ability of a science to find (its first or the next) paradigm, to become "normal" (Ibid: 5f, 24, 64f, 79, 144f).[12]

We have to be cautious enough and not equalize falsification activity and the scientific character of activity. Even astrology can be put in a form that allows falsification of its claims but it does not know a continuous practice of reshaping of falsified hypotheses into (falsifiable) hypotheses which explain more facts with better accuracy and resist serious attempts of refutation (Thagard, 1998, p. 69). Popper was well aware of the fact that many scientific hypotheses and theories have arisen from the original unverifiable and non-refutable speculations and that many interesting hypotheses cannot be empirically tested at present.[13] However, it is wise to extend the use of empirical testing on hypotheses and theories which are as scientific as possible because then they are taken more seriously as explanatory or predictive tools. This advice, however, is to be taken into consideration "with a pinch of salt", i.e. to take into consideration the development of a given science and its particularities. This demand cannot apply to all sciences uniformly, as some of them do know only rough or vague methods of empirical testing. Only after it has been defined for each science what the "empirical content" of hypotheses and theories means does it become clear what research of scientific material is, how it is defined and interpreted and how suitable methods of testing and possible falsification can be formulated and applied.

Development of the theory of science in many ways modified and completed Popper's original idea of falsification; first of all by the realization that in many cases (e.g., in the case of a complex, theoretically grounded hypothesis) the finding of counter-facts cannot refute the hypothesis (even if we know the necessary auxiliary hypotheses) but we have to have at hand another, alternative hypothesis or theory which explains all the "successes" of the former hypothesis, and its failure in the case of new counter-facts, and secondly by realizing that falsification needs inductive reasoning, and does not exist (as Popper believed) only in the logical contradiction of observed facts to the logical conclusions from the hypothesis.[14]

---

[12] I do not accept all Kuhn's theory of scientific revolutions and normal sciences, especially not his claims on the rigid, even dogmatic nature of normal science and on complete incommensurability of the different theories concurring on the status of the next paradigm, but I accept his thesis on normal science as guided by the mix of verification and falsification practices and on extraordinary science as searching for challenging alternatives to the established paradigm which evade verification and falsification. I accept the concept of a scientific paradigm as a set of patterns and models of solutions of scientific problems which implicitly guide scientific research but not as a "code of conduct", a norm or an unchallengeable authority. I also distinguish between explicit attempts to compare different scientific theories and implicit and often unconscious "labeling" of different theories which may lead to seemingly instantaneous changes of belief among scientists discussing concurrent theories. Explicit comparisons of different theories may not ever end in a final decision regarding which theory is correct or at least better but this does not mean scientists behave non-rationally when they accept one and reject another theory. There may be quite objective qualities of a theory which makes it at least very promising and worth of further research, while another may be less promising or worthy of further research. These topics need a lot of further discussion but I limit it to these few remarks.

[13] See especially the chapter "Way of Science" in Popper's Logic of Scientific Discovery (1959).

[14] As Newton-Smith has shown, Popper involuntarily assumes some pieces of inductive reasoning and support for his theory of corroboration of hypotheses. Popper tries to show that modern science has (by and large) more verisimilitude (that is, they are nearer to the truth) than previous science and that a high degree of corroboration of hypotheses (that is, hypotheses which have successfully survived many serious attempts at their falsification) leads to greater verisimilitude. Both theses can be supported only by induction from previous cases, and not, as Popper sometimes claims, by the assumption of scientific realism (Newton-Smith,

Despite all modifications and relativizations: the concept of falsification, i.e., the readiness and ability of scientists to construct falsifiable hypotheses and theories and the readiness to reject or radically reshape the dearest hypotheses and theories when proven wrong, stays a crucial sign of scientific rationality. In this sense we can still accept the Popper's claim that the existence of a systematic falsification practice divides the epistemic enterprise in empirical research which might be scientific from the enterprises which are non-scientific, or even do not want to be scientific.

If one considers the activities in the large area of research which occurs on the borders of modern sciences from this viewpoint then one can see that empirical testing and the possibility of falsification is allowed only by a smaller part of these activities. Many such activities are based on theses and theories which are purely speculative, which means they are in principle irrefutable and empirically non-testable. However, it would be false to exclude them a priori from science and to consider them as epistemically unworthy claims. Such theses and theories are sometimes scientifically interesting because they offer important heuristic starting points in serious scientific research and theorizing, or direct future research and theorizing. Some experiences that are reported by the proponents of "alternative sciences" might be scientifically relevant but the first condition here is the demand of intersubjectivity and repeatability of experiences performed in similar conditions and open readiness to construct empirically testable claims, which may follow from these theories. Let us take for example the case of acupuncture, which was the kernel of traditional Chinese medicine but today it is becoming part of modern medicine and is partly scientifically justified. Something similar is happening today with some other forms of traditional "folk" medicine, e. g. with the collections of remedies and healing drugs. They have become an important source of new pharmaceutical products and medical therapies.[15] Hypnosis, whose historical basis lies in magic and occult practices and in the method of "animal magnetism" propagated by Mesmer, has become a respected therapeutic and medical method, even if we do not know fully how it works. The "transitions" of some traditional or para-scientific practices into scientific ones usually drastically change the conceptual frames, terminology and professional methods but at the same time some traditional knowledge is being integrated into modern sciences.

We can thus assume that something similar can happen with some portions of research occurring on the borders of modern sciences. This assumption supports my claim that some

---

1981: 67-70). Eric Mc Cord put in his critique of Popper's methodology another objection to Popper's seeming deductivism. Any case of successful falsification implicitly assumes an inductivist belief, namely the Principle of Temporal Uniformity (PTU): "PTU is the unjustified assumption that the future will resemble the past. The logic of falsification assumes this principle in its assumption that a theory falsified at t will remain false at t'. That is, that because an explanation is deductively falsified by an inability to meet the facts and is thus an inadequate explanation, it will similarly be inadequate and false in the future.... Without assuming PTU, deductive falsification is impossible because it is unjustified to believe that any of the premises will be true in the future. ... Although constructing theories in such a way as to falsify them once and for all whenever a falsifying event is observed averts the premise stability problem, it causes other significant troubles. Most obviously, it seems certain that the scientists that construct theories in such a way assume PTU. If they were not to assume PTU, that nature is uniform and consistent through time, then they would not believe that the truth of a theory rests on the results of observation at any one time. They would then not falsify a theory at time t that may very well be true for all times after t. Scientists clearly prefer some unfalsified theories over others, perhaps on the sole inductive reason that one theory has survived more falsification attempts than the other. Induction certainly plays a part in even the simplest examples of theory weighing." (McCord, 1997)

[15] Matjaž Barborič et al report on the use of a plant used in traditional Indian medicine in a modern genetic remedy for neurodegenerative illnesses.

experiences and paths of knowledge cannot be excluded in advance from scientific research and theorizing merely because they do not satisfy the theoretical frames of scientific work.

7. CONCLUSION

The first question of this article asks if modern scientific research and discoveries can significantly contribute to an internal connection of scientific mind and ethical sensitivity. *Yes it can,* but only if scientists transcend their fears of responsibility and do not blind themselves with hopes of seemingly value-free research. As Russell pointed out, the cooperation of 'reason and the heart' is necessary. This cooperation also refers to "users" of science and not only to scientists. They need an even more developed awareness of their responsibility for the use of science than scientists because they further develop scientific results and apply them to a vast area of uses that are outside quite strict measures of scientific control of events. Scientists are mainly responsible for the consequences of scientific discoveries, whereas the "users" are responsible for the ethically appropriate and humane aims and means chosen for the realization of the aims.

The second question of this article asks if new emerging lines of research and scientific theories change or even dissolve the lines of division between established science and its alternatives. *Partly yes*, because new scientific research and theories may change the lines of division in cases where paradigmatic changes occur, and *partly no*, because any "normal" science (in Kuhn's sense) needs stability in its research methods and delineation of research problems. Only in some rare cases of scientific revolutions may it seem even to working scientists that "anything goes", that the lines of division between established science and its alternatives dissolve.

A complete exclusion from science should apply to the unethical practices based on explicitly misleading ideas or on manipulations of people or higher animals, their bodies and minds for the sake of profit or power over people. This exclusion follows from ethical rather than methodological or theoretical considerations. Cases of such practices done in the name of science are known[16]. Even if this work could formally be considered as scientifically valid, it would be considered non-scientific if some minimal ethical standards were accepted as constitutive for science. In recent decades, high ethical standards were introduced into all areas of scientific research, but particularly in the domains of modern biology, such as human stem cell research and experiments on animals. The articles in this book each discuss how these standards apply to a particular research topic, and pinpoint the important topics for future consideration.

[16] Manipulative or anti-human research such as cruel medical experiments on internees in Nazi Germany gave no or very poor scientific results. It seems that extreme forms of weird research tend to annul them (Annas, Grodin, 1992). It would be false, however, to conclude that this is necessarily so. Criteria with respect to the scientific soundness or weirdness of research or theorizing change over time, and it may happen that a theory which is counted as scientifically sound in its time becomes "weird" in the future (or vice versa). Consider, for example, the history of geocentric and heliocentric astronomy.

References

Annas, G. J., Grodin, M. A. (Eds.) (1992): *The Nazi Doctors and the Nuremberg Code: Human Rights in Human Experiments*. New York, Oxford: Oxford University Press.

Barnes, B., Bloor, D., Henry, J. (1996): *Scientific Knowledge. A Sociological Analysis*. Cambridge: Cambridge University Press.

Bekoff, M. (2000): Redecorating Nature: Reflections on Science, Holism, Community,

Humility, Reconciliation, Spirit, Compassion, and Love. *Human Ecology Review,* Vol. 7, No. 1, 2000: 59 – 67.

Churchland, P. M. (1981): Eliminative materialism and the propositional attitudes. *Journal of Philosophy*, 78: 67-90.

Detel, W. von (2007): Grundkurs Philosophie, Band 4: Erkenntnis und Wissenschaftstheorie. Stuttgart: Reclam

Ethics in Science and Scholarship: the Toronto Resolution (1994), publ. in *Accountability in Research*, vol3, 69-72.

Grünbaum, A. (1963): The falsifiability of theories: total or partial? A contemporary evaluation of the Duhem-Quine thesis. In M. W. Wartofsky (ed.), *Boston Studies in the Philosophy of Science, Proceedings of the Boston Colloquium for the Philosophy of Science 1961/1962*. Dordrecht-Holland: D. Reidel: 178-196.

Jonas, H. (1984): The Imperative of Responsibility. In Search of an Ethics for the Technological Age. Chicago: The University of Chicago Press.

Gustafsson, B., Ryden, L., Tibell, G. and Wallensten, P. (1984): Focus on: The Uppsala Code of Ethics for Scientists. *Journal of Peace Research*, Vol. 21, No 4, 1984.

Kitcher, P. (2001): *Science, Truth and Democracy*. Oxford, New York: Oxford Univ. Press.

Lakatos, I. (1978): *The Methodology of Scientific Research Programmes*. Cambridge: Cambridge University Press.

McCord, E. (1997): Induction-Free Science? A Critique of Popper's Falsificationist Methodology. (http://www.utdallas.edu/orgs/ntpa/McCord%20paper%20(1997).html, accessed on 20 October 2008)

Newton-Smith, W. H. (1981): *The Rationality of Science*. London, New York: Routledge.

Popper, K. (1959): *The Logic of Scientific Discovery*. London: RKP.

Popper, K. (1963): *Conjectures and Refutations*. London: Routledge & Kegan Paul.

Russell, B. (1925): *Icarus or the Future of Science*. London: Kegan Paul, Trenc, Trubner & Co.

Russell, B. (1952): *The Impact of Science on Society*. London: G. Allen and Unwin

Russell, B. (1978): *Autobiography*. London: Unwin Paperback.

Smart, J. J. C. (1981): Ethics and science. *Philosophy*, 56: 449- 465. Government Office for Science (2007): A Universal Ethical Code for Scientists. September 2007 (http://www.dius.gov.uk/policy/science_society/code.html, accessed on 16 October 2008)

Science Council of Japan (2006): Statement: Code of Conduct for Scientists. October 3. (http://www.scj.go.jp/ja/info/kohyo/pdf/kohyo-20-s3e-1.pdf, accessed on 16 October 2008).

Silberstein, M. (2002): Reduction, emergence, and explanation. In P. Machamer, M. Silberstein (eds.), *Philosophy of Science*. Oxford, Malden: Blackwell.

Thagard, P. R. (1998): Why astrology is a pseudoscience. Reprinted in E. D. Klemke, R. Hollinger, D. W. Rudge, and A. D. Kline (eds.), *Introductory Readings in the Philosophy of Science.* Amherst, New York: Prometheus Books.

UNESCO (2005): Code of Conduct for Scientists: expert group meeting report. UNESCO, March 7. (http://portal.unesco.org/shs/en/ev.php-URL_ID=8529&URL_DO =DO_TOPIC&URL_SECTION=201.html, accessed on 16 October 2008)

UNESCO (2006): Interim analysis of codes of conduct and codes of ethics. Division on Ethics of Science and Technology, September 2006 (http://unesdoc.unesco.org/images/0014/ 001473/147335E.pdf, accessed on 16 October 2008)

Weber, M. (1919): Der Sinn der 'Wertfreiheit' der soziologischen und ökonomischen Wissenschaften. *Logos*, 7: 40-88. (Reprinted as "The Meaning of 'Ethical Neutrality' in Sociology and Economics" in M. Weber, *The Methodology of the Social Sciences*, New York: The Free Press, 1949).

PART A. CONTEMPORARY SCIENCE: NEW DEVELOPMENTS

In: Philosophical Insights about Modern Science
Eds: Eva Žerovnik, Olga Markič and Andrej Ule
ISBN: 978-1-60741-373-8

Chapter 1

NEUROPLASTICITY, OR THE IMPORTANCE OF HAVING A PLASTIC BRAIN

M. Bresjanac[*1] **and G. Repovš**[*2]
[1] Laboratory for Neural Plasticity and Regeneration; Institute of Pathophysiology; Faculty of Medicine; University of Ljubljana, Ljubljana, Slovenia
[2] Mind and Brain Laboratory; Department of Psychology; Faculty of Arts; University of Ljubljana, Ljubljana, Slovenia

"To know the brain...is equivalent to ascertaining the material course of thought and will, to discovering the intimate history of life in its perpetual duel with external forces."

S. Ramon y Cajal, Recollections of My Life, 1937

1. SYNOPSIS

According to the Oxford American Dictionary the term *plasticity* denotes a quality of being easily shaped or molded. Transposed to biology, the term refers to adaptability of an organism to changes in its environment. In terms of human ability to learn, modify behavior and adapt to the ever changing environment, our nervous system, the material substrate of these abilities, is remarkably plastic. Mechanisms of nervous system plasticity span all its functional levels from molecular and cellular to the level of physiology and anatomy of the system as a whole. Importantly, plasticity is an inherent property of the nervous system and not an occasional state or condition, brought about by drastic challenges, such as injury or disease. Still, in the context of damage, which causes structural and functional neural perturbations, additional manifestations of neural plasticity can either provide basis for restoration of function through reinforcement of the existing neural connections and/or via development of new pathways (*restorative plasticity* and *rewiring*), or give rise to a worsening of the condition (*maladaptive plasticity*). Current research is unveiling intricate mechanisms underlying nervous system

[*] Zaloška 4, Ljubljana; Slovenia, EU; Ph: + 386 1 543 7033; e-mail: maja.bresjanac@lnpr.mf.uni-lj.si
[*] Aškerčeva 1, Ljubljana; Slovenia, EU

plasticity in health and disease and a comprehensive review of this rapidly growing body of work is well beyond the scope of this chapter. Therefore, a brief outline of selected examples is provided merely as an illustration of the remarkable malleability of the nervous system during development, in health and disease. In addition, selected accents are provided to demonstrate that the insight into mechanisms of neuroplasticity is contributing significantly to the understanding of how the mind evolves in the plastic brain, based on its unique genetic inheritance and the life-long experience of the environment and the self.

2. Introduction: The Mind Evolves in the Plastic Brain

Nervous system has evolved to guide animals' behavior in a way that ensures species survival. It has evolved from simple networks of a few cells that provided reflex responses to systems that support advanced sensory processing, enable complex representations of environment, and promote goal directed planning, problem solving, execution and control of behavior.

Human brain arguably presents the most advanced nervous system. Its main advantage is not in the complexity of specific processing feats but rather in its flexibility and ability to (re)shape itself based on its experiences. Many organisms are capable of highly complex behavior, which however is rigid and unchangeable, "wired-in". A change in the cues that should trigger a behavior or any changes in the demands of the environment can lead to utter failure in completing the task. Mammalian and especially human brain on the other hand is provided with very little in terms of pre-programmed behavior. Instead of a limited set of preset behaviors, it possesses key mechanisms that enable it to learn and shape its processing: mechanisms of plasticity. The ability of the nervous system to reshape itself leads to a number of significant advantages and consequences on many levels. First, it affords more powerful computational capabilities. Second, it forms the basis of learning and development of novel processing capabilities and behaviors, enabling extremely fast adaptations and responses to changes in the environment. Third, it gives rise to external transmission of knowledge and skills, promoting the development of culture and dependency on society.

The development of brain and mind enabled by plasticity progresses on two connected and interdependent levels. First, synaptic plasticity enables learning and formation of novel neuronal processing patterns and behaviors, it provides the crucial foundation for acquiring knowledge and processing abilities that enable and support cognitive control and voluntary processing. Second, once established, voluntary processing enables on-line planning and active reprogramming of behavior. It enables formulation of novel behavior, processing patterns and mental representations that are not "wired-in" but rather run as transient programs on top of a neuronal system. When practiced and maintained these processes and behaviors spur further plastic changes through which they are automated, "burnt in", making them a permanent part of the repertoire of the neuronal system. Through the interaction between these two processes human brain actively reshapes itself, rapidly developing and forming novel skills and abilities, adapting to the specific and changing challenges of the environment.

The initial indeterminacy and flexibility of human brain however presents important challenges. What is not explicitly coded, has to be learned. While most animals can move (walk, run, fly) within hours of "birth", it takes humans a year to learn how to stand and walk. And it takes much longer to perfect that activity to a proficient level. The development of things not coded specifically in the genome is now dependent on appropriate learning opportunities. The skills are not carried by genes but are rather shaped by the opportunities and challenges provided by the environment and passed from generation to generation by society and culture. While most humans are able to learn how to talk, it takes exposure to and interaction with speaking adults for a child to be able to acquire and master language - both as a communication skill as well as a system for development of concepts, representation and storage of knowledge.

The development of human brain and its inherent plasticity has led to the point where both phylogenetic as well as ontogenetic development proceeds and depends on three parallel carriers of information. First, genes code our material structure and form as well as a key set of mechanisms that enable neuronal plasticity and learning. Second, the environment with the stimuli, challenges and opportunities it presents helps shape the basic perceptual and cognitive processes and behavioral skills. And last, society and culture carries, develops and passes the complex behavioral patterns, cognitive skills and knowledge that enable a person to orient her-/himself, to cope with everyday challenges, and potentially contribute back to the growing pool of human knowledge and experience.

3. The Basics of Plasticity

Nerve cells forming specialized connections and shaping networks are the basis of our behaviors, from the simplest, such as the patellar reflex (knee-jerk or tendon stretch reflex) to the complex, such as long-term pursuit of defined goals. In invertebrates and some lower vertebrates there may be a single cell (a so called *command cell*) or a small group of cells, that can initiate complex behavioral sequences (Wiersma and Ikeda, 1964; Kupfermann and Weiss, 1978; but see also Edwards et al., 1999). In humans, however, it takes more than a few cells to generate behavior. Human nervous system is exceedingly complex and the structure and interconnections of its many parts are still not well understood. In a practical approximation, we can state that our brain relies on localized functional specialization for processing specific types of information, and on complex connectivity patterns between specialized parts for shaping and evoking representations which underlie behavioral patterns and strategies. Information for each of our senses is predominantly processed in a distinct brain region, where the afferent input typically forms a precise map of the pertinent receptor sheet on the body surface – the skin (touch), retina (sight), the basilar membrane of the cochlea (hearing), etc (Kandel, 2000). These maps are the basis for creating representations of ourselves and the world we live in. The neurons that make up these maps (sensory, motor and the interposed interneurons) fulfill specific functions because of the connections they make (Kandel, 2000). These connections, established as the brain develops, determine the behavioral function of individual cells. What makes the brain a remarkable processing machine is not so much the complex diversity of its individual neurons, but rather its many elements and the complexity of connections between them. All nerve cells have similar signalling properties. Individual neurons are able to convey unique information not because each of them is unique, but because of different and specific ways in which they are wired together and organized (Kandel, 2000).

Inevitable question follows: how, if the nervous system is wired so precisely, is behavior modified? First proposed by Ramon y Cajal at the turn of the 20th century, the plasticity hypothesis was later advanced by Konorski stating: "*The ... property by virtue of which certain permanent functional transformations arise in particular systems of neurons as a result of appropriate stimuli or their combination, we shall call plasticity and the corresponding changes plastic changes*" (Konorski, 1948).

In discussing the importance of plasticity to whole brain activity and the concept of mind, Azmitia recently evoked a statement on homeostasis by Walter Cannon: "*By an apparent contradiction, [an organism] maintains its stability only if it is excitable and capable of modifying itself according to external stimuli, and adjusting its response to the stimulation. In a sense it is stable because it is modifiable—the slight instability is the necessary condition for the true stability of the organism*". By analogy, neuroplasticity is required for a balanced brain, a framework for a stable mind (Azmitia, 2007).

In simplified terms, basic forms of nervous system plasticity (figure 1) include molecular changes taking place in neuronal terminals during synaptic activation, which change the strength of synaptic connection (long term potentiation and long term depression, or LTP and LTD), new synapse formation, structural modification of existing synapses and even elimination (pruning) of synapses between neurons already in contact, mechanisms central to the ability to learn, store the information, remember and forget. As we shall illustrate in more detail below, new neurons can also be added to the network at least in certain regions of the brain, whose normal function may critically depend on their ability to permanently remodel themselves – adult neurogenesis providing yet another dimension of the remarkable plasticity of the normal nervous system.

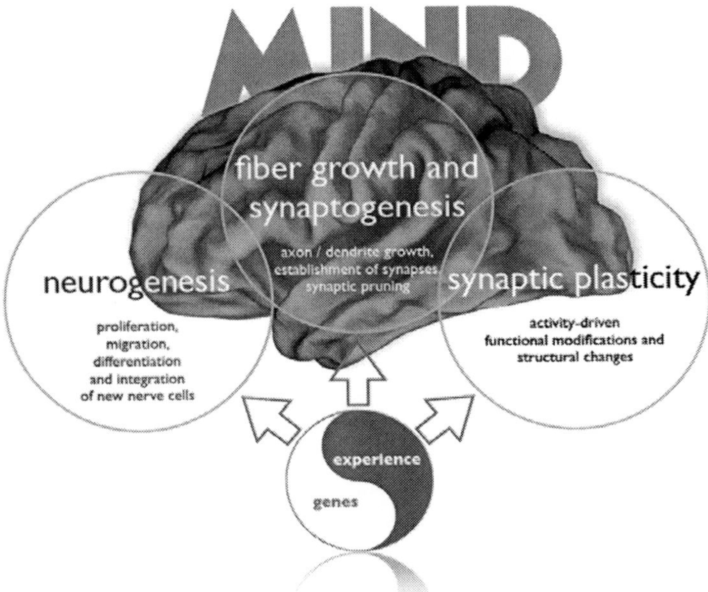

Figure 1. Mind emerges and evolves as a function of the plastic brain. (See text for more information).

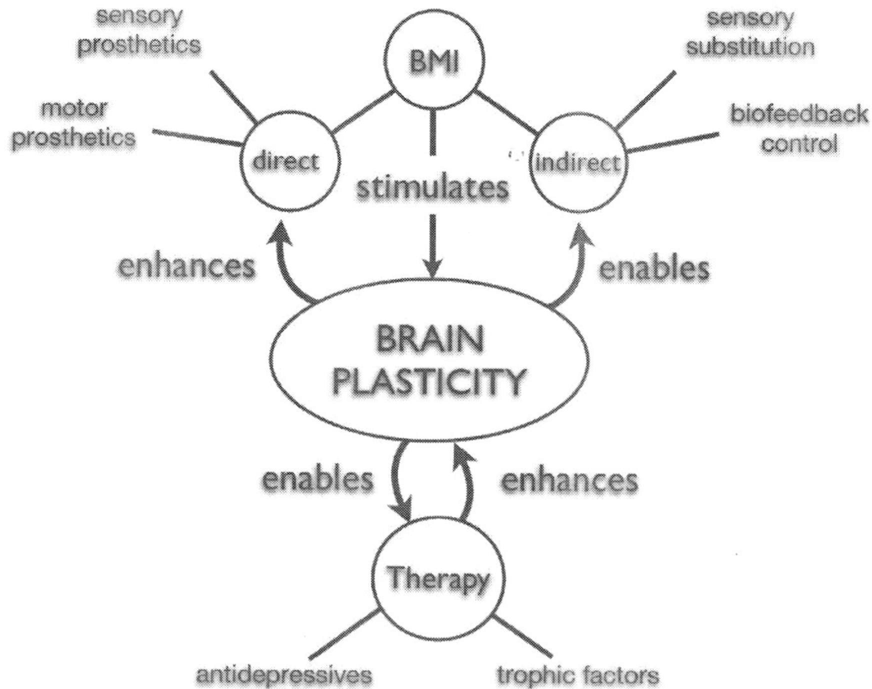

Figure 2. Harnessing brain plasticity. A number of methods aim to either repair or replace part of the human brain function following injury or disease, or extend the capabilities of healthy brain. Two examples are brain–machine interfaces (BMI) and the use of pharmacological substance therapy (Therapy). Both not only crucially depend on inherent brain plasticity but also enhance and stimulate it, opening new possibilities. (See text for more information).

Following injury, surviving parts of the nervous system respond by regenerative (enacted by the severed connections) and collateral (expressed by the surrounding uninjured connections) fiber sprouting, establishment of new synapses and rewiring within the existing connections or through the formation of new ones, as well as with enhanced neurogenesis – different forms of structural plasticity which may contribute to functional recovery of a damaged nervous system. Unfortunately, as we shall also briefly describe below, sometimes the plastic nature of the nervous system may underlie undesirable functional outcomes (*maladaptive* neuroplasticity).

4. Developmental Plasticity of the Nervous System

All the tasks performed by the nervous system, from processing of the sensory input to the control and execution of the motor output, including learning, memory and other cognitive functions, depend on appropriate and sometimes very precise connections between many nerve cells. How do the amazingly complex neural connections ever form during embryonic development?

Development of the nervous system is reminiscent of a masterfully orchestrated symphony where the expression of specific genes takes place in appropriate locations and at precise times under the conductorship of both hard-wired molecular programs and epigenetic processes. Both the embryo and the external environment contribute modifying factors that regulate neuronal differentiation and growth: the embryo produces specific humoral and cell surface factors that guide and stimulate growth of neuronal processes, as well as endogenous transcription factors, which regulate gene expression, while the environment provides nutrients, sensory stimuli and social interaction, all of which influence the basic well-being and activity pattern of nerve cells.

Neural cell *progenitors* are a uniform population of cells in the neural plate which have not yet committed to a specific lineage of nervous system cells. At a certain moment, these cells get "recruited" and start to differentiate into glia or neurons. The latter begin to migrate to their future positions and extend axons toward their target cells. When the axons reach the target cells, the establishment of synapses takes place, accompanied by a selection process during which some synapses are eliminated, while other synaptic contacts are strenghtened. The surviving synapses define the pattern of connectivity as well as the neuronal phenotype (Kandel, 2000). Neural differentiation, connectivity and functional maturation of the nervous system is achieved through its activity-derived experience.

Detailed mechanisms of determination and control of cell differentiation are coming to light through exciting research that reveals remarkable degree of evolutionary conservation in regulatory mechanisms at different levels of the nervous system development, even down to the same molecular signals (e.g., sonic hedgehog protein, bone morphogenetic protein family, retinoic acid, etc.) acting on the same receptor types to control the same developmental programs within neural cells in organisms so diverse as insects and humans. The overall outcome of the normal course of this process is a recognizable amazingly complex and "correctly" structured nervous system, which enables its owner the organism-specific repertoire of behaviors.

4.1. Science Immitating Art: Modelling a Developing Visual System *in Silico*

Plasticity plays a significant role in development of computational capabilities of the brain. The execution of any computational task within a massively parallel neuronal system depends on complex patterns of excitatory and inhibitory connections. While the basic local circuitry of the system can be relatively simple to describe and reconstruct, the engineering of the whole system represents an insurmoutable challenge. The amount of information needed to specify the connectivity of the system with estimated trillion (10^{12}) neurons connected by 10 quadrillion (10^{16}) synapses exceeds even the informational capabilities of human DNA (Boahen, 2005).

Faced with a similar challenge in an attempt to recreate early visual processing *in silico* Boahen and coleagues (Merolla and Boahen, 2003; Taba and Boahen, 2003; Taba and Boahen, 2006) realized that rather than specifing each connection in advance, they can let the system organize itself by implementing a few basic rules and properties of the system. To ensure that neighbouring ganglion cells in the silicone "retina" connect to neighbouring cells in the silicone "tectum" they used a simple rule - first proposed by Hebb in 1949 (Hebb, 1949) - stating that cells that fire together also wire together. By implementing a mechanism of promoting connections between concurently active cells in the tectum the system was able to self-organize the wiring in a way that preserved topography of input at later processing stages (Taba and Boahen, 2003). Further, to recreate spatial distribution of differential orientational sensitivity characteristic of the cells in the primary visual cortex, instead of explicitly specifying the appropriate wiring pattern, they relied on inherent local random differences in the connectivity, which – sufficiently enhanced – led to appropriate self-organization (Merolla and Boahen, 2003). Simulation showed that such system can successfully mimick the distribution of orientational sensitivity found in the visual cortex. These simulations *in silico* efficiently demostrated how complex parallel computational system can arise based on the interaction between simple mechanisms of neuroplasticty and experience.

4.2. Experience-Driven Neuroplasticity in the Development of the Nervous System

The significance of early visual experience was dramatically illustrated by Hubel and Wiesel in a series of studies of early visual development in cats (Hubel and Wiesel, 2004). Especially informative were visual deprivation studies in which deprivation of light over several months lead to marked abnormalities such as behavioral blindness, morphological changes and disruptions in cortical connectivity. It was found that the most cruical period were the first three months of life after which both the deprivation effects as well as the animals' capacity to recover were severely limited (Wiesel and Hubel, 1965; Hubel and Wiesel, 1970).

Typicaly, a light-deprivation of a single eye would lead to lack of neurons in the visual cortex that would respond to stimuli from the previously light-deprived eye. These results were initially ascribed to either failure of the connections to develop postnatally due to lack of visual experience or a deterioration of existing connections due to disuse. Such explanation, however, failed to account for the existence of visual cortex neurons that responded to stimuli from one or both eyes after both eyes were light-deprived. The latter results favored an explanation based on competition between inputs from each eye, leading to virtually all the visual cortex being driven by one eye only in a single eye deprivation and an undifferentiated response in the case of light deprivation to both eyes (Hubel and Wiesel, 2005).

Although highly influential, the studies by Hubel and Wiesel represent only a small sample of deprivation studies demonstrating the importance of early visual experience in shaping both the properties of the remarkably plastic visual system such as ocular dominance and orientation specificity, as well as its integration in visuomotor development (for an extensive review see Black and Greenough, 1986).

The extent to which functional organization of the human brain is defined by neuroplasticity and experience is also vividly demostrated by studies of early blind and early deaf human subjects. A wide range of neuroimaging studies have shown that – compared to normaly sighted participants – the visual cortex areas of the early blind subjects can be activated by Braille reading and tactile discrimination tasks (e.g., Sadato et al, 1996; Buchel et. al, 1998). Later studies using transcranial magnetic stimulation (TMS) also confirmed the functional relevance of these activations. A study by Cohen et al. (1997) for instance showed that whereas transient stimulation of visual cortex significantly impared Braille reading and identification of embossed roman letters, as well as distorted the tactile perception of blind subjects, it had no such effect on normal-sighted subjects. Similarly, studies of auditory and language processing have demonstrated activation of visual cortex during auditory localization (Collignon et al., 2008; Weeks et al., 2000), verb generation (Ofan and Zohary,

2007) and verbal memory (Amedi et. al, 2003; Burton et. al, 2002) tasks. Additionally, TMS applied to the occipital pole of the visual cortex resulted in disruption of high-level semantic verbal processing in a verb-generation task in early blind but not in normal-sighted individuals (Amedi et al., 2004; 2007).

With similar findings observed in early deaf subjects (for a recent review see Kral, 2007), these studies demonstrate the basic strategy and key mechanisms of development of the nervous system in which the function of the highly malleable cortex is shaped through its use (figure 1).

5. Adult Nervous System Plasticity

Nervous system is not remarkably malleable only during development – it remains plastic throughout lifetime conferring on its owner the life-long ability to learn and adjust to the experiences of the outside world and the self. To illustrate this, we offer a glimpse of some changes known to be involved in this proces at the levels of individual synapses and of representational maps in the brain cortex.

5.1. Synaptic Plasticity, Learning and Memory

Learning refers to the fact that behavior of an organism is not constant but changes over time as a function of experience. Changes in neuronal circuits that define the capacity to learn are the basic expressions of nervous system plasticity. The site where most of these changes occur are the synapses, the points of closest contact and communication between neurons. Learning critically depends on synaptic plasticity – ability of synapses to be created and modified (Kandel, 2001). The mechanisms involved seem to be highly preserved across species and much of what is known today about the basic neurobiology of human learning has been gathered in a molusc, *Aplysia* (Kandel, 2001). Like in vertebrate learning, memory storage in *Aplysia* has two phases: a transient memory that lasts minutes and an enduring memory that lasts much longer. And like the saying "Practice makes perfect!" suggests, conversion of short-term to long-term memory storage requires spaced repetition.

Short-term memory trace involves molecular events that result in increased neurotransmitter availability and release in the activated synapse. This process does not require protein synthesis. It depends on modification of preexisting proteins and enhancement of the transmitter release from the presynaptic terminal.

Long-term information storage and maintenance, on the other hand, calls for more pronounced changes: (a) it involves neuronal nucleus, (b) it depends on activation of memory-enhancing genes and simultaneous inactivation of memory-suppressor genes, (c) it requires synthesis of new proteins, and (d) it involves growth of new synapses.

The process of long-term memory storage starts after signalling molecules from the activated synapse reach the nucleus of the neuron, where they activate transcriptional factor CREB-1 (for cAMP response element binding protein). CREB-1 activates a set of immediate response genes, which in turn act on a cascade of downstream genes (Kandel, 2001) to enable formation of new synapses. Two interesting details stand out: (i) existence of memory-suppressor genes, and (ii) reversibility of synapse formation. The former may provide a "checkpoint" for memory storage, possibly ensuring that only salient features are learned. The latter probably facilitates forgetting: over time, as the memory fades, the synaptic connections retract (for an exhaustive review see Kandel, 2001 and the references therein).

A puzzling problem arises from the arrangement described above: if long-term memory trace formation and maintenance involves the nucleus and is therefore a cell-wide phenomenon, what (if any) mechanisms ensure actual synapse specificity of long-term facilitation? Indeed, despite involvement of the nucleus and protein synthesis, mechanisms have evolved to guarantee that long-term augmentation of synaptic function and structure are restricted to the activated synapse (Kandel, 2001). This happens through specific marking (*tagging*) of the synapses that go through repeated short-term activation. Marking of the active synapses relies on the protein synthesis that takes place at the site – in the presynaptic terminal (see the excellent paper by Barborič et al., Chapter 2, part A of this volume for details on mRNA presence and function in dendrites). Local protein synthesis in the presynaptic terminal of the activated synapse produces the key proteins needed locally for the synaptic structure and function, resulting in their stabilization (Kandel, 2001; Barborič et al, Chapter 2, Part A).

Learning at the neuronal level results from changes in the strength of synaptic connections between precisely interconnected cells. The genetic program of an organism instructs the establishment of connections between cells, but it does not specify their precise strength. Rather, experience alters the strength and effectiveness of these preexisting chemical connections (figure 1). Synaptic plasticity as an information storage mechanism is built into the very molecular architecture of chemical synapses.

5.2. Maps in the Brain

As we experience ourselves and the world we live in, representational maps of our knowledge are formed in our brain. Sensory cortical areas represent their respective peripheral receptive fields in a topographic manner. In the somatosensory cortex, maps of the body surface are somatotopic in the sense that neighboring cortical regions respond to adjacent skin site stimulation. Similarly, the auditory and visual cortex are organized according to tonotopic and retinotopic coordinates, respectively. In the last quarter of a century research has shown that even in the adults these maps are not static. Rather, they undergo plastic changes in response to both peripheral manipulations and behaviorally important experience throughout life.

Brain mapping studies have revealed that sensory and motor cortical representation of the body is fluid. Clark and coworkers (1988) posited that cortical maps are formed by the selection of a subset of a large number of inputs and proposed that inputs are selected on the basis of temporal correlation. They tested this idea by altering the correlation of inputs from two adjacent digits on the adult owl monkey hand by surgically connecting the skin surfaces of the two fingers, which increased the correlation of inputs from skin surfaces of adjacent fingers. The characteristic discontinuity between the zones of representation of adjacent digits found in the somatosensory cortex under normal sensory input disappeared, supporting the hypothesis that the topography of the body-surface map in the adult cortex is influenced by the temporal correlations of afferent inputs.

Taking this reseach a step further and a notch up the evolutionary ladder, Mogilner and coworkers (1993) used TMS to reveal changes in cortical representation of the hand and finger area in patients who had undergone successful surgical separation of fingers congenitally fused together in a condition known as syndactily. Prior to surgery, they found abnormal topography of the patients' cortical maps of the deformed hands. One week after surgery, however, this pattern was already substituted by a new, distinct representation of individual fingers, reflecting a new functional status of the hand, which was now able to mobilize each finger independently, and of the patients, who now perceived each finger as a separate entity for the first time in their lives.

Not all stimuli that an individual is exposed to are relevant for survival or merit attention. The frequency of occurence of stimuli or their temporal correlation alone is not sufficient for representational remodelling. Rather it is the behavioral value of stimuli that has been shown to play a significant modulating role in representational plasticity of the mammalian cortex. For example, in a study on the auditory cortex plasticity in rats, Kilgard and coworkers (1998) showed that the nucleus basalis, whose cholinergic neurons are activated as a function of the behavioral significance of stimuli, modulates the direction and extent of cortical representational plasticity. Nucleus basalis is a good candidate subcortical structure for such modulatory effect on cortical representational plasticity as a function of salience of the incoming sensory signals, because it receives input from the limbic and paralimbic structures and projects to the entire cortex (Mesulam et al., 1983; Rye et al.,1984). Thus cortical self-organizational capability is not simply a passive response to the increased pressure of sensory input, but a result of an active selection of behaviorally relevant stimuli that warrant changes in the subject's representation of the external world and self.

Learning of motor skills entails organizing muscle synergies into effective movement sequences (Monfils et al., 2005). Motor cortex representational map topography reflects skilled movement capacity and suggests that maps can adapt in response to motor learning. Indeed, several studies have demonstrated that motor training can induce changes in motor map organization that reflect the nature of the acquired skill. Animal studies done by Nudo and colleagues in squirrel monkeys trained in a skilled digit manipulation task showed that this training caused an expansion of digit representations (Nudo et al., 1996). Interestingly, the expansion was reverted during skill extinction.

Another set of insightful studies on plasticity in skill learning was reported by Pascual-Leone (2001; Pascual-Leone et. al, 1995). Using focal TMS Pascual-Leone and colleagues were able to map the extent of motor cortical area that targets finger muscles and observe its changes due to daily 2 hour practice of playing a predefined fingering sequence on a musical keyboard. The authors observed two types of changes in the size and threshold of the mapped cortical area. First, both the size was increased and threshold reduced significantly when measured 20 to 30 minutes after each practice session, with changes dissipating and returning to baseline when measured the next day before practice. The extent of change peaked during first two weeks of practice and then slowly diminished over weeks 3 to 5. Second, much slower and discrete increases in area size and decreases of threshold measured as baseline each week before practice were observed as well. The changes were minimal in the first three weeks and more pronunced in weeks four and five.

The authors proposed that the observed changes reflect two complementary mechanism of plasticity. The fast, extensive, but transient change in cortical representation can be explained best by recruitment or unmasking of pre-existing synaptic connections. The underlying mechanism most probably being LTP – the already mentioned transient functional change in the excitability of the postsynaptic membrane. The slow and persistent change on the other hand most probably reflects more permanent structural changes brought about either by strengthening of existing synapses, or by sprouting and establishment of new connections.

The most surprising results, however, came in a subsequent study in which the participants were asked to visualize the movement of the fingers rather than actually performing it (Pascual-Leone et. al, 1995). Even though the participants were performing mental practice only, their cortical output maps as measured by TMS showed changes comparable to physical practice group. Albeit to a lesser degree than the physical practice group, the performance of the mental practice group improved significantly both in speed and accuracy compared to no-practice group. Moreover, the performance of mental practice group caught up with the performance of physical training group a single 2hr session at the end of day five, resulting in a jump in performance equal to two days of physical practice.

Put together the studies by Pascual-Leone and colleagues present an insightful demonstration of plasticity in action. Their results are convincing in showing that plastic reorganization of cortex, driven by two separable processes, is cruical for acquisition of novel skills. Additionaly, the results also show that changes in the motor cortex can be brought about by mental practice only.

In summary, adult mammalian cerebral cortex is a highly sophisticated self-organizing system (Singer, 1986), which shapes its effective local connections and responses in accordance with alterations in central and peripheral input, and in response to behavior (Buonomano and Merzenich, 1998). This capacity for reorganization likely accounts for certain forms of perceptual and motor learning. Studies of neuronal mechanisms of cortical reorganization and especially explorations of the relationship between synaptic plasticity and plasticity of representational maps are among the fastest growing areas of research. Specific receptor-dependent forms of synaptic plasticity (LTP and LTD) are likely synaptic mechanisms underlying the experience-dependent regulation of synaptic strength required for reshaping and fine-tuning of the cortical maps. However, a direct causality between synaptic plasticity and cortical map plasticity has only been demonstrated in a few cases (Daw et al., 2007). There are even studies showing a dissociation between LTP/LTD and receptive field plasticity. It is obvious that there is not a simple relationship between synaptic plasticity and cortical map plasticity and more work will be required to determine the precise relationship between these synaptic and systems-level phenomena and the exact mechanisms involved (Daw et al., 2007).

5.3. Adult Neurogenesis

Most tissues of the human organism undergo life-long turnover of their cellular populations. *Neurogenesis* denotes a process of generating neurons from multipotent neural progenitor cells (the same cell population neurons derive from during development), which proliferate, migrate and differentiate into specific neuronal phenotypes. Although neurogenesis has been highly conserved through evolution and occurs in adult nervous systems of both invertebrates and vertebrates (e.g., Altman and Das, 1965), it was long believed that new neurons are not generated in a mature human brain.

A decade ago, however, it was demonstrated that neural progenitor cells give rise to new neurons also in the adult human brain (Eriksson et al., 1998). Adult neurogenesis has been conclusively demonstrated in the dentate gyrus of the human hippocampus (Eriksson, 1998) and it may possibly also occur in the subventricular zone (Curtis, 2007; Sanai et al., 2004). Thus far studies had not revealed any evidence of neurogenesis in the adult human cortex (reviewed in Zhao et al., 2008).

Adult hippocampal neurogenesis is strongly influenced by experience. For example, voluntary exercise and enrichment of animal living environment, which are known to improve the performance of young as well as aged mice in tests of spatial learning and memory, have been shown to increase hippocampal neurogenesis and to promote the survival of new, immature neurons (Kee et al., 2007; Tashiro et al., 2007). Maturation of the newborn neurons has distinct stages and both their survival as well as their full functional development depend on their proper integration and sensory input.

Importantly, it takes specific learning experience and not simply exposure to stimuli to enhance neurogenesis (figure 1). Thus, in the rodent subventricular zone, which "feeds" new neurons into the olfactory bulb, the neurogenesis is regulated by the animal's olfactory experience (reviewed by Lledo et al., 2006). More young neurons survived in the olfactory bulb of mice who learned an odor discrimination task (Alonso et al., 2006) than in the olfactory bulb of those mice who were simply exposed to odors. In other words, similar to the case of cortical representational map plasticity mentioned above, it is the behavioral value of the stimuli that modulates the expression of the inherent neurogenic potential of an adult brain.

In contrast to stimulatory effects of voluntary physical exercise and enriched environment, hippocampal neurogenesis can also be suppressed, most notably by aging and stress (reviewed in Zhao et al., 2008), revealing the reversible regulatory effects of external stimuli on this form of neural plasticity.

6. Lesions of the Nervous System: Challenge to Neural Plasticity

Adult mammalian brain's intricate circuitry was long considered to be fixed as specified by genome and to remain stable throughout life (Ramachandran and Hirstein, 1998). It was believed that no new neural cells can be added and no new connections can be formed beyond early development (fetal period and certain critical periods of early infancy). Indeed, even the "father of modern neuroscience", Ramon y Cajal – who posited the concept of synaptic plasticity as the basis of learning – concluded in his later years based on his studies of central nervous system (CNS) response to injury: "*...once the development has ceased, the springs of growth and regeneration irrevocably dry out. In the adult central nervous system the neural connections are fixed, final and immutable. Everything can die. Nothing can be renewed. It will be up to science to reverse this cruel verdict, if possible.*" (Santiago Ramon y Cajal, 1928).

However, a rapidly expanding body of research in the recent decades has revealed an impressive potential of the mammalian brain for structural plasticity, expressed in phenomena of regenerative growth, rewiring of the existing connections and adult neurogenesis – all of these having a potential to affect functional recovery after injury or in disease.

6.1. Sprouting, Regeneration and Rewiring in an Injured Mammalian Brain

If a peripheral nerve is injured and the continuity of its axons is interrupted, it can be repaired. If the proximal and distal stump of the nerve are apposed, a number of axonal sprouts from the proximal stump traverse the gap and lead the axonal regeneration into and through the distal nerve stump toward the denervated targets. They can grow at a speed measured in milimeters per day. In doing so, they are assisted by the components of the peripheral nerve micro-environment: the proliferating Schwann cells (peripheral glia) produce chemoattractant molecule gradient, the basal lamina tubes and Schwann cell surfaces offer growth promoting adhesive cues, while the non-permissive perineurium restricts axonal growth outside the nerve sheath. In this manner, the targets are often reinnervated and function is restored to a variable, yet often quite impressive degree. The process is not perfect: in order to successfully restore lost function, sufficient number of functional synapses need to form between regenerating fibers and their targets. Some injured neurons die, and many axons end up misguided while crossing the gap into the distal stump leading to suboptimal functional recovery. Nevertheless, peripheral nervous system demonstrates remarkable capacity for structural and functional plasticity and repair.

In stark contrast, injury to the neural pathways of the adult mammalian CNS gives rise to little spontaneous restoration, leaving the victim of a CNS injury with little hope for recovery.

From the early experiments of Tello (1911, cited by Ramon y Cajal, 1928) and later work of David and Aguayo (1981) it has been known that the transected central nervous system axons sprout, grow long distances and even establish new functioning synapses with their targets when peripheral nerve segments were used as "bridges" to bypass the natural environment of the mammalian CNS. The *milieu* provided to the axonal sprouts by the CNS seemed far less permissive than the peripheral nerve bridges. Exploring the differences between the central and peripheral nervous system environment further, Schwab and Thoenen (1985) opened the way to discovery of several potent growth-inhibitory components of the CNS myelin. It seemed as if the intricate complexity of the CNS connections established during development needs to be maintained at a cost of the frustrated or abortive capacity for the system to repair itself.

Recent results from human and animal studies suggest that under certain conditions CNS is capable of significant spontaneous reorganization (Nudo, 1999; Napieralski et al., 1996; Carmichael et al., 2005). Extensive cortical rewiring involving axonal sprouting near the site of injury and establishment of novel connections within a distal target had been demonstrated in adult squirrel monkey brain following an ischemic injury (Dancause et al., 2005). Most significantly, human data are now becoming available through the use of new techniques, such as diffusion tensor imaging (DTI), which enables studying the density, diameter and geometry of myelinated fibers in the CNS. Employing DTI, Voss and coworkers (2006) found changes in intracortical and cerebellar white matter connectivity suggestive of extensive white matter reorganization over the course of 18 months in a patient who regained consciousness and recovered reliable communication after 19 years in a minimally conscious state.

Together these findings point to the need for prospective and longitudinal studies of neuroplasticity following CNS injuries and insults, and they provide hope that functional recovery can be attributed to and possibly facilitated by careful enhancement of brain rewiring.

6.2. Neurogenesis in the Context of Injury or Disease of the Nervous System

Many neurological disorders are characterized by neuronal death. As noted above, it had been demonstrated that neurogenesis spontaneously occurs in certain regions of a normal adult mammalian brain. A growing body of research findings suggest that changes in the nervous system parenchyma brought about by injury or disease are likely to influence formation of new neurons in the affected brain. This in turn may affect the course and functional outcome of the disorder.

Increased neurogenesis has been seen in rodent models of stroke (Zhao et al., 2008) and a similar process appears to happen in human stroke patients (Shen et al., 2008), but to date no studies have shown the degree to which this may influence the functional recovery.

Epileptic seizures induce neurogenesis. Interestingly, hippocampal neurogenesis may be increased for weeks after a seizure, but eventually declines to basal state levels or lower (Zhao et al., 2008). Also, many newborn neurons fail to mature properly and do not successfully integrate into the neural circuitry.

Alzheimer disease – despite its prominent loss of neurons overall – has also been associated with increased hippocampal neurogenesis (Jin et al., 2004). These findings are a good example that enhancement of neural plasticity need not have functional significance if the new neurons are not exposed to favorable environment and successfully integrated into the neural circuitry.

Neuroplasticity expressed as adult neurogenesis, from birth of new neurons to their successful functional integration, is strongly influenced by external factors. The newborn neurons likely play a role in behavior in health and disease, but the full extent of their role remains to be determined.

6.3. Maladaptive Plasticity

Under certain pathological circumstances molecular and structural mechanisms of rapid learning and information storage may be "hijacked" by processes resulting in disfunctional patterns of neural activation and maladaptive behavior – as seen in the case of *phantom limb pain*.

Following a peripheral nerve transection and reconnection, or following a localized peripheral nerve crush, the cortical representation of the body part innervated by that nerve initially becomes silent and then grows responsive to the input originating in the body surface area supplied by adjacent uninjured peripheral nerves, which may sprout and innervate the denervated body surface area - a phenomenon known as *collateral sprouting*. If the crushed or reconnected peripheral nerve successfully regenerates and reinnervates its appropriate peripheral targets, the previously deafferented cortical maps slowly revert to normal. Regenerative sprouting, axon elongation, and appropriate target reinnervation are more successful after a nerve crush than following transection (or other lesions causing structural discontinuity of the nerve) also leading to a better reconstruction of the normal cortical representation.

Following amputation of an extremity (or a body part), when transected nerves have no targets to grow to and reinnervate, a *phantom* of the missing extremity can be perceived by the amputee. The phantom can be very vivid: it can assume odd positions in respect to the body, it can itch, but above all it may be excruciatingly painful. The incidence of severe phantom pain poses a clinical problem, as 70% of phantoms have been found to remain painful 25 years post-amputation. The patient recognizes that the sensations are not veridical – he/she is not dellusional. They know they are experiencing an illusion (Ramachandran and Hirstein, 1998). Phantom limb pain has beleaguered its victims and intrigued clinicians and scientists for ages. Lord Nelson (1758 – 1805), who lost his right arm in battle and suffered from a painful phantom arm, saw in this ghostly sensation a "direct proof" for the existence of the soul: "*If an arm can survive physical anihilation, why not the whole person*?" (Ramachandran and Rogers-Ramachandran, 2000).

Phantom limb sensations have been loosely attributed to hypersensitive transected axons forming a blindly ending stump neuroma, or to activation changes in the "diffuse neural matrix" (Melzack, 1990). However, more recently, amputation of an extremity has been shown to lead to massive reorganization in the adult human somatosensory cortex (Ramachandran, 1993). Thus, for example, after amputation of an arm, sensory input from the face was shown to activate the hand area of the primary somatosensory cortex. Sensory input from the face went to two different cortical areas: the original "face area" and the area that earlier received information from the arm. In terms of space, this reorganization of the cortical map was found to cover a distance of 2 – 3 cm. Most interestingly, the referred sensations were modality-specific: hot, cold, vibration, rubbing, metal, or massage were felt as hot, cold, vibration, rubbing, metal, and massage at precisely localized points on the phantom limb (Ramachandran, 1993). These results pointed to the possibility that sensations referred to the phantom limb from the facial "trigger area" emerged from the changes in cortical representational topography following amputation — an idea referred to by the author as the "remapping hypothesis" (Ramachandran, 1993), a prominent example of *maladaptive neuroplasticity*. On the cellular level, it was attributed largely to unmasking of preexisting connections in the somatosensory cortex rather than sprouting and formation of new connections based on the observation that modality-specific referral from the face to the phantom limb can occur even just a few hours after amputation (Borsook rt sl., 1998). Mechanisms, which could explain such rapid maladaptive plasticity could involve acute injury-related changes in neuronal membrane excitability, removal of local inhibition, and/or various forms of short- or long-term synaptic plasticity. A neuropharmacological study in healthy volunteers, where deafferentation effects of amputation were simulated by forearm ischemic nerve block (INB) combined with low-frequency repetitive transcranial magnetic stimulation (rTMS), provided evidence that acute increase in excitability of the motor cortex following INB with rTMS involves rapid removal of gamma-aminobutiric acid (GABA)-related cortical inhibition (allowing weak existing synapses to become disinhibited) and short-term changes in synaptic efficacy, dependent on Na^+ and Ca_2^+ channels (Ziemann et al., 1998). Although "remapping hypothesis" does not succeed in explaining all aspects of the phantom limb phenomena, it offers the best explanation to date of the central aspects of the phantom limb pain.

Interestingly, the plastic nature of the nervous system provides a remarkable degree of reversibility even in this setting. An ingenious technique, employing a vertical mirror in a box, which reflected the image of the healthy arm into the space of the missing other limb, enabled a proportion of the forearm amputees to use visual sensations to learn to relieve the painful spasms in the phantom limb (Ramachandran and Hirstein, 1998).

7. The Shape of Things to Come: Harnessing Neural Plasticity?

7.1. Manipulating Neuroplasticity for Treatment of Neurological and Psychiatric Disorders

The evidence that adult neurogenesis can be augmented by external factors from exercise to antidepressant medication (Malberg et al, 2000) suggests a possibility for creating specific strategies by which this process can be modified *in vivo* to restore neuronal population lost through disease or injury. Similar conclusion applies to harnessing other forms of plasticity, such as neural connectivity, rewiring and regenerative growth in attempts to relieve conditions spanning from phantom limb pain to trauma-related long term reduced consciousness (e.g., use of deep brain stimulation of the thalamus by Schiff and co-workers to enable a patient regain communication after 6 years in minimally conscious state (2007)). Several clinical trials of exogenous trophic factor delivery into brains of patients with neurodegenerative diseases, with the aim to enhance their inherent neuroplasticity and restorative potential had been attempted in the last two decades. To date none of these attempts resulted in a desired clinical improvement, but it appears that this may be due primarily to the issues relating to compound delivery and potential side effects than to the failure of the premise that carefully and selectively stimulated neuroplasticity can lead to functional improvement in humans, as has been demonstrated in animals (Fallon et al., 2000; Nakatomi et al., 2002). The search for successful strategies to best exploit the neuroplastic potential of the human brain are currently focusing on controlling molecular switches that regulate this process at one end of the neuroplasticity research field (i.e., molecular neuroscience), and on patterns of experience that can reproducibly favorably modify the phenotype (e.g., in the course of neuro-rehabilitation) at the other end.

7.2. Brain-Machine Interface

Both in science and in fiction, an interest in replacing lost sensory and motor abilities as well as extending the existing ones has a long tradition. Under the title of neuroprosthetics, brain-computer interfaces (BCI) or brain-machine interfaces (BMI), a number of research teams worldwide are actively developing methods of providing nervous system with alternative sources of sensory input and interpreting neural activity to guide computers and machines. Both of these goals depend crucialy on the plasticity of the human nervous system, which on one hand opens exciting possibilities and on the other presents significant challenges.

There are many different varieties of BMI and approaches to their development. For the purpose of this chapter we will arbitrarily divide them into direct and indirect interfaces, taking into account that many of them fall somewhere in between. As direct interfaces we are cosidering those devices that try to directly produce or interpret the relevant neuronal code. In case of visual neuroprostheses the device is supposed to replace the disfunctional part of the visual system by providing direct stimulation of the nervous system further down the visual pathway, such as stimulating retinal ganglion cells or input layer of the primary visual cortex (for a review see Merabet et. al, 2005). In the motor domain, the aim of the interface is to record the firing pattern of neurons in primary motor cortex, interpret the intended motion and drive the prosthetic device (e.g., Carmena et. al, 2003, Velliste et. al, 2008). Similarly, fMRI have been recently successfully used to deduce motor intentions from brain activity using complex pattern matching alghorhythms (e.g., Kamitani and Tong, 2006).

In both described cases the intention is to adjust the interface to the normal neural system and directly and seamlesly replace the missing or disfunctional sensory or motor system. For that reason such devices should be only minimaly dependent on neuroplasticity. In reality, however, significant technical difficulties still stand in the way of long term robust measurment of the activity of sufficient number of individual neurons or precise excitation of the relevant neurons in the sensory pathways. In both cases the experience has shown that subjects through process of neuroplasticity actively adjust either the firing of motor neurons or interpretation of stimulated input and optimize the performance of the interface. Empirical results have for instance shown, that despite the reduced number of functional recordings, the precision of the use of a motor BMI significantly improved over time (Carmena et. al, 2003). The improvement was accompanied by functional reorganization of multiple cortical areas, suggesting that BMI is being incorporated into sensory and motor cortical representations.

Direct interfaces can on the other hand be significantly impaired by neuroplasticity. As described earlier, in the absence of relevant input, neuroplasticity leads to recruitment of deafferented cortex by other sensory modalities and cognitive processes. In these cases, direct stimulation of the cortex would not elicit the desired experience but can rather lead to perceptions in other sensory modalities or disturbances in cognitive processing that are now being represented in that part of the cortex, as shown by TMS studies of the early blind described above (Amedi et. al, 2004; Cohen et. al, 1997). In these cases the brain has to re-learn the appropriate processing and interpretation of the incoming information. Such re-learning can be supported by training based on providing relevant feedback or concordant information in other modalities (for more information see Merabet et. al, 2005).

Due to the immense plasticity of the human brain our options in developing BMI are however not limited to reproducing and interpreting the original neural code in developing BMI. A number of systems - which we call indirect interfaces - have been developed that depend on the ability of the human brain for cross-modal plasticity. On the input side, such systems build on the ability to reinterpret stimuli from one modality in another modality (sensory substitution), whereas on the output side, they depend on the ability to recode desired motor action into another, more easily detectable neural response. In both cases the systems heavily depend on brain plasticity and training suported by immediate feedback.

As indirect systems are less invasive, they have been more extensively studied, explored and used in human subjects. In the sensory domain, pioneered by Bach-y-Rita (e.g., Bach-y-Rita et al., 1969), a wide range of systems employing sensory substitution were developed. Most common are systems recoding visual information into tactile (for a review see Bach-y-

Rita, 2004) or auditory (Amedi et. al, 2007; Meijer, 1992) modality, but they were also successfuly employed in remapping tactile information to restore peripheral sensation (Bach-Y-Rita, 1999) and providing alternative input of vestibular information (Tyler, Danilov & Bach-Y-Rita, 2003).

In the ouput domain significant progress has been made with systems for recording large scale neuronal activity such as electroencephalography (EEG; Iversen et al., 2008), magnetoencephalography (MEG; Buch et al., 2008; Mellinger et. al, 2007) and functional magnetic resonance imaging (fMRI; Sitaram et. al, 2007). In such cases subjects, provided with relevant feedback, are asked to consciously manipulate some aspect of neuronal processing such as modulating MEG sensorimotor *mu* and *beta* rhythms (Buch et. al, 2007; Mellinger et. al, 2007) or amplitude of slow EEG cortical potentials (Iversen et. al, 2008) to adjust the position of a cursor on a computer screen, or to manipulate the activity of a specific brain region (for a review see Sitaram et. al, 2007).

Be it direct or indirect systems, animal studies as well as human trials repeatedly show that brain machine interfaces are relatively quickly and smoothly integrated in the subjects' motor body schema, sensory representations and subjective experience. In this way they present yet one more compelling demonstration of the ability and potential of the brain to dynamically mold itself to the information and challenges presented by the environment.

7. 3. Brain Enhancement

When the aim of a procedure or a drug is to correct or alleviate a deficit or specific pathology, the intervention is considered *therapeutic*. If the intervention is not repairing a damage or treating a deficit, but is aimed at improving a functional system, it can be considered an *enhancement*. Modern society faced with population aging in parallel with ever increasing demands of the accelerated technological development generates a growing need for strategies that would maintain and possibly improve its members' ability to cope and be efficient.

Cognition is a summary word for processes an organism uses to organize information. It includes acquiring, selecting, representing (aspects of learning) and retaining (memory) information, and using it to guide behavior (reasoning and coordination of motor outputs). As illustrated in the examples listed in this chapter, human cognitve function hinges on neural plasticity, and improving cognition would imply better utilization of the brain's plastic potential.

As a principle, cognitive function gradually declines with age, with pathological decrease in cognition being most prevalent in the people over 60 years. Interventions to improve cognitive function may be directed at any one of the core faculties involved in cognition and seem to be of interest to all age groups, as they may significantly improve all aspects of life (e.g., scholastic achievement, employment and social success of an individual). There are numerous ways in which cognition of an adult organism has been shown to be improved from laboratory animals to humans, starting from care for general health, education and enrichment of the living environment, mental training, drugs, brain stimulation with transcranial magnetic pulses, brain-machine interaction, etc.

Advances in neuroscience and better insight into possibilities to "mold the brain in order to shape the mind" accelerate the potential for cognitive enhancement to become an ordinary and everyday thing, but like wealth it would not be distributed equally in society. Better access to effective cognitive enhancement in certain socio-economic groups can thus be expected to increase inequalities and raise social tensions. In addition, manipulating our brains to modify our cognitive abilities, personality traits and subjective experiences raises moral and ethical concerns that modern society will yet have to address. Recent birth of *neuroethics* as a separate field at the intersection of bioethics and neuroscience devoted to research and discussion of ethical issues associated with mind and behavior (Illes and Bird, 2006) instills hope that both the neuroscientists and the society are aware of the challenges lying ahead.

REFERENCES

Alonso, M., Viollet, C., Gabellec, M.M., Meas-Yedid, V., Olivo-Marin, J.C. and Lledo, P.M. (2006). Olfactory discrimination learning increases the survival of adult-born neurons in the olfactory bulb. *J. Neurosci.,* 26, 10508–10513.

Altman, J. and Das, G.D. (1965). Autoradiographic and histological evidence of postnatal hippocampal neurogenesis in rats. *J. Comp. Neurol.,* 124, 319–335.

Amedi, A., Floel, A., Knecht, S., Zohary, E. & Cohen, L. (2004). Transcranial magnetic stimulation of the occipital pole interferes with verbal processing in blind subjects. *Nature Neuroscience,* 7 (11), 1266-70.

Amedi, A., Raz, N., Pianka, P., Malach, R. & Zohary, E. (2003). Early 'visual' cortex activation correlates with superior verbal memory performance in the blind. *Nature Neuroscience,* 6 (7), 758-66.

Amedi, A., Stern, W., Camprodon, J., Bermpohl, F., Merabet, L., Rotman, S. et al. (2007). Shape conveyed by visual-to-auditory sensory substitution activates the lateral occipital complex. *Nature Neuroscience,* 10 (6), 687-9.

Azmitia, E.C. (2007). Cajal and brain plasticity: insights relevant to emerging concepts of mind. *Brain Res. Rev.,* 55 (2), 395-405.

Bach-Y-Rita, P. (2004). Tactile sensory substitution studies. Annals of the New York Academy of Sciences, 1013, 83-91.

Bach-y-Rita, P. (1999). Theoretical aspects of sensory substitution and of neurotransmission-related reorganization in spinal cord injury. *Spinal cord : the official journal of the International Medical Society of Paraplegia,* 37 (7), 465-74.

Bach-y-Rita, P., Collins, C., Saunders, F., White, B. & Scadden, L. (1969). Vision substitution by tactile image projection. *Nature,* 221 (5184), 963-4.

Barborič, M., Lenasi, T., Lovšin, N., Ule, J. (2009). *The genes and the junk: recent advances in the studies of gene regulation?* (this volume)

Black, J.E. & Greenough, W.T. (1986). Induction of pattern in neural structure by experience: Implications for cognitive development. In M.E. Lamb, A.L. Brown and B Rugoff (Eds.) Advances in Developmental Psychology Volume 4. Lawrence Erlbaum Associates, Hillsdale.

Boahen, K. (2005). Neuromorphic Microchips. *Scientific American,* 292 (5), 56-63.

Borsook, D., Becerra, L., Fishman, S., Edwards, A., Jennings, C.L., Stojanovic, M., Papinicolas, L., Ramachandran, V.S., Gonzalez, R.G., Breiter, H. (1998). Acute plasticity in the human somatosensory cortex following amputation. *NeuroReport,* 9, 1013–1017.

Buch, E., Weber, C., Cohen, L., Braun, C., Dimyan, M., Ard, T. et al. (2008). Think to move: a neuromagnetic brain-computer interface (BCI) system for chronic stroke. *Stroke; a journal of cerebral circulation,* 39 (3), 910-7.

Büchel, C., Price, C., Frackowiak, R.S., Friston, K. (1998). Different activation patterns in the visual cortex of late and congenitally blind subjects. *Brain.* 121(Pt 3), 409-19.

Buonomano, D.V., Merzenich, M.M. (1998). Cortical plasticity: from synapses to maps. *Annu. Rev. Neurosci.,* 21, 149-86.

Burton, H., Snyder, A., Diamond, J. & Raichle, M. (2002). Adaptive changes in early and late blind: a FMRI study of verb generation to heard nouns. *Journal of neurophysiology,* 88 (6), 3359-71.

Carmena, J., Lebedev, M., Crist, R., O'doherty, J., Santucci, D., Dimitrov, D. et al. (2003). Learning to control a brain-machine interface for reaching and grasping by primates. *PLoS biology,* 1 (2), E42.

Carmichael, S.T. et al. (2005). Growth-associated gene expression after stroke: evidence for a growth promoting region in peri-infarct cortex. *Exp. Neurol.*, 193, 291–311.

Clark, S.A., Allard, T., Jenkins, W.M., Merzenich, M.M. (1988). Receptive fields in the body-surface map in adult cortex defined by temporally correlated inputs. *Nature,* 332, 444–445.

Cohen, L., Celnik, P., Pascual-Leone, A., Corwell, B., Falz, L., Dambrosia, J. et al. (1997). Functional relevance of cross-modal plasticity in blind humans. *Nature,* 389 (6647), 180-3.

Collignon, O., Voss, P., Lassonde, M. & Lepore, F. (2008, Sep 2). Cross-modal plasticity for the spatial processing of sounds in visually deprived subjects. *Experimental Brain Research.*

Curtis, M.A., Kam, M., Nannmark, U., Anderson, M.F., Axell, M.Z., Wikkelso, C., Holtas, S., van Roon-Mom, W.M., Bjork-Eriksson, T., Nordborg, C. et al. (2007). Human neuroblasts migrate to the olfactory bulb via a lateral ventricular extension. *Science,* 315, 1243–1249.

Dancause, N. et al. (2005). Extensive cortical rewiring after brain injury. *J. Neurosci.,* 25, 10167–10179.

David, S., Aguayo, A.J. (1981). Axonal elongation into peripheral nervous system bridges after CNS injury in adult rats. *Science,* 214, 931-3.

Daw, M.I., Scott, H.L., Isaac, J.T.R. (2007). Developmental Synaptic Plasticity at the Thalamocortical Input to Barrel Cortex: Mechanisms and Roles. *Mol. Cell Neurosci.,* 34(4), 493–502.

Edwards, D. H., Heitler, W. J. & Krasne, F. B. (1999). Fifty years of command neurons: the neurobiology of escape behavior in the crayfish. *Trends Neurosci.,* 22, 153–161.

Eriksson, P.S., Perfilieva, E., Björk-Eriksson, T., Alborn, A.M., Nordborg, C., Peterson, D.A., Gage, F.H. (1998) Neurogenesis in the adult human hippocampus. *Nat. Med.,* 4(11), 1313-7.

Fallon, J., Reid, S., Kinyamu, R., Opole, I., Opole, R., et al. (2000). In vivo induction of massive proliferation, directed migration, and differentiation of neural cells in the adult mammalian brain. *Proc. Natl. Acad. Sci. USA.* 97:14686–91.

Hebb, D.O. (1949). The Organization of Behavior. Wiley, New York.

Hubel, D. & Wiesel, T. (1970). The period of susceptibility to the physiological effects of unilateral eye closure in kittens. *The Journal of Physiology,* 206 (2), 419-36.

Hubel, D.H. and Wiesel, T.N. (2005). Brain and Visual Perception. Oxford University Press, New York.

Illes, J. and Bird, S. J. (2006). Neuroethics: a modern context for ethics in neuroscience. *Trends Neurosci.* 29(9),511-7.

Iversen, I., Ghanayim, N., Kübler, A., Neumann, N., Birbaumer, N., & Kaiser, J. (2008). A brain-computer interface tool to assess cognitive functions in completely paralyzed patients with amyotrophic lateral sclerosis. *Clinical neurophysiology : official journal of the International Federation of Clinical Neurophysiology,* 119 (10), 2214-23.

Kamitani, Y., & Tong, F. (2006). Decoding seen and attended motion directions from activity in the human visual cortex. *Current biology* : CB, 16 (11), 1096-102.

Kandel, E. R., Schwartz, J.H., Jessel, T.M. (2000). Principles of Neural Science. McGraw-Hill Medical, 4th ed.

Kandel, E. R. (2001). The molecular biology of memory storage: a dialogue between genes and synapses. *Science,* 294, 1030-1038.

Kee, N., Teixeira, C.M., Wang, A.H., and Frankland, P.W. (2007). Preferential incorporation of adult-generated granule cells into spatial memory networks in the dentate gyrus. *Nat. Neurosci.,* 10, 355–362.

Kilgard, M.R., Merzenich, M.M. (1998). Cortical Map Reorganization Enabled by Nucleus Basalis Activity. *Science,* 279, 1714 – 8.

Konorski, J. (1948). Conditioned reflexes and neuron organization. Cambridge University Press, Cambridge.

Kral, A. (2007). Unimodal and cross-modal plasticity in the 'deaf' auditory cortex. *International Journal of Audiology,* 46 (9), 479-93.

Kupfermann, I. and Weiss, K. R. (1978). The command neuron concept. *Behav. Brain Sci.,* 1, 3–39.

Lledo, P.M., Alonso, M. and Grubb, M.S. (2006). Adult neurogenesis and functional plasticity in neuronal circuits. *Nat. Rev. Neurosci.*, 7, 179–193.

Malberg, J.E., Eisch, A.J., Nestler, E.J., and Duman, R.S. (2000). Chronic Antidepressant Treatment Increases Neurogenesis in Adult Rat Hippocampus. *J. Neurosci.* 20(24):9104-9110.

Meijer, P. (1992). An experimental system for auditory image representations. *IEEE transactions on bio-medical engineering,* 39 (2), 112-21.

Mellinger, J., Schalk, G., Braun, C., Preissl, H., Rosenstiel, W., Birbaumer, N. et al. (2007). An MEG-based brain-computer interface (BCI). *NeuroImage,* 36 (3), 581-93.

Melzack, R. (1990). Phantom limbs and the concept of a neuromatrix. *Trends Neurosci.*,13, 88-92.

Merabet, L., Rizzo, J., Amedi, A., Somers, D. & Pascual-Leone, A. (2005). What blindness can tell us about seeing again: merging neuroplasticity and neuroprostheses. *Nature reviews Neuroscience,* 6 (1), 71-7.

Merolla, P., & Boahen, K. (2003, Jan 1). A recurrent model of orientation maps with simple and complex cells. Advances in neural information processing systems, proceedings of Neural Information Processing Systems Conference.

Mesulam, M.M., Mufson, E.J., Wainer, B.H., Levey, A.I. (1983). Central holinergic pathways in the rat: an overview based on an alternative nomenclature (Ch1-Ch6). *Neuroscience,* 10(4), 1185-201.

Mogilner, A., Grossman, J.A.I., Ribary, U., Joliot, M., Volkmann, J., Rapaport, D., Beasley, R.W., Llinas, R.R. (1993). Somatosensory cortical plasticity in adult humans revealed by magnetoencephalography. *Proc. Natl. Acad. Sci. U.S.A.* 90, 3593-7.

Monfils, M.-H., Plautz, E.J., Kleim, J.A. (2005). In Search of the Motor Engram: Motor Map Plasticity as a Mechanism for Encoding Motor Experience. *Neuroscientist,* 11, 471-83.

Nakatomi H, Kuriu T, Okabe S, Yamamoto S, Hatano O, et al. (2002). Regeneration of hippocampal pyramidal neurons after ischemic brain injury by recruitment of endogenous neural progenitors. *Cell.* 110:429–41.

Napieralski, J.A., Butler, A.K. and Chesselet, M.F. (1996). Anatomical and functional evidence for lesion–specific sprouting of corticostriatal input in the adult rat. *J. Comp. Neurol.*, 373, 484–497.

Nudo, R.J., Milliken, G.W., Jenkins, W.M., Merzenich, M.M. (1996). Use dependent alterations of movement representations in primary motor cortex of adult squirrel monkeys. *J. Neurosci.,* 16, 785–807.

Nudo, R.J. (1999) Recovery after damage to motor cortical areas. *Curr. Opin. Neurobiol,* 9, 740 –747.

Ofan, R. & Zohary, E. (2007). Visual cortex activation in bilingual blind individuals during use of native and second language. *Cerebral cortex,* 17 (6), 1249-59.

Pascual-Leone, A. (2001). The brain that plays music and is changed by it. Annals of the New York Academy of Sciences, 930, 315-29.

Pascual-Leone, A., Nguyet, D., Cohen, L., Brasil-Neto, J., Cammarota, A., & Hallett, M. (1995). Modulation of muscle responses evoked by transcranial magnetic stimulation during the acquisition of new fine motor skills. *Journal of neurophysiology,* 74 (3), 1037-45.

Ramachandran, V.S. (1993) Behavioral and MEG correlates of neural plasticity in the adult human brain. *Proc. Natl. Acad. Sci. U. S. A.* 90, 10413-10420.

Ramachandran, V.S. and Hirstein, W. (1998). The perception of phantom limbs. The D. O. Hebb lecture. *Brain,* 121(9), 1603-1630.

Ramachandran, V.S., Rogers-Ramachandran D. (2000). Phantom limbs and neural plasticity. *Arch. Neurol.*, 57, 317-320.

Ramon y Cajal, S. (1928). Degeneration and Regeneration of the Nervous System. Translated and Edited by Raoul M. May. (1959). Hafner Publishing Co, NY.

Rye, D.B., Wainer, B.H., Mesulam, M.M., Mufson, E.J., Saper, C.B. (1984). Cortical projections arising from the basal forebrain: a study of cholinergic and noncholinergic components employing combined retrograde tracing and immunohistochemical localization of choline acetyltransferase. *Neuroscience,* 13(3), 627-43.

Sadato, N., Pascual-Leone, A., Grafman, J., Ibañez, V., Deiber, M., Dold, G. et al. (1996). Activation of the primary visual cortex by Braille reading in blind subjects. *Nature,* 380 (6574), 526-8.

Sanai, N., Tramontin, A.D., Quinones-Hinojosa, A., Barbaro, N.M., Gupta, N., Kunwar, S., Lawton, M.T., McDermott, M.W., Parsa, A.T., Manuel-Garcia Verdugo, J. et al. (2004). Unique astrocyte ribbon in adult human brain contains neural stem cells but lacks chain migration. *Nature*, 427, 740–744.

Schiff, N.D., Giacino, J.T., Kalmar, K., Victor, J.D., Baker, K., Gerber, M., Fritz, B. et al. (2007). Behavioural improvements with thalamic stimulation after severe traumatic brain injury. *Nature.* 448(7153):600-3.

Schwab, M.E. and Thoenen, H. (1985). Dissociated neurons regenerate into sciatic but not optic nerve explants in culture irrespective of neurotrophic factors. *J. Neurosci,* 5, 2415-23.

Shen, J., Xie, L., Mao, X., Zhou, Y., Zhan, R., Greenberg, D.A., Jin, K. (2008). Neurogenesis after primary intracerebral hemorrhage in adult human brain. *J. Cereb. Blood Flow Metab.,* 28(8), 1460-8.

Singer, W. (1986). The brain as a self-organizing system. *Eur. Arch. Psych. Neurol. Sci.*, 4, 236.

Sitaram, R., Caria, A., Veit, R., Gaber, T., Rota, G., Kuebler, A. et al. (2007, Jan 1). FMRI brain-computer interface: a tool for neuroscientific research and treatment. Computational intelligence and neuroscience: *CIN,* 25487.

Taba, B. & Boahen, K. (2006, Jan 1). Silicon growth cones map silicon retina. Advances in neural information processing systems, proceedings of Neural Information Processing Systems Conference.

Taba, B. & Boahen, K. (2003, Jan 1). Topographic Map Formation by Silicon Growth Cones. Advances in neural information processing systems, proceedings of Neural Information Processing Systems Conference.

Tashiro, A., Makino, H., and Gage, F.H. (2007). Experience-specific functional modification of the dentate gyrus through adult neurogenesis: a critical period during an immature stage. *J. Neurosci.,* 27, 3252–3259.

Tyler, M., Danilov, Y. & Bach-Y-Rita, P. (2003). Closing an open-loop control system: vestibular substitution through the tongue. *Journal of integrative neuroscience*, 2 (2), 159-64.

Velliste, M., Perel, S., Spalding, M., Whitford, A. & Schwartz, A. (2008). Cortical control of a prosthetic arm for self-feeding. *Nature,* 453 (7198), 1098-101.

Voss, H.U. et al. (2006). Possible axonal regrowth in late recovery from the minimally conscious state. *J. Clin. Invest.,* 116, 2005–2011.

Weeks, R., Horwitz, Aziz-Sultan, A., Tian, B., Wessinger, C., Cohen, L. et al. (2000). A positron emission tomographic study of auditory localization in the congenitally blind. The Journal of neuroscience : *the official journal of the Society for Neuroscience,* 20 (7), 2664-72.

Wiersma, C.A.G. and Ikeda, K. (1964). Interneurons commanding swimmeret movements in the crayfish, Procambarus clarkii (Girard). *Comparative Biochemistry and Physiology,* 12, 509-525.

Wiesel, T. & Hubel, D. (1965). Extent of recovery from the effects of visual deprivation in kittens. *Journal of neurophysiology,* 28 (6), 1060-72.

Zhao, C., Deng, W., Gage, F.H. (2008). Mechanisms and functional implications of adult neurogenesis. *Cell,* 132, 645–660.

Ziemann, U., Hallett, M., Cohen, L.G. (1998). Mechanisms of Deafferentation-Induced Plasticity in Human Motor Cortex. *J. Neurosci,* 18(17), 7000-7007.

In: Philosophical Insights about Modern Science
Eds: Eva Žerovnik, Olga Markič and Andrej Ule

ISBN: 978-1-60741-373-8

Chapter 2

THE GENES AND THE JUNK: RECENT ADVANCES IN THE STUDIES OF GENE REGULATION

Matjaž Barborič[1], Tina Lenasi[1], Nika Lovšin[2] and Jernej Ule[3]

[1] Departments of Medicine, Microbiology, and Immunology, Rosalind Russell Medical Research Center, University of California at San Francisco, San Francisco, California 94143-0703, USA

[2] Department of Biochemistry, Faculty of Chemistry and Chemical Technology, University of Ljubljana, Ljubljana, Slovenia

[3] MRC Laboratory of Molecular Biology, Cambridge, United Kingdom

SYNOPSIS

The rapid progress of DNA sequencing technologies in last decade has led to sequencing of many genomes including that of our own species, which greatly changes our view of genes and their evolution. It was found that only 1.5% of our DNA encodes proteins, the cellular machines and structural components. Vast regions of the DNA have at first sight no apparent function, and were often referred to as "junk". However, detailed analyses of the genome have in recent years revealed hidden codes in these DNA regions, which serve to control expression of nearby genes, contain self-replicating genetic elements or express RNA that does not encode proteins, but rather regulates expression of other genes. While the repetitive genomic elements selfishly replicate themselves, they shuffle around pieces of genes and thereby increase the speed of evolution. These elements use similar mechanisms of self-replication as viruses such as HIV, which are able to circumvent our immune system by hiding themselves inside our genome. Interestingly, our genome produces a large number of diverse RNA molecules that do not encode proteins, so large that they are sometimes referred to as the 'dark matter' of molecular biology. In neurons, transport of RNA molecules to dendrites lays a foundation for the mechanism underlying memory. Finally, differential control of gene expression appears to be the primary factor distinguishing chimpanzee and human genomes. Understanding the way genes are expressed in our cells thus uncovers a wealth of knowledge about ourselves that lays hidden within our genome.

INTRODUCTION

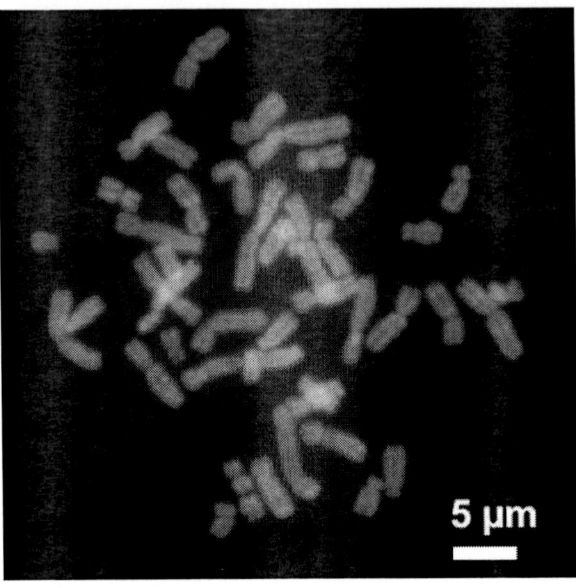

Picture. Human chromosomes. In the early stages of cell division, the chromosomal DNA ceases to transcribe into RNA and instead condenses to allow its own transport into the two daughter cells. In this compact form each chromosome is visible with an optical microscope as a four-arm structure. (from http://en.wikipedia.org/wiki/Image:HumanChromosomesChromomycinA3.jpg).

A century ago Thomas Hunt Morgan provided us with the basis of how to imagine our genome by showing that genes reside on chromosomes (Morgan, 1911). However, the question of what a gene was made of remained unanswered until 1944, when DNA was established as the molecule encoding the gene (Avery et al., 1944). Since then, the molecular biology continued to reveal the secrets of the genes with an ever increasing speed, the first being the discovery that DNA encodes genetic information via the sequence of four different nucleotides in a long double-stranded helical molecule that is condensed into a chromosome (Watson and Crick, 1953).

The common definition of a gene is a genomic region producing an mRNA that encodes a protein. From this point of view, proteins are considered the major actors in all cell processes. However, this view has changed dramatically with the recent sequencing of the human genome (Lander et al., 2001). It is now known that only 1.5% of the DNA in our genome encodes proteins. The remaining vast regions of DNA were often referred to as "junk DNA" due to their apparent lack of function.

The completion of several genome projects has revealed that the number of genes does not justify the complexity of an organism. Namely, humans and nematode worm have similar number of genes (22726 vs. 20060 genes, www.ensembl.org), despite the differences in organismal complexity. However, the size of human genome is 30 fold larger than that of the nematode worm. This has raised the question whether the "junk" DNA might hide the clues to the evolution and function of higher organisms.

This article will review how the discoveries of the last decade have changed our understanding of the DNA regions that used to be referred to as "junk". The first chapter will show how detailed analysis of the genome can reveal hidden codes in the DNA regions that control transcription of genes, and how these regulatory codes contributed to the complexity of multicellular organisms. The second chapter will discuss the repetitive genomic elements that initially appeared to play no other role in the genome but to selfishly replicate themselves. These apparent "genomic parasites" now turned out to be important players in the evolution of our genome by shuffling around pieces of genes. Studies of another type of cellular parasites, the viruses, have also taught us a lot about the unusual ways of gene regulation. Therefore, the third chapter will discuss the mechanisms whereby HIV circumvents our immune system by hiding itself inside our genome. In the fourth chapter, we will present the roles RNA molecules play in regulating and processing genetic information, with particular emphasis on the brain. In the final two chapters, we will discuss how these discoveries have changes our view of evolution, and how the ability to determine the sequence of human genome offers new possibilities for medicines tailored to the individual, but also pose new legal challenges.

Transcriptional Regulation and Organismal Complexity

Detailed analyses of whole-genome sequence assemblies of several organisms of varying complexities (such as yeast and human) have proposed a novel model that could explain the overwhelming diversity of organisms. According to this model, organismal complexity may correlate well with the intricacies and likely number of gene expression patterns exhibited in the course of animal's life cycle. But which parameters determine the nature of expression patterns and how do these elements change from simple to more complex organisms? To better understand these processes, we need to start with the basics of transcriptional control.

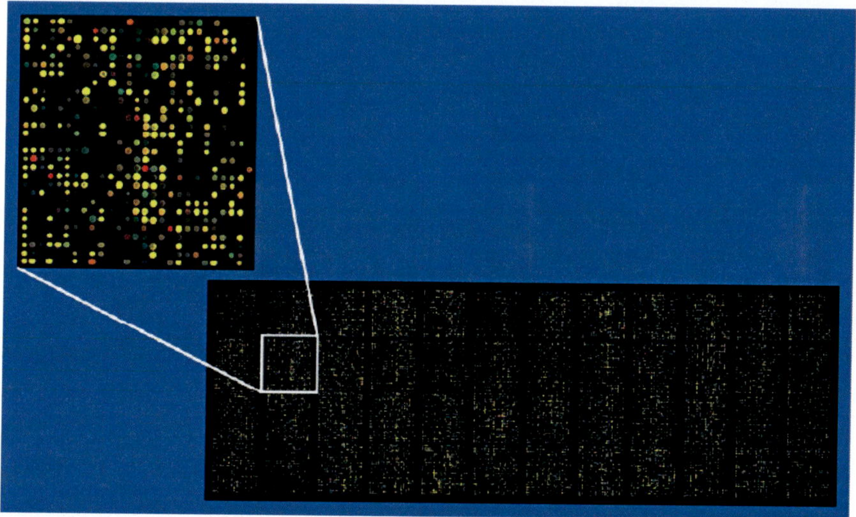

Picture. DNA microarray. Each spot represents one DNA sequence complementary to a human mRNA. The color of the spots corresponds to the amount of the bound mRNAs in the biological samples. (from http://en.wikipedia.org/wiki/Image:Microarray2.gif).

Basics of Transcriptional Regulation

Transcription of the genetic information from DNA to RNA by diverse sets of multi-protein RNA polymerase machines is a highly regulated cellular process that ensures normal development, growth and survival of an organism. Transcription of the protein-coding genes by RNA polymerase II (RNAPII) has been studied most intensely. It can be regulated at multiple levels, including the recruitment of RNAPII with pre-initiation complex to DNA, transcription initiation, elongation, and termination (Saunders et al., 2006). This regulation is predominantly achieved through the interplay between the proteins called transcription factors and a plethora of DNA regulatory elements. General transcription factors, such as those present in the multi-subunit TFIID complex, bind RNAPII as well as the DNA sequences of the promoter, the region close to the transcription start site, and thereby recruit RNAPII to the DNA. Additionally, tissue-specific transcription factors recognize and bind to a series of DNA regulatory elements near or far from promoters, and in this way regulate transcription of specific genes.

Importantly, DNA is not naked in the cell nucleus. It is neatly wrapped around the histone proteins, constituting the chromatin, which represents a barrier to active transcription (Kouzarides, 2007). To overcome this chromatin block, DNA-binding transcription factors or chromatin itself recruit recently discovered chromatin remodeling and modification complexes, thereby helping the transcribing RNAPII to initiate gene expression and effectively navigate through chromatin.

Several recent studies indicate that RNAPII often pauses subsequently to transcription initiation (Guenther et al., 2007; Muse et al., 2007), thus challenging the once predominantly held view that recruitment of RNAPII to promoters is the rate-limiting event. This type of control has been well documented earlier while studying the regulations of HIV gene expression (Barboric and Peterlin, 2005). In this way, an already engaged RNAPII can rapidly resume transcription in response to diverse sets of stimuli, including those during embryonic development or of the immune system. Overall, a complex crosstalk between chromatin, RNA polymerases, transcription factors and DNA regulatory elements determines gene expression.

Expanding the Complexity of Transcriptional Regulation

Having the genome sequences of evolutionary diverse organisms at hand, investigators next asked some simple questions. Do the ratios and absolute numbers of transcription factors per genome correlate with the organismal complexity? Is, for example, the structure of regulatory DNA sequences more complex in the fruitfly than in yeast? Indeed, the answers to both questions were unequivocal. A staggering 5-10% of the total protein-coding genes in multicellular organisms is dedicated to proteins that regulate transcription. Our genome codes for as many as 3000 transcription factors (Lander et al., 2001), whereas the yeast genome contains only 300 (Wyrick and Young, 2002). Yeast genome contains one transcription factor per 20 genes, while human genome contains approximately one per 10 genes. Given the combinatorial nature of gene expression, even this two-fold difference could account for a striking increase in regulatory complexity.

What is more, the number and types of DNA regulatory elements vary greatly among different species. For example, a typical yeast gene contains a promoter and immediately adjacent upstream activating sequences, which bind one or two different transcription factors. In contrast, expression of genes in multicellular organisms is controlled by a much greater variety of regulatory elements. In humans, this distance reaches 100 kilobases, and several reports indicate that regulatory elements on one chromosome could even regulate expression of a gene present on a different chromosome (Dekker, 2008). Importantly, these regulatory elements come in various flavours. Whereas enhancers stimulate the expression of a target gene, silencers repress it. In addition, insulators, which are present at the boundaries of transcription units, ensure that a certain enhancer or silencer regulates only its dedicated gene or a group of genes, preventing inappropriate expressions of neighboring genes. Of note, one regulatory element can bind several different transcription factors, each one of them responding to a distinct signal from the environment. Having all this in mind, one can easily imagine increased degrees of complexities in organisms with expanded sets of transcription factors and DNA regulatory elements. Hence, one can perceive a difference between yeast and man.

Rewiring the Transcription Network

Scientists have taken an interesting approach to study the role of the transcriptional regulatory complexity as an evolutionary force. In experiments using the queen of prokaryotic genetics, a bacterium *E. coli*, they switched promoters from one gene to another on a large genome-wide scale, thereby creating new promoter-gene connections (Isalan et al., 2008). This systematic rewiring of the transcription network was well tolerated by the cells. The vast majority of the approximately 600 new connections showed no apparent deleterious effects, and some strains even grew better than the original. These findings suggest that organisms can evolve by changing the architecture of their genetic networks. As a result, it seems highly intuitive that as the number of genes and regulatory elements expands, the possibilities for rewiring increase as well. Here, transposable elements (TEs) that we discuss in the next chapter can do the "switching" job and collaborate with the expanded regulatory complexity to rewire genetic networks and thereby contribute to organismal complexity (Feschotte, 2008).

Technology Shifts Gears

Significant technological progress has been made in the past decade and scientist can now observe the interplay between transcription factors and the genetic material. The development of novel mass spectrometry and crystallography methods has allowed to identify complexes of transcription factors and to determine their three-dimensional structure. For example, the breakthrough structure of RNAPII allowed us to understand how this multi-subunit enzyme catalyzes the synthesis of RNA molecules and how it couples RNA transcription to the later steps of RNA processing (Cramer et al., 2001).

Analysis of transcriptional changes on a genome-wide scale has been made possible by the advent of DNA microarrays. They consist of an arrayed series of millions of spots attached to a surface such as glass or a silicon chip, each containing a specific DNA sequence, corresponding to a gene or a DNA regulatory element. Thus, all genes of an organism or portions of chromosomes can be represented on DNA microarrays, and the spots are used as probes to which RNA transcripts or pieces of DNA can bind under high-stringency conditions. This technology allows scientists to observe gene expression patterns of specific cell types or whole organisms under varying environmental conditions, greatly advancing the speed and quality of experiments.

Once transcriptional changes are found by a microarray, it is necessary to find transcription factors that might regulate these changes. This can be achieved using a method that finds binding sites of a particular transcriptional factor on a genome-wide scale. This method, called ChIP on chip, combines chromatin immunoprecipitation (ChIP) assay with the DNA microarray technology (Ren et al., 2000). A protein is captured together with its bound DNA, the DNA is labeled and hybridized to the DNA microarray. This allows identification and quantification of binding sites of a particular DNA-binding protein along the whole genome. It was the ChIP on chip approach that revealed that RNAPII pauses shortly after transcription initiation, waiting for signals from the environment to restart gene expression.

Lastly, a recent exciting technique called chromosome conformation capture (3C) has shed light on the gene regulation in the third dimension (Dekker, 2008). This approach demonstrated that regulatory elements, which are dispersed over great distances in higher eukaryotes such as humans or mice, physically interact with the target genes by the virtue of long-range chromosome looping events. For example, these transient trans-interactions have been determined between enhancers and the cognate olfactory receptor genes. It is believed that active rather than passive processes, possibly via the nuclear actin and myosin machines, guide different genetic loci toward each other, constituting a "transcription factory", responding to the environmental demands. Overall, it will be exciting to watch how these powerful technologies will further "unweave the rainbow" of the inner workings of cells.

Genomic Parasites

The sequence of the human genome confirmed that most of our DNA constitutes of repetitive sequences (Lander et al., 2001). The majority of repetitive DNA sequences are transposable elements (TE), which used to be dismissed as selfish, or 'junk' DNA (Doolittle and Sapienza, 1980). TEs are selfish sequences, in the sense that they replicate and move along the genome, using extensive cellular resources but without any apparent benefit for the cell. Therefore they were regarded as the intracellular genetic parasites, which exploit cellular machineries for their own survival (Doolittle and Sapienza, 1980). TEs jump from one part of the genome to another either by copying themselves via an RNA intermediate (retrotransposons) or via cut and paste mechanisms (DNA transposons). They are found in all currently known eukaryotic genomes where they can form different portions of the genomes. Multiple copies of the same TE are scattered throughout genomes. For instance, in humans, half a million copies of LINE-1 retrotransposon comprise 17% of the genome (Lander et al., 2001).

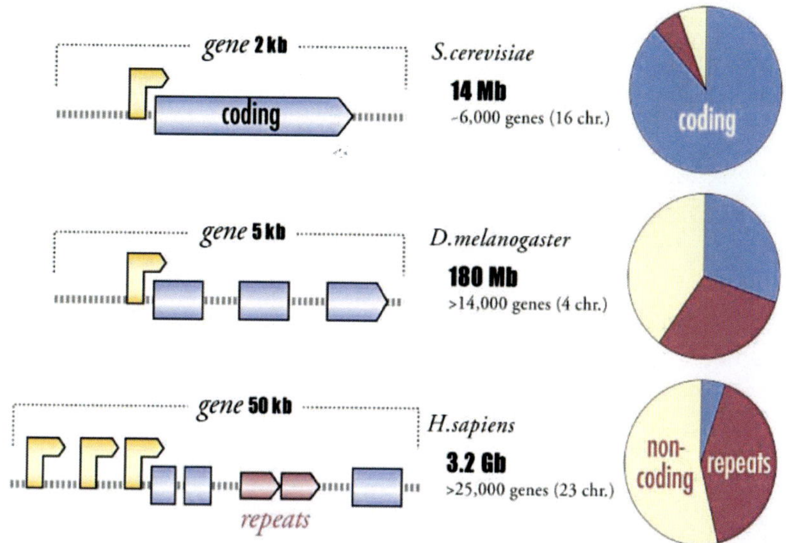

Picture. Size and composition of genome of baking yeast (S. cerevisiae), fruitfly (D. melanogaster) and human (H. sapiens). The increase in genome size correlates with the expansion of non-protein coding (grey lines showing intronic and intergenic sequences) and repeat DNA (red arrows repeats, such as transposable elements, TEs) sequences. An increase in size and complexity of transcriptional regulatory units (yellow arrows) is also evident. (from http://en.wikipedia.org/wiki/Genomic_organization).

Usefulness of Junk

The genomes of different species maintain different types and amounts of TEs. Why have these repetitive sequences been maintained in mammalian genomes in species-specific patterns? One possible explanation might lie in their usefulness for regulating gene expression. For example, the predominant human SINE (short interspersed nuclear element) in human genome is Alu, a type of TEs specific for primate genomes, which is related to the B1 and B2 SINEs in the mouse genome. Alu element was shown to express a non-protein coding RNA (ncRNA) that acts as a transcription regulator directly impacting RNAPII. When cells are exposed to heat, expression of Alu increases by 100 fold, leading to transcriptional repression (Mariner et al., 2008). Similar role was also demonstrated for mouse B2 elements (Allen et al., 2004). Interestingly, Alu element contains parts of both B1 and B2 elements, thereby combining two functions required to bind two RNAPII molecules.

A lot of complex biological processes have to be regulated immediately and at the same time so that a number of genes should be regulated simultaneously. Perhaps different SINE transcripts serve as master regulators of gene expression by targeting many different steps. Another striking observation was that high level of retrotransposon expression in embryonic stem cells gives rise to numerous pseudogenes. Recently a pseudogene was demonstrated to serve as a template in the RNA interference pathway of gene regulation, confirming the importance of TEs in the evolution of regulatory networks (Tam et al., 2008). However, the usefulness of TEs does not end with genomic regulation and evolution. TEs were demonstrated to contribute to the repair of DNA double stranded brakes, and thereby maintaining the non-mutated state of our DNA (Morrish et al., 2002). Taken together, it appears that the TEs and our genome have co-evolved into a state of mutual dependence, and therefore the repetitive elements are after all not all "junk".

Ultraconserved Sequences

Ultraconserved elements in the human genome are defined as stretches of at least 200 base pairs of DNA that match identically with corresponding regions in the mouse and rat genomes. In addition, most of these elements have been evolutionarily conserved since mammal and bird ancestors diverged over 300 million years ago. The reason for this extreme conservation remains a mystery. Interestingly, even though the conserved regions of the genome normally encode proteins, most ultraconserved sequences do not encode a protein sequence, and are in fact under an even stronger evolutionary selection than the protein-coding sequences (Katzman et al., 2007). Even though it is not yet clear if this is the reason behind their strong conservation, some of the ultraconserved sequences were found to play a role in regulating transcription or alternative splicing (Bejerano et al., 2004; Lareau et al., 2007; Pennacchio et al., 2006). Interestingly, one ultraconserved sequence has already been found to be a relic of a transposable element (Bejerano et al., 2006). Moreover, it was shown that TEs played a role in mammalian brain evolution. Transgenic mouse model demonstrated that highly conserved mammalian AmnSINE1 TEs function as distal transcriptional enhancer elements in developing mouse embryos (Sasaki et al., 2008). The hope is that further studies of TEs might provide further explanation to the riddle of ultraconservation.

Our Mosaic Genome

The impact of TEs can be passed to the next generation only when new insertions happen in germ cells or embryonic stem (ES) cells. In the development of new therapies based on ES cells it is therefore important to bear in mind the role that TEs play in the stability of genomes. In the majority of somatic cells, the DNA regions encoding TEs are methylated, and as such not transcribed. However, recent studies with transgenic mice have shown that transgenic LINE-1 element under the control of endogenous promoter can be expressed and retrotransposed in neural stem cells suggesting that newly generated LINE-1 insertions in neural progenitor cells can affect gene expression and neural fate (Muotri et al., 2005). Such LINE-1 retrotransposon insertions lead to somatic mosaicism. Expression of TEs can be induced by stress; therefore the extent of mosaicism in organisms may be influenced by environmental conditions. It remains to be examined how such genetic mosaicism contributes to individual differences, from cognition to disease predisposition (Muotri et al., 2007).

Control of Transposition

Although TEs contribute to the origin of new genes and sophisticated regulatory network system, their overexpression and activity can be harmful for their hosts and several diseases are caused by TE insertions into genes. The overexpression of TEs can also lead to tumor development in somatic cells (Ostertag and Kazazian, 2001; Sciamanna et al., 2005). For this purpose, several mechanisms have evolved to keep TE in check. These include transcriptional silencing via DNA methylation (Walsh et al., 1998), posttranscriptional silencing via RNA interference (RNAi) (Kanellopoulou et al., 2005) and cytidine deamination via the APOBEC3 family of proteins (Esnault et al., 2005). During embryonic development, TEs are hypomethylated therefore their transposition needs to be controlled by other mechanisms, such as RNAi and editing by APOBEC3 proteins. Small RNA based control of TEs has been discovered in different organisms, which associate with a specialized family of Piwi proteins that activate mechanisms to silence TE expression (Girard and Hannon, 2008). The third group of proteins that restricts mammalian retrotransposons are APOBEC3 cytidine deaminases. APOBEC3 proteins only inhibit retrotransposition of elements that undergo reverse transcription (Esnault et al., 2005). The exact mechanism of their action is unclear. Although mice lacking APOBEC3 protein were more susceptible to infection of mouse mammary tumor virus, no increase in retrotransposition was observed (Okeoma et al., 2007).

In spite of the many mechanisms genomes use to stem the expansion of TEs, TEs have found their own ways to evade these mechanisms. In fact, it seems that in some genomes, TE won the battle and highly expanded during evolution, as has been observed for a type of Alu sequences in the human genome (see the fifth chapter). In fact, such expansions of TEs may have played crucial roles in the evolution of species-specific traits, such as for example the human brain.

Hide and Seek of the HIV

Picture. The Mamala plant produces prostratin, which provided a new venue for the treatment of HIV (from http://www.lightyearsip.net/ip_samoa_mamala.shtml).

We have so far discussed the "selfish" genetic elements that have been replicating as part of our genome for ages. Now we will move on to similarly selfish genome invaders - viruses, the particles that contain their own genetic material, but require a host cell for their replication. One of the most famous and thoroughly studied viruses is human immunodeficiency virus (HIV), which belongs to the family of retroviruses. HIV uses RNA as its genome, but replicates via a DNA intermediate. In the course of its life cycle, HIV integrates its DNA intermediate into the host genome and can thereby enter a latent status, where it can remain hidden for many years.

Problem of the Latent HIV

HIV was discovered in 1984 as a cause of the acquired immunodeficiency syndrome (AIDS). Great advances were made in the last 25 years in the search for a cure for AIDS. Currently, the most commonly used treatment is known as highly active antiretroviral therapy (HAART), and is successful in temporarily suppressing viral expression. However, this treatment does not eliminate the virus from the body, since a small pool (~a million) of cells contains HIV in a latent form (Chun et al., 1995). These cells contain the DNA encoding viral genome, but because the DNA is not transcribed, the cells do not express viral proteins and can thus not be targeted by the immune system or the HAART therapy, which both target only viral proteins. The latent reservoir consists predominantly of resting immune T cells, in which various cellular processes are slowed down. Upon activation of these cells, the virus can reemerge and contribute to the development of AIDS. Since the viral DNA can persist in the genome for years, it is estimated that at least 70 years of HAART treatment would be required for the eradication of the latent viruses (Finzi et al., 1999).

Viral Pathway towards Latency

HIV has to surmount numerous barriers before it can integrate its genome into the genome of the host cell and start its replication cycle. To accomplish this task, the virus hijacks the cellular machinery for its own purpose. First, the virus attaches to the cell surface via interaction with receptors on the cellular membrane, which normally serve other purposes to the cell. In the cellular cytoplasm, the virus partially disassembles and its enzyme reverse transcribes the viral RNA into a linear double-stranded DNA. Next, viral DNA is imported to the nucleus together with viral proteins, which recruit the cellular machinery to enable integration of the viral DNA into the host genome. After integration, the viral DNA is indistinguishable from the rest of cellular genes, and is referred to as a 'provirus'. Provirus can either be transcribed into RNA by the cellular RNAPII, or remain silent for many years.

Maintenance of the Latent Provirus

After integrating into the cellular genome, the fate of provirus depends on several features. The three major factors affecting transcriptional activity of the provirus and thus establishing latency are transcriptional interference, the chromatin environment and the activity of transcription factors (Contreras et al., 2006).

Transcription of the viral DNA can be inhibited as a result of transcription of a proximal gene, a phenomenon termed transcriptional interference (Adhya and Gottesman, 1982). In this situation, RNAPII transcribing the host gene antagonizes transcription originating from the viral promoter. Thus, when viral DNA integrates into an actively transcribed host gene, TI suppresses transcription of the HIV genome (Lenasi et al., 2008). Because more than 90% of proviruses reside in actively transcribed host genes (Han et al., 2004; Lewinski et al., 2005), transcriptional interference is very likely a key cause of latency. Another feature of provirus is that the cellular histone proteins that coat the DNA of viral promoter are modified in a way that inhibits transcription (Steger and Workman, 1997). In other words, the viral promoter has a repressed chromatin status, where the histone proteins assume a structure that prevents association with the transcription factors (Williams et al., 2006). Finally, HIV often infects resting immune T cells, where levels of transcription factors occupying the viral promoter are often too low for efficient transcription of the provirus.

The different mechanisms that contribute to proviral latency are mutually connected. For example, transcription factors vital for HIV gene expression are present in limiting amounts in resting cells, therefore transcription of the proviral DNA is not efficient. Furthermore, repressed chromatin can be associated with actively transcribed genes, in spite of the prevailing notion that actively transcribed genes have de-repressed chromatin. For instance, recent studies in yeast have shown that transcribing RNAPII is preceded by de-repressed chromatin and followed by rapid repression of chromatin (Kaplan et al., 2003; Li et al., 2007). This mechanism normally prevents any cryptic promoter-like sequences within the transcribed genes from functioning as transcription start sites. Therefore, the viral promoters residing within actively transcribed host genes may be repressed by the same mechanism that normally operates to silence cryptic promoters within genes. Taken together, multiple mechanisms operating at the level of gene expression act cooperatively to establish and maintain latency of the HIV provirus.

How to Eradicate the Latent Viral Reservoir?

It is now clear that the current approach to HIV therapy, which inhibits viral replication and spreading, can not completely eradicate the virus from the human body. Perhaps the only way to completely get rid of the HIV is by targeting the latent viral reservoir. This could be achieved by activating expression of all latent proviruses, which can then be eliminated by the immune system and existing therapies.

One approach to induce expression of the latent provirus is to overcome transcriptional interference by inhibiting transcription from the host gene. This could then be combined with agents that stimulate the recruitment of key transcription factors to the viral promoter. Unfortunately, HIV genome randomly integrates into various host genes, each of which has different levels of transcription, therefore their collective inhibition is not practical. However, the second approach is more effective because it activates viral transcription regardless of the viral integration sites. Often, transcription factors also cause de-repression of chromatin due to recruitment of chromatin remodeling complexes (Williams et al., 2007; Zhong et al.,

1998). Currently, scientists are still looking for an ideal approach for viral activation that is not harmful to the host cells.

New Hope for HIV Eradication

It has long been thought that diverse plants in the different habitats of the Earth synthesize an enormous variety of chemicals, which may hold cures for many diseases. Two such chemicals have recently proved to be promising for the treatment of HIV. One of these is prostratin, found in the Mamala plant that grows in the Samoan rainforest.

However, it would be hard to obtain enough of the compounds from the plants for therapeutic use, and heavy harvesting of the wild plants could cause ecological damage. Prostratin, used in combination with an immune system stimulator, managed to flush out and eliminate approximately 80 percent of the dormant virus. But with HIV, 80-percent efficiency is not enough.

The Mamala plant produces prostratin for its own reasons, so this chemical is certainly not fully optimized for treatment of HIV. Therefore, scientists have devised ways to produce synthetic prostratin, which allows them to modify the structure and related function of the compound in a way to maximize its ability to induce proviral expression while avoiding any side affects (Wender et al., 2008).

Prostratin has long been used by traditional Samoan healers without their patients experiencing acute side effects. The Samoan healers willingly shared their knowledge with Paul Cox, an ethnobotanist who saw them prescribing a tea made from Mamala bark for patients with hepatitis-like symptoms. Cox, in turn, sent samples to the National Institutes of Health in US, in hopes that the bark might have antiviral properties useful in fighting some cancers. Researchers at NIH then analyzed the bark and isolated prostratin, and while no effect on cancer was found, they eventually found its effect on HIV.

We can learn from this study that today's global community could bring forth many more such discoveries, as long as the diversity of species and the knowledge of traditional communities is preserved for the future generations.

RNA and the Brain

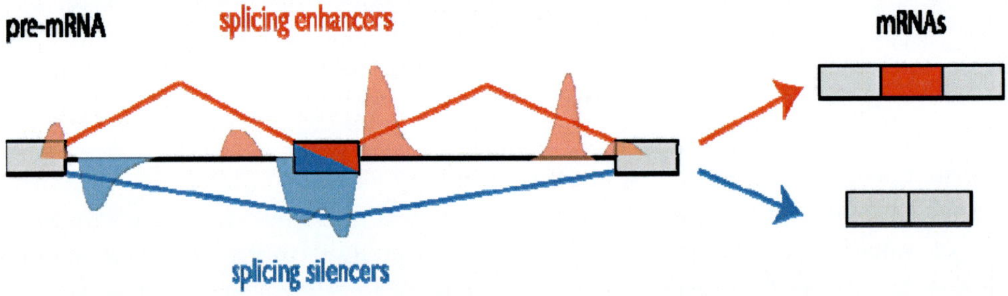

Picture. The RNA map of genome-wide location of Nova-dependent splicing regulatory elements. The red peaks show the locations on pre-mRNA where Nova binds to promote inclusion of the alternative exon, and the blue peaks the locations when Nova silences exon inclusion. (adapted from (Ule et al., 2006)).

The first clue that neurons have a unique system for regulating gene expression at the level of RNA came from studies at the intersection of cancer cells and neurons. Comparative gene expression studies of a thyroid tumor found a unique transcript of the calcitonin gene that was expressed in the tumor cell line, but was normally restricted to the brain (Rosenfeld et al., 1983). This raised the possibility that many other such transcripts might be present in the brain, which was confirmed in the recent years by a number of genome-wide studies.

The Importance of Being Alternative

Most genes encode proteins in small pieces (called exons), which are separated by longer non-coding pieces (called introns). In some genes, the exons combine in different ways allowing production of several alternative mRNA isoforms from a single gene, in a process referred to as alternative splicing. In the early 90's, alternative splicing was considered a rare event and its importance was not appreciated. However, the large-scale sequencing of genomes and mRNAs in the last decade revealed that over 60% of human genes produce at least two different mRNA isoforms (Sharov et al., 2005). Alternative splicing is particularly prevalent in the brain, which contains the highest number of tissue-specific isoforms, and its own set of factors that regulate alternative splicing (Ule et al., 2005). There are three main reasons for the importance of alternative splicing in the brain.

Alternative splicing enables the limited number of existing genes to generate the proteomic diversity required for development and function of neuronal circuits. For example, each of the three neurexin genes contains multiple alternative exons and an alternative promoter, allowing for the generation of over a thousand different isoforms (Ullrich et al., 1995). Specific alternative neurexin isoforms at the pre-synaptic neuron preferentially bind to a specific post-synaptic neuroligin isoform, and the diversity of isoforms together with isoform-specific interactions contributes to proper development and wiring of excitatory and inhibitory synapses (Chih et al., 2006).

Alternative splicing contributes to the diversity of neuronal sub-populations (Lipscombe, 2005). For instance, the sensory neurons that perceive pain express an alternative RNA isoform encoding a calcium channel that remains open for longer and forms a higher density of functional channels than the isoform expressed in other neuronal types. This RNA isoform in sensory neurons is specifically required to mediate pain response to heat (Altier et al., 2007).

Finally, alternative splicing expands the spectrum of responses to neural activity. Splicing of numerous alternative exons has already been found to respond to neuronal activity. One such example is the activity-dependent splicing of the RNA encoding the NMDA receptor 1 (NR1) subunit, a core component of the signaling pathway that leads to synaptic plasticity. Splicing of NR1 exon 20 affects trafficking of the NMDA receptor from the endoplasmic reticulum to the synapse (Mu et al., 2003).

Splicing for the Brain

To unravel the mechanisms that allow the brain to produce its own set of alternative mRNAs, similar genome-wide methods were used as those developed for transcription factors (see previous chapter). To find the sites on the neuronal RNAs where an RNA-binding protein binds, a method was developed that relies on the ability of UV light to crosslink proteins to the target RNA sites, termed UV crosslinking and immunoprecipitation (CLIP). This method was first used in the first attempt to understand the role of a brain-specific alternative-splicing regulator Nova (Ule et al., 2003). To find the changes in alternative splicing that occur in the absence of Nova, a DNA microarray was designed to recognize alternative exon-exon junctions (Ule et al., 2005).

The genome-wide studies of Nova RNA targets led to two intriguing observations. Firstly, most of the Nova-regulated RNAs encode proteins that primarily function at the neuronal synapse. This finding suggested that a neuronal splicing regulation can reconfigure the genes that normally play other roles in the rest of the body for the purpose of neuronal communication (Ule and Darnell, 2007). Secondly, a bioinformatic analysis of data generated by CLIP and splicing microarray revealed a precise set of positions where Nova can bind the pre-mRNA to silence or enhance exon inclusion, which were termed 'RNA map'. It came as a surprise to the splicing community that the RNA map could define the activity of Nova on a genome-wide scale (Ule et al., 2006).

RNA and Memory

The last decade has uncovered a large diversity of mechanisms by which different types of RNA molecules allow the cells to precisely control the time and amount of proteins to be synthesized at different locations in the cell. Nowhere is the role of RNA more apparent than in its contribution to neuronal capacity to encode memory. Memory depends on changes in neuronal circuits that occur in response to a sensory experience. Most of these changes occur at the synapses, the contacts between neurons. The ability of synapses to be modified or created during the process of synaptic plasticity is therefore crucial for memory (Kandel, 2001).

Studies in various organisms have shown that most aspects of synaptic plasticity require translation of new proteins (Kandel, 2001). Not only do new proteins need to be synthesized, but the neurons need to precisely determine the synapse that will produce new protein, and the time and amount of protein translation (Ule and Darnell, 2006). This precision is achieved by regulation of gene expression at the level of DNA transcription (Hong et al., 2005; Kandel, 2001) and RNA processing, localization and translation (Kiebler and Bassell, 2006).

The first clues to the role of RNA in memory came during the earliest stages of molecular neurobiology. In 1982, electron microscopy evidence indicated that ribosomes are present in the neuronal dendrites (Steward and Levy, 1982). Later, translation initiation factors and a number of different mRNAs were documented in dendrites (Kiebler and Bassell, 2006). These discoveries have raised an important question; why should the neurons invest energy and resources into post-transcriptional RNA regulation, and its transport into dendrites? To address this question, we need to explore the different benefits of RNA regulation for synaptic plasticity and memory.

RNA and the Synapse

The size and shape of a synapse is determined by its own 'micro-skeleton', extended filaments formed by the protein beta-actin. Interestingly, the mRNA encoding this protein is present in neuronal dendrites. The transport of beta-actin mRNA to the dendrite requires binding of an RNA-binding protein, zipcode-binding protein 1 (ZBP1) to the 3' untranslated region (3'UTR) of the beta-actin mRNA (Gu et al., 2002). In addition to promoting the transport of the mRNA to the dendrite, ZBP1 also inhibits its translation, thereby preventing premature synthesis of the protein (Huttelmaier et al., 2005). Preliminary analysis in cultured fibroblasts suggests that when the mRNA arrives to the synapse, ZBP1 is modified by phosphorylation, which releases beta-actin m2□□□and allows synthesis of the protein. Therefore, ZBP1 appears to couple mRNA localization to the translational inhibition, ensuring that only the mRNA that arrives to an activated synapse is free to be translated into protein.

One of the biggest stories of the last decade relates to a new type of ncRNAs that are generally only ~21 nucleotides long and are commonly referred to as microRNAs (miRNA). Each miRNA is able to bind to many different mRNAs via partial complementarity, and recruit a number of proteins that lead to mRNA degradation or translational inhibition. MicroRNAs have already been shown to play roles in tumor development and differentiation of tissues (Stefani and Slack, 2008). In addition, miRNAs can regulate mRNA translation within neuronal dendrites. Brain-specific miRNA134 was found localized to neuronal dendrites where it silences translation of mRNA encoding LimK1, a regulator of actin filament dynamics (Schratt et al., 2006). When a neuron is stimulated, miRNA134 inhibition is relieved, allowing the synapses to change their shape during neuronal activity, and thereby encode memory.

Protein Synthesis at the Synapse

As mentioned earlier, the ability of synapses to be modified or created during the process of synaptic plasticity is crucial for memory, and synaptic modification is heavily dependent on translation of protein locally at the activated synapse (Kandel, 2001). A small subset of neuronal mRNAs are localized to the neuronal dendrites. These mRNAs are transported in a translationally silent state, but are translated once they arrive to an activated synapse. Presence of mRNA at the synapse allows protein production to be restricted only to the activated synapse, and not the rest of synapses of the same cell (Kiebler and Bassell, 2006).

Interestingly, the mRNAs that have so far been found in the dendrite encode the proteins that are most abundant at the synapse. Production of these proteins probably represents the time-limiting step in synaptic plasticity. Presence of mRNAs for these proteins at the synapse avoids the delay that would occur if new protein would have to be transported from cell bodies, which are often at a considerable distance away from the synapse. Taken together, synaptic mRNA translation increases the precision of information processing by neurons, in the dimensions of both space and time (Ule and Darnell, 2006).

One of the most definitive pieces of evidence that mRNA regulation in neuronal dendrites plays a role in synaptic plasticity was found by analysis of mRNA encoding CaMKIIα. CaMKIIα is one of the most abundant proteins in neuronal synapses, is required for most forms of synaptic plasticity and for proper spatial memory (Silva et al., 1992). The mRNA encoding CaMKIIα is abundant in the neuronal dendrites. To study the role of the localized mRNA, transgenic mice lacking the untranslated region of the CaMKIIα were generated, which prevents the mRNA to localize to neuronal dendrites. Even though the mice synthesize a normal amount of CaMKIIα protein, they show a severe reduction in synaptic plasticity and long-term spatial memory (Miller et al., 2002). Thus, the protein synthesized in the cell body can not compensate for the protein synthesized at the activated synapse from the dendritically localized CaMKIIα mRNA.

The mRNAs located at the synapses allow fast and spatially restricted protein translation. In fact, local RNA translation might be the crucial mechanism allowing neuronal synapses to act as independent information processing units. Since synapses are much smaller than neurons, their independence greatly expands the brain capacity to process and store information.

RNA Therapy

The discoveries of the role of RNA in neuronal dendrites came at the same time as the therapeutic potential of RNA began to be explored. Antisense RNA oligonucleotides against specific molecular targets are already becoming important reagents for the therapy of cancer and other diseases (Crooke, 2004). Studies are underway to test if neurons could be induced to uptake therapeutic RNAs into the synapse in a way that would modulate translation of proteins that are important for memory. Such an RNA therapy could provide useful to treat Alzheimer and other neurodegenerative diseases that affect memory.

EVOLUTION OF GENOMES

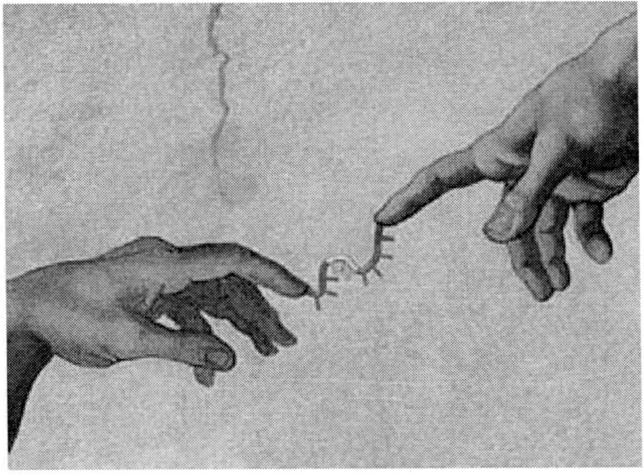

Picture. The June 14 2007 cover of the Economist magazine was dedicated to the RNA and the magazine's leader explained, under a 'The RNA revolution, Biology's Big Bang' heading, that biology will be to the 21st century what physics was to the 20th. (from: http://www.economist.com/opinion/displaystory.cfm?story_id=9339752).

Evolution Is Driven by Gene Regulation

In 1973, Russian biologist Theodozius Dobzhansky famously stated: "Nothing in biology makes sense except in the light of evolution" (Dobzhansky, 1973). Similarly, the sequence of human genome is of little use if one cannot interpret it in the light of evolution. Only a tiny fraction of the 3 billion letters of the human genome could be easily interpreted in 2001 when its sequence was released. Since then, 21 vertebrate genomes have been sequenced to date, allowing precise analyses of the conserved regions that regulate and encode expression of RNAs.

Using a number of high-throughput experimental and bioinformatic methods, a consortium of 35 groups of researchers at 80 institutions in 11 nations has undertaken a detailed study of the genome as part of the Encyclopedia of DNA Elements (ENCODE) project (Birney et al., 2007). These researchers have spent four years sifting through more than 400 million letters of chromosome 9 to make sense of just one percent of the human genome by using the bioinformatics and experimental approaches described in the first chapter.

Scrutinizing a small portion of our DNA has shown that the genome hosts a lot more activity than was previously seen. In fact, the study found that most of the genome is being transcribed in one form or another. This happens, in part, because each gene is often transcribed along with a surprisingly large number of non-protein-coding sequences to produce some extraordinarily long RNA fragments. The results also indicate that a single gene can be transcribed into many different RNA fragments of varying lengths such that each gene is represented, on average, by more than five transcripts that share overlapping sequences. In addition, the study found many new ncRNA (noncoding RNA) molecules transcribed from regions of the genome that were previously relegated to the "junk" DNA. Majority of these transcripts have unknown functions and are sometimes referred to as the "dark matter" of molecular biology (Johnson et al., 2005).

The overwhelming presence of ncRNAs, and their unusual genomic organization are challenging the notion of the gene as a distinct region of the genome. The overlapping transcripts, sometimes transcribed from opposite DNA strands, and regulatory elements that affect transcription of multiple unrelated transcripts make it difficult to divide the genome into independent portions of genes as pearls on a string. Instead, it appears that the genome functions more like a control deck of an airplane, where the instruments are crammed up all around the driver's seat.

The ENCODE project has also identified new regions of the genome that control the transcriptional program of the genes. It is the changes in the DNA or RNA elements that bind to the regulatory proteins, rather than changes in the proteins themselves, that appear to play the main role in the evolution of vertebrate gene regulation (Jelen et al., 2007; Odom et al., 2007). Mutations in these regions are often associated with complex diseases, such as heart disease and diabetes. By mapping out where exactly the regulatory regions lie and how they work, the ENCODE project will allow to understand how the disease-causing mutations lead to faulty expression of the distal genes.

Transposable Elements in Evolution

The origin of TEs remains a matter of debates. Retrotransposons possibly trace back to the hypothetical RNA world during the origins of life. It is postulated that life evolved from RNA-like molecules, since RNA is the only modern molecule of life that is capable of both encoding genetic information as well as catalyzing enzymatic reactions (Joyce, 2002). In modern life, RNA plays the largest variety of roles in the cell; it can act as a carrier of genetic information to the translation machinery or as a catalyst of translation within the ribosome. RNA instructs the processing of precursor messenger RNAs during splicing and editing, and regulates translation and degradation of mRNAs via miRNA silencing mechanisms. The basic components of RNA (nucleotides) are used to derive important signaling molecules and participate in most central metabolic reactions. Taken together, this versatility of RNA suggests that an RNA-like molecule could have evolved first during origins of life, and retrotransposons might be remnants of such self-replicating molecules.

TEs must have somehow benefited their hosts to survive in the genomes for such a long period of time. Interestingly, when Barbara McClintock discovered TEs 60 years ago, she recognized them as "controlling" elements due to their affect on expression of neighboring genes (McClintock, 1956). However, it was only after the availability of large-scale genomic data that the roles of TEs became fully apparent. By moving pieces of one gene to another, TEs often lead to creation of new genes that combine functions of two unrelated genes. In addition, TEs can move elements that regulate gene expression, and can thereby modulate expression of the genes in different tissues, or lead to creation of new alternative mRNA isoforms from an existing gene (Deininger et al., 2003). As active genomic components they promote recombination and provide ready to use motifs, even protein coding sequences. Indeed, almost 600 genes, including some short regulatory RNAs (such as miRNAs) originated from transposable elements in humans (Iwashita et al., 2003).

One of the most interesting TEs in humans are Alu elements, since they comprise approximately 10 percent of the genome. Comparison of a sub-family of Alu elements known as the AluYb lineage in the genomes of human, chimpanzees, bonobos, gorillas, orangutans, gibbons and siamangs showed an impressive expansion of these elements in the human genome. The AluYb sub-family had a 20-million-year general inactivity in the primate genomes with exception of human genome, where it underwent a major expansion during the past 3-4 million years. In fact, massive expansion of Alu sequences to over one million copies in hominid species and recombination between them was probably a major force in human evolution. Some AluYb elements are still actively mobilizing in the human genome, causing insertion mutations that cause about half a percent of all human genetic disorders, including hemophilia and some cancers (Han et al., 2005). Importantly, comparison of RNA editing in mice and humans revealed that editing of Alu transcripts is highly increased in humans, especially in brains. We already pointed out that metabolism of RNA in brains importantly contributes to functioning of our nervous system. Predominance of Alu elements in humans may therefore very well be connected with the evolution of cognitive functions and not just an accidental remains of history. This is just one of many examples indicating that Alu and other TEs have greatly contributed to the evolutionary jumps of higher vertebrates.

A Personal Genome Sequence

The $1000 Genome

When James Watson and Francis Crick fit together cardboard cut-outs to propose the double-helical structure of DNA 55 years ago (Watson and Crick, 1953) they did not expect that the first author would get his own personal genome sequenced in his lifetime. This is indeed what was recently achieved using massively parallel DNA sequencing (Wheeler et al., 2008). 3 billion base pairs present in the human genome were sequenced in just four months. The massive parallelization is achieved by attaching single DNA molecules to synthetic beads, trapping these in tiny water droplets within a water–oil emulsion and then amplifying these minute islets of DNA. In a later step, during which optical measurements are used to collect the actual sequencing data, each bead is confined to a picolitre-scale well etched into the end of a glass fibre within a fibre-optic bundle.

The actual practice of personalized medicine based on genome sequences will have to wait until we are able to predict what particular variations in the genome mean. At present, we have little ability to do so. One prerequisite for such analysis would be to have a large number of individual genome sequences available for comparative studies. The cost of Watson's genome sequencing was $2 million, but the hope is that once the cost decreases to 10,000, many people will decide to have their genomes sequenced, allowing their doctors to use the data for more-personalized medical advice.

In an attempt to speed up the progress toward personalized medicine, the National Institute of Health (NIH) is planning to sequence the genomes of 100 different people in the next two years. In 2007 the X Prize Foundation announced a $10 million award for the first privately funded team that can sequence 100 human genomes in 10 days.

Indeed, there is a flurry of activity and new approaches developed almost every month. A team at the University of California sequences DNA by forcing it through tiny pores and logging each nucleotide as it passes. A firm based in New York (Reveo) runs knives that measure only a few atoms along the surfaces of DNA molecules. Since each nucleotide has a different shape, the result can be used to decode their order. A company in Cambridge, Massachusetts (Helicos) attaches billion of DNA molecules to a surface of ten square centimeters, and then sequences each individual molecule by shining a laser at them and taking many high resolution images while the DNA polymerase is synthesizing the complementary strand of the DNA molecule.

Analysis of Watson's genome revealed about 3,300,000 single nucleotide polymorphisms (SNPs) — simple substitutions of a common base for another at a particular site in the genome. While we can't predict the functional impact of vast majority of these variations, even our limited current knowledge is enough to reveal some potentially important sites. For instance, Watson carries at least 10 mutations that have previously been associated with severe diseases in humans (in most cases he only carries one copy of a mutation, where two would be required to cause disease). Some of these mutations that boost the risk of cancer, including one linked to breast cancer, which might explain why Watson's sister had suffered from a serious breast cancer. This suggests that each of us may carry quite a large number of mutations that could potentially result in serious disease in our children, should we be unlucky enough to mate with someone carrying mutations in the same gene.

Genetics and Privacy

The sequencing technologies continue to improve with an incredible speed, and will soon allow sequencing of individual genomes for a reasonable cost. Someday in the future, every human might have their DNA sequenced at birth. This most personal information will allow more rational therapies tailored to the individual. It is important, though, that the ethical and legal implications of this new wealth of knowledge about ourselves are taken into account. Many diseases are still untreatable, and knowing that one contains a mutation that predisposes them for the disease can therefore be of little help. For example, Watson has a family history of Alzheimer's, but after his genome was sequenced, he did not want to know if he carries a specific variation known as APOE4 that's linked to increased risk of the disease.

One of the public's biggest worries about personal genomics is the possibility of genetic discrimination when applying for a job or trying to get health insurance. To address this concern, the US Senate has passed the Genetic Information Non-discrimination Act (GINA) on April 24th 2008. GINA will help prevent companies from using genetic information in deciding whether to employ someone. It will also forbid insurers from discriminating against individuals because of genetic predispositions.

PERSPECTIVES

The genome sequencing has allowed development of high throughput methods that have remarkably changed our view of the genome. Rather than the number of genes, it appears that the increased complexity of gene expression regulation has been the driving force of evolution. There seems to be an intricate battle between transposable elements and genomic defense mechanisms that keep their numbers in check, between viruses and the immune system that tries to eradicate them from the body. These battles appear to have played a major role in speeding up the evolution of genomes. Furthermore, recent years have placed RNA molecule at the forefront of genome regulation, with the discoveries in the fields of pre-mRNA alternative splicing, miRNAs and other types of non-protein coding RNAs. What all these different topics hold in common is that they all address the function of genomic sequences that do not encode proteins, and were therefore sometimes erroneously referred to as ‘junk DNA’.

These discoveries will provide fertile ground for important advances in medicine. They will allow more rational treatments of human diseases tailored to individual genomes. Whereas most current medicines are a product of random screening of thousands of chemicals, the medicines of the future may often use nucleic acid analogues of RNA to correct expression of specific mis-regulated genes. Sequencing of individual genomes will allow development of such tailored medicines, but at the same will lead to new challenges. It will be important to develop legal mechanisms to prevent misuse of the most private information, the sequence of our own DNA.

Finally, we can wonder about the relationship between the genome and the selfish elements that pervade it. In the seventies, Richard Dawkins has proposed that our genes are the units of evolution, whereas we serve just as their carriers (Dawkins, 1976). Since then, genomic analyses have shown that genes do not act as independent units. Instead, regulatory elements in the genome, rather than the coding pieces of genes have been found to be at the core of evolution. However, Dawkins did show some vision by naming his book 'Selfish genes' long before the role of TEs in genome evolution became appreciated. The delicate balance between the genomic defense mechanisms and TEs attempts at expansion may have played crucial roles in our own evolution. The coding information in genes may be split into pieces just so that TEs and other processes of DNA recombination can move these pieces around, and so that new patterns of alternative splicing can evolve. Our genome seems to have built within itself the potentiality for the new evolutionary jumps in the unforeseen future. It has made its pact with the Mephistopheles of Alu elements long before Goethe wrote his Faust.

Acknowledgment

We would like to thank Tina Bregant, Andrej Ule, Eva Žerovnik and Luka Avsec for reading the article and giving helpful advice. Also, we thank Matija Peterlin for instilling in us the love for genome regulation.

References

Adhya, S., and Gottesman, M. (1982). Promoter occlusion: transcription through a promoter may inhibit its activity. *Cell. 29*, 939-944.

Allen, T. A., Von Kaenel, S., Goodrich, J. A., and Kugel, J. F. (2004). The SINE-encoded mouse B2 RNA represses mRNA transcription in response to heat shock. *Nat. Struct. Mol. Biol. 11*, 816-821.

Altier, C., Dale, C. S., Kisilevsky, A. E., Chapman, K., Castiglioni, A. J., Matthews, E. A., Evans, R. M., Dickenson, A. H., Lipscombe, D., Vergnolle, N., and Zamponi, G. W. (2007). Differential role of N-type calcium channel splice isoforms in pain. *J. Neurosci. 27*, 6363-6373.

Avery, O. T., MacLeod, C. M., and McCarty, M. (1944). Studies on the Chemical Nature of the Substance Inducing Transformation of Pneumococcal Types: Induction of Transformation by a Desoxyribonucleic Acid Fraction Isolated from Pneumococcus. *J. Exp. Med. 79*, 137-158.

Barboric, M., and Peterlin, B. M. (2005). A new paradigm in eukaryotic biology: HIV Tat and the control of transcriptional elongation. *PLoS Biol. 3*, e76.

Bejerano, G., Lowe, C. B., Ahituv, N., King, B., Siepel, A., Salama, S. R., Rubin, E. M., Kent, W. J., and Haussler, D. (2006). A distal enhancer and an ultraconserved exon are derived from a novel retroposon. *Nature. 441*, 87-90.

Bejerano, G., Pheasant, M., Makunin, I., Stephen, S., Kent, W. J., Mattick, J. S., and Haussler, D. (2004). Ultraconserved elements in the human genome. *Science. 304*, 1321-1325.

Birney, E., Stamatoyannopoulos, J. A., Dutta, A., Guigo, R., Gingeras, T. R., Margulies, E. H., Weng, Z., Snyder, M., Dermitzakis, E. T., Thurman, R. E., et al.. (2007). Identification and analysis of functional elements in 1% of the human genome by the ENCODE pilot project. *Nature. 447*, 799-816.

Chih, B., Gollan, L., and Scheiffele, P. (2006). Alternative splicing controls selective trans-synaptic interactions of the neuroligin-neurexin complex. *Neuron. 51*, 171-178.

Chun, T. W., Finzi, D., Margolick, J., Chadwick, K., Schwartz, D., and Siliciano, R. F. (1995). In vivo fate of HIV-1-infected T cells: quantitative analysis of the transition to stable latency. *Nat. Med. 1*, 1284-1290.

Contreras, X., Lenasi, T., and Peterlin, B. M. (2006). HIV latency: present knowledge and future directions. *Future Virol. 1*, 733-745.

Cramer, P., Bushnell, D. A., and Kornberg, R. D. (2001). Structural basis of transcription: RNA polymerase II at 2.8 angstrom resolution. *Science. 292*, 1863-1876.

Crooke, S. T. (2004). Progress in antisense technology. *Annu. Rev. Med. 55*, 61-95.

Dawkins, R. (1976). The Selfish Gene: Oxford University Press).

Deininger, P. L., Moran, J. V., Batzer, M. A., and Kazazian, H. H., Jr. (2003). Mobile elements and mammalian genome evolution. *Curr. Opin. Genet. Dev. 13*, 651-658.

Dekker, J. (2008). Gene regulation in the third dimension. *Science. 319*, 1793-1794.

Dobzhansky, T. (1973). *The American Biology Teacher. 35*, 125-129.

Doolittle, W. F., and Sapienza, C. (1980). Selfish genes, the phenotype paradigm and genome evolution. *Nature. 284*, 601-603.

Esnault, C., Heidmann, O., Delebecque, F., Dewannieux, M., Ribet, D., Hance, A. J., Heidmann, T., and Schwartz, O. (2005). APOBEC3G cytidine deaminase inhibits retrotransposition of endogenous retroviruses. *Nature. 433*, 430-433.

Feschotte, C. (2008). Transposable elements and the evolution of regulatory networks. *Nat. Rev. Genet. 9*, 397-405.

Finzi, D., Blankson, J., Siliciano, J. D., Margolick, J. B., Chadwick, K., Pierson, T., Smith, K., Lisziewicz, J., Lori, F., Flexner, C., et al.. (1999). Latent infection of CD4+ T cells provides a mechanism for lifelong persistence of HIV-1, even in patients on effective combination therapy. *Nat. Med. 5*, 512-517.

Girard, A., and Hannon, G. J. (2008). Conserved themes in small-RNA-mediated transposon control. *Trends Cell Biol. 18*, 136-148.

Gu, W., Pan, F., Zhang, H., Bassell, G. J., and Singer, R. H. (2002). A predominantly nuclear protein affecting cytoplasmic localization of beta-actin mRNA in fibroblasts and neurons. *J. Cell Biol. 156*, 41-51.

Guenther, M. G., Levine, S. S., Boyer, L. A., Jaenisch, R., and Young, R. A. (2007). A chromatin landmark and transcription initiation at most promoters in human cells. *Cell. 130*, 77-88.

Han, K., Xing, J., Wang, H., Hedges, D. J., Garber, R. K., Cordaux, R., and Batzer, M. A. (2005). Under the genomic radar: the stealth model of Alu amplification. *Genome Res. 15*, 655-664.

Han, Y., Lassen, K., Monie, D., Sedaghat, A. R., Shimoji, S., Liu, X., Pierson, T. C., Margolick, J. B., Siliciano, R. F., and Siliciano, J. D. (2004). Resting CD4+ T cells from human immunodeficiency virus type 1 (HIV-1)-infected individuals carry integrated HIV-1 genomes within actively transcribed host genes. *J. Virol. 78*, 6122-6133.

Hong, E. J., West, A. E., and Greenberg, M. E. (2005). Transcriptional control of cognitive development. *Curr. Opin. Neurobiol. 15*, 21-28.

Huttelmaier, S., Zenklusen, D., Lederer, M., Dictenberg, J., Lorenz, M., Meng, X., Bassell, G. J., Condeelis, J., and Singer, R. H. (2005). Spatial regulation of beta-actin translation by Src-dependent phosphorylation of ZBP1. *Nature. 438*, 512-515.

Isalan, M., Lemerle, C., Michalodimitrakis, K., Horn, C., Beltrao, P., Raineri, E., Garriga-Canut, M., and Serrano, L. (2008). Evolvability and hierarchy in rewired bacterial gene networks. *Nature. 452*, 840-845.

Iwashita, S., Osada, N., Itoh, T., Sezaki, M., Oshima, K., Hashimoto, E., Kitagawa-Arita, Y., Takahashi, I., Masui, T., Hashimoto, K., and Makalowski, W. (2003). A transposable element-mediated gene divergence that directly produces a novel type bovine Bcnt protein including the endonuclease domain of RTE-1. *Mol. Biol. Evol. 20*, 1556-1563.

Jelen, N., Ule, J., Zivin, M., and Darnell, R. B. (2007). Evolution of Nova-dependent splicing regulation in the brain. *PLoS Genet. 3*, 1838-1847.

Johnson, J. M., Edwards, S., Shoemaker, D., and Schadt, E. E. (2005). Dark matter in the genome: evidence of widespread transcription detected by microarray tiling experiments. *Trends Genet. 21*, 93-102.

Joyce, G. F. (2002). The antiquity of RNA-based evolution. *Nature. 418*, 214-221.

Kandel, E. R. (2001). The molecular biology of memory storage: a dialogue between genes and synapses. *Science. 294*, 1030-1038.

Kanellopoulou, C., Muljo, S. A., Kung, A. L., Ganesan, S., Drapkin, R., Jenuwein, T., Livingston, D. M., and Rajewsky, K. (2005). Dicer-deficient mouse embryonic stem cells are defective in differentiation and centromeric silencing. *Genes Dev. 19*, 489-501.

Kaplan, C. D., Laprade, L., and Winston, F. (2003). Transcription elongation factors repress transcription initiation from cryptic sites. *Science. 301*, 1096-1099.

Katzman, S., Kern, A. D., Bejerano, G., Fewell, G., Fulton, L., Wilson, R. K., Salama, S. R., and Haussler, D. (2007). Human genome ultraconserved elements are ultraselected. *Science. 317*, 915.

Kiebler, M. A., and Bassell, G. J. (2006). Neuronal RNA granules: movers and makers. *Neuron. 51*, 685-690.

Kouzarides, T. (2007). Chromatin modifications and their function. *Cell. 128*, 693-705.

Lander, E. S., Linton, L. M., Birren, B., Nusbaum, C., Zody, M. C., Baldwin, J., Devon, K., Dewar, K., Doyle, M., FitzHugh, W., et al.. (2001). Initial sequencing and analysis of the human genome. *Nature. 409*, 860-921.

Lareau, L. F., Inada, M., Green, R. E., Wengrod, J. C., and Brenner, S. E. (2007). Unproductive splicing of SR genes associated with highly conserved and ultraconserved DNA elements. *Nature. 446*, 926-929.

Lenasi, T., Contreras, X., and Peterlin, B. M. (2008). Transcriptional Interference Antagonizes Proviral Gene Expression to Promote HIV Latency. *Cell Host Microbe. in press*.

Lewinski, M. K., Bisgrove, D., Shinn, P., Chen, H., Hoffmann, C., Hannenhalli, S., Verdin, E., Berry, C. C., Ecker, J. R., and Bushman, F. D. (2005). Genome-wide analysis of chromosomal features repressing human immunodeficiency virus transcription. *J. Virol. 79*, 6610-6619.

Li, B., Carey, M., and Workman, J. L. (2007). The role of chromatin during transcription. *Cell. 128*, 707-719.

Lipscombe, D. (2005). Neuronal proteins custom designed by alternative splicing. *Curr. Opin. Neurobiol. 15*, 358-363.

Mariner, P. D., Walters, R. D., Espinoza, C. A., Drullinger, L. F., Wagner, S. D., Kugel, J. F., and Goodrich, J. A. (2008). Human Alu RNA is a modular transacting repressor of mRNA transcription during heat shock. *Mol. Cell. 29*, 499-509.

McClintock, B. (1956). Controlling elements and the gene. *Cold Spring Harb. Symp. Quant. Biol. 21*, 197-216.

Miller, S., Yasuda, M., Coats, J. K., Jones, Y., Martone, M. E., and Mayford, M. (2002). Disruption of dendritic translation of CaMKIIalpha impairs stabilization of synaptic plasticity and memory consolidation. *Neuron. 36*, 507-519.

Morgan, T. H. (1911). The Origin Of Five Mutations In Eye Color In Drosophila And Their Modes Of Inheritance. *Science. 33*, 534-537.

Morrish, T. A., Gilbert, N., Myers, J. S., Vincent, B. J., Stamato, T. D., Taccioli, G. E., Batzer, M. A., and Moran, J. V. (2002). DNA repair mediated by endonuclease-independent LINE-1 retrotransposition. *Nat. Genet. 31*, 159-165.

Mu, Y., Otsuka, T., Horton, A. C., Scott, D. B., and Ehlers, M. D. (2003). Activity-dependent mRNA splicing controls ER export and synaptic delivery of NMDA receptors. *Neuron. 40*, 581-594.

Muotri, A. R., Chu, V. T., Marchetto, M. C., Deng, W., Moran, J. V., and Gage, F. H. (2005). Somatic mosaicism in neuronal precursor cells mediated by L1 retrotransposition. *Nature. 435*, 903-910.

Muotri, A. R., Marchetto, M. C., Coufal, N. G., and Gage, F. H. (2007). The necessary junk: new functions for transposable elements. *Hum. Mol. Genet. 16 Spec No. 2*, R159-167.

Muse, G. W., Gilchrist, D. A., Nechaev, S., Shah, R., Parker, J. S., Grissom, S. F., Zeitlinger, J., and Adelman, K. (2007). RNA polymerase is poised for activation across the genome. *Nat. Genet. 39*, 1507-1511.

Odom, D. T., Dowell, R. D., Jacobsen, E. S., Gordon, W., Danford, T. W., MacIsaac, K. D., Rolfe, P. A., Conboy, C. M., Gifford, D. K., and Fraenkel, E. (2007). Tissue-specific transcriptional regulation has diverged significantly between human and mouse. *Nat. Genet. 39*, 730-732.

Okeoma, C. M., Lovsin, N., Peterlin, B. M., and Ross, S. R. (2007). APOBEC3 inhibits mouse mammary tumour virus replication in vivo. *Nature. 445*, 927-930.

Ostertag, E. M., and Kazazian, H. H., Jr. (2001). Biology of mammalian L1 retrotransposons. *Annu. Rev. Genet. 35*, 501-538.

Pennacchio, L. A., Ahituv, N., Moses, A. M., Prabhakar, S., Nobrega, M. A., Shoukry, M., Minovitsky, S., Dubchak, I., Holt, A., Lewis, K. D., et al.. (2006). In vivo enhancer analysis of human conserved non-coding sequences. *Nature. 444*, 499-502.

Ren, B., Robert, F., Wyrick, J. J., Aparicio, O., Jennings, E. G., Simon, I., Zeitlinger, J., Schreiber, J., Hannett, N., Kanin, E., et al. (2000). Genome-wide location and function of DNA binding proteins. *Science. 290*, 2306-2309.

Rosenfeld, M. G., Mermod, J. J., Amara, S. G., Swanson, L. W., Sawchenko, P. E., Rivier, J., Vale, W. W., and Evans, R. M. (1983). Production of a novel neuropeptide encoded by the calcitonin gene via tissue-specific RNA processing. *Nature. 304*, 129-135.

Sasaki, T., Nishihara, H., Hirakawa, M., Fujimura, K., Tanaka, M., Kokubo, N., Kimura-Yoshida, C., Matsuo, I., Sumiyama, K., Saitou, N., et al. (2008). Possible involvement of SINEs in mammalian-specific brain formation. *Proc. Natl. Acad. Sci. U. S. A. 105*, 4220-4225.

Saunders, A., Core, L. J., and Lis, J. T. (2006). Breaking barriers to transcription elongation. *Nat. Rev. Mol. Cell Biol. 7*, 557-567.

Schratt, G. M., Tuebing, F., Nigh, E. A., Kane, C. G., Sabatini, M. E., Kiebler, M., and Greenberg, M. E. (2006). A brain-specific microRNA regulates dendritic spine development. *Nature. 439*, 283-289.

Sciamanna, I., Landriscina, M., Pittoggi, C., Quirino, M., Mearelli, C., Beraldi, R., Mattei, E., Serafino, A., Cassano, A., Sinibaldi-Vallebona, P., et al. (2005). Inhibition of endogenous reverse transcriptase antagonizes human tumor growth. *Oncogene. 24*, 3923-3931.

Sharov, A. A., Dudekula, D. B., and Ko, M. S. (2005). Genome-wide assembly and analysis of alternative transcripts in mouse. *Genome Res. 15*, 748-754.

Silva, A. J., Paylor, R., Wehner, J. M., and Tonegawa, S. (1992). Impaired spatial learning in alpha-calcium-calmodulin kinase II mutant mice. *Science. 257*, 206-211.

Stefani, G., and Slack, F. J. (2008). Small non-coding RNAs in animal development. *Nat. Rev. Mol. Cell Biol. 9*, 219-230.

Steger, D. J., and Workman, J. L. (1997). Stable co-occupancy of transcription factors and histones at the HIV-1 enhancer. *Embo J. 16*, 2463-2472.

Steward, O., and Levy, W. B. (1982). Preferential localization of polyribosomes under the base of dendritic spines in granule cells of the dentate gyrus. *J. Neurosci. 2*, 284-291.

Tam, O. H., Aravin, A. A., Stein, P., Girard, A., Murchison, E. P., Cheloufi, S., Hodges, E., Anger, M., Sachidanandam, R., Schultz, R. M., and Hannon, G. J. (2008). Pseudogene-derived small interfering RNAs regulate gene expression in mouse oocytes. *Nature.*

Ule, J., and Darnell, R. B. (2006). RNA binding proteins and the regulation of neuronal synaptic plasticity. *Curr. Opin. Neurobiol. 16*, 102-110.

Ule, J., and Darnell, R. B. (2007). Functional and Mechanistic Insights From Genome□Wide Studies of Splicing Regulation in the Brain, Vol 623: Springer).

Ule, J., Jensen, K. B., Ruggiu, M., Mele, A., Ule, A., and Darnell, R. B. (2003). CLIP identifies Nova-regulated RNA networks in the brain. *Science. 302*, 1212-1215.

Ule, J., Stefani, G., Mele, A., Ruggiu, M., Wang, X., Taneri, B., Gaasterland, T., Blencowe, B. J., and Darnell, R. B. (2006). An RNA map predicting Nova-dependent splicing regulation. *Nature. 444*, 580-586.

Ule, J., Ule, A., Spencer, J., Williams, A., Hu, J. S., Cline, M., Wang, H., Clark, T., Fraser, C., Ruggiu, M., et al. (2005). Nova regulates brain-specific splicing to shape the synapse. *Nat. Genet. 37*, 844-852.

Ullrich, B., Ushkaryov, Y. A., and Sudhof, T. C. (1995). Cartography of neurexins: more than 1000 isoforms generated by alternative splicing and expressed in distinct subsets of neurons. *Neuron. 14*, 497-507.

Walsh, C. P., Chaillet, J. R., and Bestor, T. H. (1998). Transcription of IAP endogenous retroviruses is constrained by cytosine methylation. *Nat. Genet. 20*, 116-117.

Watson, J. D., and Crick, F. H. (1953). Molecular structure of nucleic acids; a structure for deoxyribose nucleic acid. *Nature. 171*, 737-738.

Wender, P. A., Kee, J. M., and Warrington, J. M. (2008). Practical synthesis of prostratin, DPP, and their analogs, adjuvant leads against latent HIV. *Science. 320*, 649-652.

Wheeler, D. A., Srinivasan, M., Egholm, M., Shen, Y., Chen, L., McGuire, A., He, W., Chen, Y. J., Makhijani, V., Roth, G. T., et al. (2008). The complete genome of an individual by massively parallel DNA sequencing. *Nature. 452*, 872-876.

Williams, S. A., Chen, L. F., Kwon, H., Ruiz-Jarabo, C. M., Verdin, E., and Greene, W. C. (2006). NF-kappaB p50 promotes HIV latency through HDAC recruitment and repression of transcriptional initiation. *Embo J. 25*, 139-149.

Williams, S. A., Kwon, H., Chen, L. F., and Greene, W. C. (2007). Sustained induction of NF-kappa B is required for efficient expression of latent human immunodeficiency virus type 1. *J. Virol. 81*, 6043-6056.

Wyrick, J. J., and Young, R. A. (2002). Deciphering gene expression regulatory networks. *Curr. Opin. Genet. Dev. 12*, 130-136.

Zhong, H., Voll, R. E., and Ghosh, S. (1998). Phosphorylation of NF-kappa B p65 by PKA stimulates transcriptional activity by promoting a novel bivalent interaction with the coactivator CBP/p300. Mol. Cell. *1*, 661-671.

In: Philosophical Insights about Modern Science
Eds: Eva Žerovnik, Olga Markič and Andrej Ule
ISBN: 978-1-60741-373-8

Chapter 3

UNDERSTANDING MOLECULAR BACKGROUND OF ALZHEIMER'S DISEASE: IN SEARCH FOR A CURE

Eva Žerovnik*

Department of Biochemistry, Molecular and Structural Biology,
Jožef Stefan Institute, Jamova 39, 1000 Ljubljana, Slovenia

SYNOPSIS

Many neurodegenerative diseases, such as Alzheimer's, Parkinson's and prion disease might be a result of similar molecular defects. So far there is little evidence that a bacterial or viral infection contribute to these diseases. Instead, the Prusiner's "protein only" hypothesis proposes that a misfolded protein alone might be the infective agent, at least for prion diseases. In the case of the so called "conformational disorders", among them systemic amyloidoses, various dementias and other neurodegenerative diseases, the patient's own misfolded proteins seem to present the underlying (primary) cause.

To repeat the basics: proteins are biological macromolecules composed of 20 kinds of amino-acids, where usually more than 100 such amino acids make a one-dimensional chain (primary sequence), which bends and twists into alpha helices, beta turns and beta strands (secondary structure), which then fold into regular and unique, three-dimensional structure (tertiary structure). An accepted dogma of molecular biology states that one gene (DNA) codes for one protein with defined structure and function, and that the information is one way. However, the protein structure (usually but not necessarily at the energetic minimum) can exert different functions, depending on gene splicing, post-translational modifications, oligomeric state, context and localization in the cell. Proteins are enzymes, transporting molecules, receptors for various ligands (receptors are usually imbedded in cell membranes), and cytoskeleton building constituents. More and more evidence is gained that multiprotein complexes, such as ubiquitin proteasome system or microtubular transport system, perform complex tasks in the cell.

Sometimes, especially when cellular conditions change from normal, which can happen in strong immune response or after intoxication by heavy metals such as Al^{3+}, the proteins can get misfolded and tend to accumulate in the cell. After infection causing fever, when heat shock proteins get overexpressed in order to fight the battle, the misfolded proteins do not get degraded fast enough (proteasome overload), which adds to

* e-mail: eva.zerovnik@ijs.si

protein aggregation. The so called aggresomes (equivalent to inclusion bodies in prokaryotes) composed of aggregated protein – usually of one kind – are believed to sequester the dangerous material, comparted at the centriole by the nucleus. They eventually get cleared by autophagy.

When the cell defense system fails due to continuous production of a mutated protein or other damages to the cell such as oxidative stress or in normal aging, familial or sporadic neurodegenerative diseases develop. Initially – for years - they are silent with no symptoms. Among most well known neurodegenerative diseases are Huntington's (inherited), Parkinson's (usually sporadic but also familial) and Alzheimer's disease (both familial and sporadic). In Parkinson's disease intra-cytoplasmic Lewy bodies, made of aggregated protein alpha-synuclein, and in Huntington's disease intranuclear inclusions made of the protein huntingtin with extended poly-glutamine repeats, accumulate. In Alzheimer's disease (AD), extracellular plaques from amyloid-beta (A-beta) peptide and intracellular tangles of aggregated Tau protein, are observed.

It could also be that protein aggregates would not be the cause of neurodegeneration but rather a response to some other trigger or even a means of defense (Howlett, 2003). However, evidence from inherited cases shows the opposite to be the case. Evidence has accumulated that soluble oligomers rather than mature amyloid fibrils (Haas & Selkoe, 2007) cause a cascade of detrimental events in the cell. The »channel hypothesis« of AD states that amyloid oligomers interact with cellular membrane, causing an influx of Ca^{2+} ions in neurons, which is an early sign of pathology and is responsible for uncontrolled neurotransmission. Another possible cause, especially for sporadic cases, could be metal ions, such as Zn^{2+}, Cu^{2+}, Fe^{3+}, Al^{3+}... The delicate balance of these in the brain is important to prevent neurodegenerative changes.

In this chapter we will limit ourselves to Alzheimer's disease (AD).

We will describe 1) diagnosis options 2) risk factors 3) preventive measures and symptomatic treatments 4) advances in molecular and cellular studies, which should provide broader understanding and 5) new therapies and perspective for a cure. As this book is meant to raise philosophical and ethical questions (not necessarily to give the answers), two last sub-section titles are devoted to such issues.

ABBREVIATIONS

A-beta	amyloid-beta peptide
AD	Alzheimer's disease,
PD	Parkinson's disease,
HD	Huntington's disease,
CJD	Creutzfeldt-Jakob disease,
TSE	transmissible spongiform encephalopathy

INTRODUCTION

Alzheimer's disease is going to become an enormous economical burden if it is not stopped by successful treatments and preventive measures. For example, health care costs for the 4.5 million Americans with Alzheimer's disease exceeded \$100 billion in the year 2003, according to the Alzheimer's Association. In other Western countries the spread of the disease is similar, in contrast to India where AD is much rarer.

As age increases, so does the risk of getting AD. For each five-year age group beyond 65, the percentage of people with AD doubles and nearly half of those over the age of 85 acquire the disease. A small number get an "early-onset Alzheimer's," which strikes people in their 30s to 50's, but most AD cases are among older people. A person with AD lives on average eight years after the onset of symptoms, but some live as long as 20 years.

AD is a disease of the brain. It occurs gradually, starting with mild memory loss, changes in personality and behavior, and a decline in thinking abilities (cognition). It progresses to loss of speech and movement, then total incapacitation and eventually death. It is normal for memory to decline and the ability to absorb complex information to slow as people get older, but AD is not a part of normal aging.

The causes of AD are not totally clear. Researchers know that people with the disease have an abundance of two abnormal structures in the brain: plaques and tangles. Plaques are dense, rather sticky patches made of amyloid-beta (A-beta), a 40 to 42 amino acid long fragment of amyloid-precursor protein (APP). Tangles are twisted paired filaments (fibers) made from the protein called *tau* after it has changed its conformation from a normally folded state to an amyloidogenic precursor state. The A-beta plaques reside in the spaces between the neurons in the brain, and the neurofibrillary tangles accumulate inside the neurons. It also has been observed that transport along the microtubules, which is essential in the long neuron's axon, is slowed down or blocked in AD. This leads to blockage of the communication between the neurons, which affects mental processes, such as thinking, memory, talking and movement. As AD progresses, nerve cells die, the brain shrinks, and the ability to function deteriorates even further.

Role of metal ions imbalance, similarly to Parkinson's disease, is claimed by some researchers as a possible cause for neurodegeneration also in Alzheimer's disease. Increased concentration of metal ions such as zinc and copper can potenciate Alzheimer's disease by participating in the aggregation of normal cellular proteins and in the generation of reactive oxygen species. Aluminum and copper can also propagate an inflammatory response in the aging brain.

Neurodegenerative diseases arise either sporadically with age or they run in families (inherited) with an early onset. Prion diseases can start in 3 ways: sporadic, inherited or infectious. The infectivity apparently distinguishes prion diseases from the rest of conformational disorders (Soto, 1999). However, similarly to prions, for which Soto and co-workers (Soto et al., 2005) have designed a propagating assay, amyloid protofibrils can seed fibril formation if the source of the amyloidogenic protein is provided.

1. Diagnosis Options

Diagnosing Alzheimer's

Scientists are uncovering clues to better diagnose the disease and to determine who is at risk. What is needed are both diagnostic and prognostic tests.

At the moment, AD can be diagnosed conclusively only by examining the brain post-mortem. However, diagnosis on living patients is possible by taking a complete medical history, administering neurological and psychological tests, and doing a physical exam, blood and urine laboratory tests, and a brain-imaging scan. Once symptoms begin, the disease can be diagnosed with up to 90 percent accuracy.

Research suggests that the illness may predate clinical symptoms by years and maybe decades. People show what is known as a mild cognitive decline much before the AD actually bursts out, therefore, the means for early diagnosis followed by selected preventive treatments of those at high risk, are highly desirable.

Advances in neuroimaging, taking pictures of the living brain by magnetic resonance imaging (MRI) and by positron electron tomography (PET), may allow researchers to see the accumulation of plaques and tangles at various points of disease development and thus allow early diagnosis. Neuroimaging may one day prove useful in monitoring the progression of the disease and assessing people's responses to drug treatment.[1]

It also is becoming possible to arrive at an early diagnosis and to confirm the disease from spinal fluid analysis. By tracking the levels in the spinal fluid of A-beta, tau and some other marker proteins[2] it will be possible to trace changes over time from the person's baseline level and make prognosis (start early treatment) in people at risk. It seems likely that spinal fluid testing will become a valuable routine diagnostic tool in the future.

2. Risk Factors

Risk factors for Alzheimer's

AD is multi-factorial, i.e., factors from environment and genes work in combination. Disregarding for a moment genetics, the main risk factors for getting AD are age, exposure to metabolic and environmental risk factors: including life style, food and heavy metals exposure. It is becoming clear that risk factors for AD are shared with other Western society illnesses such as diabetes, obesity, and coronary artery disease.

However, the genetic background of an individual is also important. Diseases such as cystic fibrosis, muscular dystrophy, and Huntington's disease are single-gene disorders. If one inherits the gene that causes one of these disorders, he will usually get the disease. AD, on the other hand, is not caused by a single gene. In proven familial cases more than one gene mutation can cause AD, and genes on multiple chromosomes are involved.

---

[1] New development has been reported at: http://www.medicalnewstoday.com/articles/98241.php

A team of researchers at University College London, has shown that MRI scans of patients with Alzheimer's can be distinguished from those of healthy individuals and patients with other forms of dementia. Computers can identify the characteristic damage of Alzheimer's disease with an accuracy as high as 96%.

The new method works by teaching a standard computer the differences between brain scans from patients with proven Alzheimer's disease and people with no signs of the disease at all. The two conditions can be distinguished with a high degree of accuracy on a single clinical MRI scan. The new method makes an objective diagnosis without the need for human intervention.

[2] Protein markers of disease states, among them neurodegenerative diseases, are getting accessible by new developments of proteomics. Samples from normal blood serum (spinal fluid) and the AD patients serum (spinal fluid) are compared (appr. 200 proteins can be separated and characterised at once). Those which are significantly changed (increased or decreased) are possible markers of the disease. If the same is done for some other similar pathology, characteristic changes of proteome for AD will be identified.

Genes Responsible for Early Onset AD

Scientists have identified several genes that play a role in early-onset AD, a rare form of the disease that strikes people in their 40s. Three genes were identified that, when mutated, cause AD: the *Aβ amyloid precursor protein gene* (*APP*), the *presenilin 1 gene* (*PSEN1*) and the *presenilin 2 gene* (*PSEN2*) (see, Ghidoni et al., 2007). Together, these mutations are responsible for 30 to 50% of autosomal dominant AD cases. It is likely that there are additional genes that can cause Early Onset Familial AD that have not yet been identified. Although mutations in the known genes are a rare cause of AD (about 0.5% of AD cases), they are important for presymptomatic diagnostics of patients. Also, the identification of the genes involved has been extremely important to the recent progress in the understanding of the biology of AD.

Genetic testing to look for mutations in the *PSEN1*, *PSEN2*, and *APP* genes is available for individuals who have a family history of Early Onset Familial AD. Such testing may inform a person whether or not they have inherited a mutation in one of these three genes. This type of testing will not always be informative because not all families have mutations in one of these three genes. Even if one of the mutations is confirmed no prognosis can be made of when the disease will break out and how severe it will be; as this varies from individual to individual.

Genes Responsible for Late-Onset Disease

For late-onset AD, defined as the disease showing symptoms after the age of 65, a gene that produces a protein called apolipoprotein E (ApoE) appears to play an important role. The gene codes for a protein that transports cholesterol[3] in the bloodstream.

The gene comes in several forms, or alleles. Having the ApoE ε4 allele increases life style risks for getting AD (Kivipelto et al., 2008), especially so in combination with K-variant of butyrylcholinesterase gene (Lane et al., 2008). About 40 percent of people with AD have the ApoE ε4 allele, but inheriting it doesn't mean a person will definitely get AD. Some people with the gene never get the disease, and some without it, do.

Another risk factor for late onset AD might be cystatin C gene (Finckh et al., 2000) which also is connected to brain hemorrhage in cerebral amyloid angiopathy and stroke (Levy et al., 2007). An allelic form of cystatin C gene leads to impaired secretion of the protein (Benussi et al, 2003).

Certainly, other causative and risk genes are involved in AD and need to be identified to fully elucidate the etiology of AD. Ultimately, this will lead to the development of effective therapies for this major disease.

There is a question, also bearing on ethics (see, heading Ethical concerns), if people at risk should be TESTED of gene mutations at all.

---

[3] There is evidence for a strong link between AD and ateriosclerosis. In both cases impaired cholesterol metabolism is a risk factor. It is well known, however, that aterioselerosis can efficiently be reduced by exercise and appropriate diet.

3. Preventive Measures and Symptomatic Treatments

Delaying the Disease

Scientists continue to search for treatments to slow the progress of AD and prevent the outbreak of the disease for as long as possible. Cholesterol-lowering drugs, anti-inflammatory drugs, antioxidants, and estrogen are some of the substances that are claimed to be of some use in slowing down AD progression, but study results have been conflicting. Even some alternative drugs or, better said - food additives, such as curcumin (Yang et al., 2005, Zhang et al., 2006), red wine and green tee polyphenols have been claimed as beneficial

That nicotine is enhancing cognitive abilities has also been declared. Unfortunately, its regular use brings other detrimental consequences, especially, the lung cancer. Notwithstanding the problem of addiction, smoking cannabis might be less harmful than smoking nicotine cigarettes and could be of some use for patients with diagnosed AD. Interestingly, it has been proved that the active component of marijuana, Δ^9-tetrahydrocannabinol (THC), competively inhibits the enzyme acetylcholinesterase (AChE), thus increasing the acetylcholine levels, as well as prevents amyloid-beta peptide aggregation, the key pathological marker of Alzheimer's disease (Eubanks et al., 2006).

Healthy lifestyle behaviors such as exercising, healthy food, avoiding too much physical and especially psychological stress, may also be of value in protecting people from AD. Some studies have shown that participating in mentally stimulating activities, such as reading books, solving crossword puzzles, playing chess, etc., may be associated with a reduced risk of AD. Similarly would be expected for intellectually challenging jobs, such as doing scientific research or teaching.

Lowering cholesterol: studies have shown a link between high blood pressure, high cholesterol levels and diets high in saturated fats with an increased risk for AD. There is also evidence that an elevated level of homocysteine, an amino acid in the blood, presents a risk for both heart disease and AD. Further, taking cholesterol-lowering drugs (statins) is associated with a lower occurrence of AD. In addition to statins, substances being tested for slowing and preventing AD are folate and vitamins B6 and B12, which lower homocysteine levels.

Anti-diabetes drugs: Knowing that diabetes and high blood pressure represent risk factors, preventing these correlates positively with lowering a chance to get AD. One such approved antidiabetic drug is Rosiglitazone (RSG), a peroxisome proliferator-activated receptor g (PPARg) agonist. It is an insulin sensitizing agent that allows the body to use endogenous insulin more efficiently, maintain normal physiological feedback mechanisms and produce anti-inflammatory response (Risner et al., 2006).[4]

Anti-inflammatory drugs, such as aspirin and ibuprofen, may in addition to reduce joint inflammation and pain, also reduce the inflammation in the brain associated with AD. None of the studies performed with the anti-inflammatory drugs to date is definite.

---

[4] There is a firm correlation between the obesity, type 2 diabetes and AD, which all could be prevented by life style changes (physical and mental life) and healthy food. In a Special Issue: Metals in Alzheimer's Disease: http://www.j-alz.com/press/2006/20061117.html : Solfrizzi V and co-authors review the possible role of macronutrients and the basic elements of carbohydrates, proteins, and fat in the development of AD. They suggest that healthy diets, antioxidant supplements, and the prevention of nutritional deficiencies or exposure to foods and water with high content of metals could be considered the first line of defense against the development and progression of cognitive decline.

Anti-oxidants: researchers are also looking at antioxidants, such as vitamin E, vitamin C and carotene (the most powerful anti-oxidant is produced from the bark of Mediterranean pine tree – known under the commercial name "Pycnogenol" and Maritime Pine Plus[5]), to possibly prevent cognitive decline. Antioxidants help by breaking down free radicals, which accumulate in AD brain. Why is this happening is not totally clear, it was reported that the plaques (extracellular aggregates of A-beta peptide) interact with metal ions, leading to free radicals (reactive oxygen species – ROS) release, whereas inside the cell the metal balance is disturbed and free radicals are consequently not cleared. The natural defense by enzymes superoxide dismutase and nitrogen oxide synthetase protect cells against free radicals, however, these protective mechanisms decline as a person ages. It should be mentioned that studies did not always confirm beneficial effects of anti-oxidants and there is no definite conclusion.

Estrogen: some studies have linked the female hormone estrogen to improved memory and possible delay or prevention of AD in women. But a large, long-term clinical trial sponsored by the NIH has provided evidence to the contrary! The study, published in *JAMA*[6], found that the hormone combination of estrogen and progestin given to women over 65 of age did not protect against AD (quite the contrary) and even not against the mild cognitive impairment.

Testosterone: positive effects of testosterone supplementation on cognition of elderly, otherwise healthy men, have been reported in some studies (Driscoll and Resnick, 2007). It was found that spatial memory is mediated through an androgen pathway, whereas verbal memory is mediated through an estrogen-dependent pathway. It should be noted that only a range of concentrations was beneficial (not too much not too low), and that the effect might not prove the same in men affected by AD. It could also be imperative, as proved with the estrogen supplementation study in 65 years old women (see, footnote 6) who deteriorated rather than improved, to start such intervention early enough to prevent irreversible loss of memory. The authors also warn that much more research is needed before recommendations for clinical practice can be made.

Balancing metal ions: A group from Melbourne led by Bush AI has developed chelating drugs, which bind metals released from amyloid plaques made of A-beta (or proteinaceous aggregates of α-synuclein) and in this way help to disintegrate the aggregate (Bush, 2001; Crouch et al., 2007). More on this issue will be explained under Part 4) Advances in Molecular and Cellular Studies.

Treating the Symptoms

At the moment, there is no cure for AD, so treatments are more or less symptomatic. There are four approved prescription drugs for people with mild-to-moderate AD: tacrine, donepezil, rivastigmine and galantamine. All of them work by the same mechanism, i.e., they increase the level in the brain of acetylcholine - a chemical – neurotransmitter, which is deficient in AD. The drugs work by inhibiting an enzyme called cholinesterase that breaks

---

[5] double-blind research studies relating to Maritime Pine Plus® 's have shown effectiveness for such conditions as Alzheimer's disease and immune system dysfunction. Also, an internationally-respected Midwest psychiatric center has started a landmark research study to document the compound's efficacy for ADD/ADHD.

[6] Shumaker SA et al. 2003 : Estrogen Plus Progestin and the Incidence of Dementia and Mild Cognitive Impairment in Postmenopausal Women: The Women's Health Initiative Memory Study: A Randomized Controlled Trial. JAMA 289: 2651 - 2662.

down the acetylcholine and thus increase its levels in the brain. The cholinesterase inhibitors have an effect on the symptoms, but there is no evidence that they have any effect on the underlying progression of the disease. Another drug, memantine, is approved to treat people with moderate-to-severe Alzheimer's disease. This drug blocks the action of glutamate, another neurotransmitter. Glutamate is thought to be increased in people with AD, which produces intoxication and neural death. Thus, blocking it's receptors is a way to spare neurons.

One can conclude that regular drugs mainly act on improving the cognition, at least initially, by increasing brain choline levels and in advanced stage to prevent glutamate neurotoxicity. However, in later stages other symptoms, such as psychiatric, have to be treated, which is more difficult as the patients are very sensitive to deleterious side effects of most antipsychotics, even the newest atypical ones – such as quetiapine.[7]

4. Advances in Molecular and Cellular Studies

If researchers want to improve the treatment options and the design of new drugs they need to know more details of the molecular processes leading to the disease. Basic science has provided us with several important clues in recent years.

Hypotheses for the Molecular Origin of AD

"Amyloid Cascade Hypothesis"

The first hypothesis (Hardy & Higgins, 1992) was that AD started with the accumulation of amyloid plaques, and that limiting this accumulation would change the progress of AD. This did not prove to be the case as correlation between amyloid plaques load and symptoms of the disease was not very high. Next to look at was *tau*, a microtubule binding protein. Indeed, in neurons from AD brain there were neurofibrillary tangles made of hyperphosphorylated *tau*. However, there were no genetic cases reported in the *tau* gene making a strong argument that this is not the primary cause.

Familial forms of AD are in one way or another connected to amyloid precursor protein processing and the resulting A-beta peptides production. These peptides, especially of the length 1-42, oligomerise and aggregate in the cell, before they get expelled into interneural space, where they form plaques. As said, the plaques load does not correlate with severity of the symptoms.

A modified amyloid (cascade) hypothesis was therefore formulated (Wirths, Multhaup, Bayer, 2004), which related intraneuronal A-beta generation to its pathological function in cell culture, transgenic AD mouse models and post mortem brain tissue of AD patients, as well as its connection to oxidative stress and apoptosis.

More and more evidence has been gained that soluble, prefibrillar A-beta oligomers might be harmful to cells, rather then extracellular fibrillar deposits. The oligomers of

---

[7] Zhong KX (2007). Quetiapine to treat agitation in dementia: a randomized, double-blind, placebo-controlled study. Curr Alzheimer Res. 4, 81-93. Rainer et al., (2007). Quetiapine versus risperidone in elderly patients with behavioural and psychological symptoms of dementia: efficacy, safety and cognitive function. Eur Psychiatry 22, 395-403.

pathological or non-pathological amyloidogenic proteins (Stefani et al., 2003, Čeru et al., 2008) were shown toxic in cellular studies. If A-beta oligomers derived from AD tissue were injected in the brain of mice, the animals temporarily performed worse in tasks connected to memory (Walsh et al., 2002; Cleary et al., 2005, Walsh & Selkoe, 2007).

"Channel Hypothesis"

The "channel" hypothesis of AD (Kagan et al., 2002) states that the so called "amyloid toxins" – they are now believed to be annular oligomers rather than mature fibrils - perturb cellular membranes or even make pores (Lansbury& Lashuel, 2005). Apart from A-beta, at least six other amyloidogenic peptides were shown to make pores into membranes (Kawahara et al., 2000). The amyloid/membrane interaction or perforation most likely triggers entry of Ca^{2+} and changes cell signaling. If the mitochondrial membrane is affected, the energy of the cell drops and eventually the cell undergoes apoptosis (Stefani & Dobson, 2003; Bucciantini et al., 2005). The hypothesis still needs experimental evidence for human neurodegenerative pathologies.

"Cholesterol / Lipids / Amyloid Interactions"

Except from direct perforation, other more specific membrane interactions may take place. For example, gangliosides bind A-beta and change its conformation to more β-sheet (Matsuzaki, 2000). There are several reports indicating that A-beta initially accumulates in the fractions with the lipid composition similar to that of lipid rafts (Tun et al., 2002; Ehehalt et al., 2003). It has been found that binding of A-beta to ganglioside GM1 is significantly accelerated in cholesterol-rich domains of membranes. Several lines of evidence suggest that apolipoprotein E modulates the distribution and metabolism of cholesterol in neuronal membranes in an allele-dependent manner (Kivipelto et al., 2008).

"Metal Hypothesis" and Oxidative Stress

There is evidence that imbalance in metal ions is taking place in all neurodegenerative diseases, which in turn increases free radical levels and impairs intracellular signaling, etc, ending in neuron death. The "metal" hypothesis of AD, proposed in 1990 by A.I. Bush and C. Masters from the University of Melbourne, sees the disease as a result of abnormal metal metabolism in the brain (Maynard et al., 2005). This hypothesis proposes that copper ions released from neurons bind with A-beta to catalyse the production of hydrogen peroxide; as the hydrogen peroxide breaks down chemically, it generates free radicals that damage proteins and nucleic acids in neural tissue.

In a special double issue of the *Journal of Alzheimer's Disease* published in November 2006, edited by Miu and Benga , the role of metals in the biochemistry and physiology of AD has been described.[8] The articles cover six major categories: Comprehensive historical

[8] Taken from http://www.j-alz.com/press/2006/20061117.html : Savory J, Herman MM and Ghribi O review the controversial outlooks on aluminum neurotoxicity, examining data on the possible cellular mechanisms underlying aluminum neurotoxicity and potential neuroprotective strategies against that. In the next review, Adlard PA and Bush AI discuss how metal ions such as zinc and copper can potentiate Alzheimer's disease by participating in the aggregation of normal cellular proteins and in the generation of reactive oxygen species. In the third review article, Campbell A focuses on how aluminum and copper can initiate or propagate an inflammatory response in the aging brain. Exley C reviews in-vitro studies of metals found in plaque cores in AD brains and concludes that aluminum and iron could cause oxidative damage but copper and zinc likely do not. The movement of metals across the blood-

reviews, methodological perspectives, a topical review, integrative genetic and epigenetic reports, and a review of risk factors. Lots of work has been put in developing drugs which bind metals from amyloid plaques and balance metals intraneurally (Bush, 2001; Crouch et al., 2007). Recently, new successful drugs of this type have been announced.[9]

It is of note that prion and APP and even α-synuclein all bind copper. Copper balance may thus be of particular importance in preventing neurodegenerative disorders.

Some researchers claim that oxidative damage together with lowered protein synthesis and degradation potential in aged brains, might be the initial trigger for AD development (Ding et al., 2008). They have found that in persons with mild cognitive impairment (MCI) there was an increased level of oxidative stress. This damages proteins, which aggregate more and do not get cleared by proteasome but rather by autophagy.

"Axonal Transport " Hypothesis

There exist even more hypotheses of AD pathology origin, among them the "Axonal transport" one, which claims that components of APP, especially the C-terminal peptide, make integral part of a multi-protein complex regulating microtubular transport along the axon (Stokin et al., 2005) and that this transport gets slowed down in AD much before any plaques accumulate.

Why are the cholinergic neurons most affected in AD and the dopaminergic ones in Parkinson's, still needs a better explanation. Perhaps, the differences between various sites / kind of neurons / of neurodegeneration in different neurodegenerative diseases arise from »loss of function« of a particular affected gene/protein, whereas the more general symptoms may be due to a gain in »toxic function« exerted by the toxic aggregates or/ and accompanying oxidative stress, perturbed axonal transport, chaperones system and ubiquitin proteasome system overload.

Protein Aggregation Studies

Normal proteins under stressful conditions for the cell or at over expression and most of all pathological mutants or fragments can undergo misfolding and aggregation. The aggregates make cells less viable, therefore, mechanisms exist to sequester them and make them less harmful. After Kopito (Kopito, 2000, Kopito & Sitia, 2000) the inclusions, similar

brain barrier is reviewed by Yokel RA. A number of transporters are described that could mediate metal transport into and out of the brain. He reviews the role of these transporters in moving aluminum, manganese, iron and other metals across the blood-brain barrier. Dolev I and Michaelson DM write about the apoE4 isoform of apolipoprotein E. Their study of the nucleation, growth and reversibility of A-beta deposition in mice should shed new light on this genetic risk factor for AD.

[9] Cited from: http://www.biotechnews.com.au/index.php/id;1827236977 "In a seminal experiment in 1996, A. Bush showed that clioquinol, an old 1960s remedy for traveller's diarrhoea, rapidly dissolves amyloid plaques by removing the copper and zinc ions that cause them to aggregate - the metal ions pin the peptide fragments together. Last year Prana (a Biotechnology Co., Melbourne) decided to move to a second-generation compound, based on the structural theme of clioquinol - PBT2 showed much more potent activity than clioquinol in all pre-clinical tests, including its ability to disaggregate purified amyloid plaque in vitro. Tested in AD mice, it proved significantly superior in reducing amyloid plaque concentrations in the brain.

to inclusion bodies in prokaryotes, which have been observed in eukaryotic cells, are termed "aggresomes".

Assuming that the basic physics and chemistry are the same *in vitro* and in the cell, in order to understand the molecular backgrounds of neurodegeneration (Goedert et al., 1998; Žerovnik, 2002, Ross & Poirier, 2004), studying the mechanisms of amyloid fibril formation and amyloid induced toxicity is meaningful. Any protein seems to be able to form amyloid-like fibrils and therefore model systems are being used *in vitro*. Understanding the process at molecular level seems important to be able to stop it at a particular stage. The toxic oligomers could be stopped from forming or, alternatively, their removal could be accelerated. Antibodies directed against a common structural epitope shared by the prefibrillar oligomers (Kayed et al., 2003) could serve such a role. Indeed they proved to reduce toxicity when applied together with an "amyloid toxin".

In Figure 1, are shown various morphologies of the prefibrillar aggregates and fibrils - as observed *in vitro*.

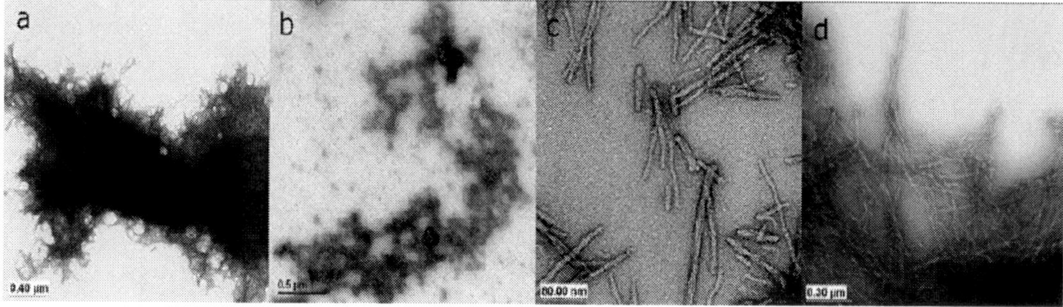

Figure 1. Different types of protein aggregates: amorphous and granular (a,b) protofibrils (c) and fibrils (d).

Common structural characteristics of amyloid fibrils are: predominantly β-sheet secondary structure, increased Thioflavin T and Congo Red binding and characteristic pattern seen by X-ray diffraction (Figure 2A,B). Electron microscopy images reveal unbranched fibrils with some repeats, which can come from twisting of the constituent protofilaments.

Oligomers of certain size are believed to interfere with membrane permeability as a means for toxicity. Amyloid-toxins may bind from within to intra-cellular membranes, such as mitochondrial or they can bind from the blood-stream to the plasma membrane. There are different possibilities, how they could permeate the membrane: by directly making pores (channels) (Kawahara et al., 2000, Kagan et al. 2002), or, by binding to gangliosides at lipid rafts (Tun et al., 2002, Ehehalt et al., 2003), resulting in changed intracellular Ca^{2+} concentration and/or affecting intracellular signaling.

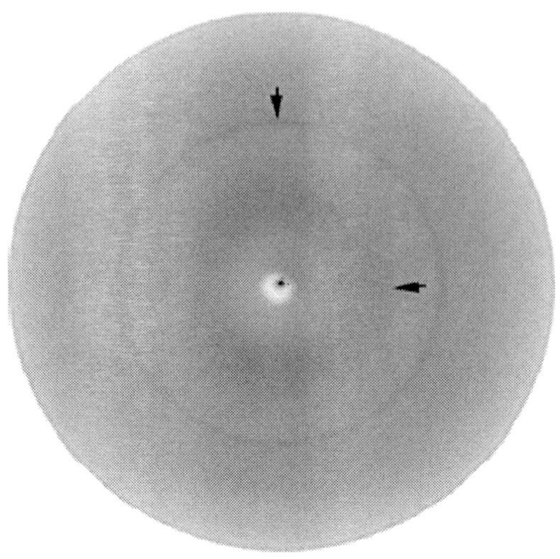

Figure 2 A. X-ray diffraction pattern of stefin B fibrils. The two distances are at 4.7 Å and at ~10 Å.

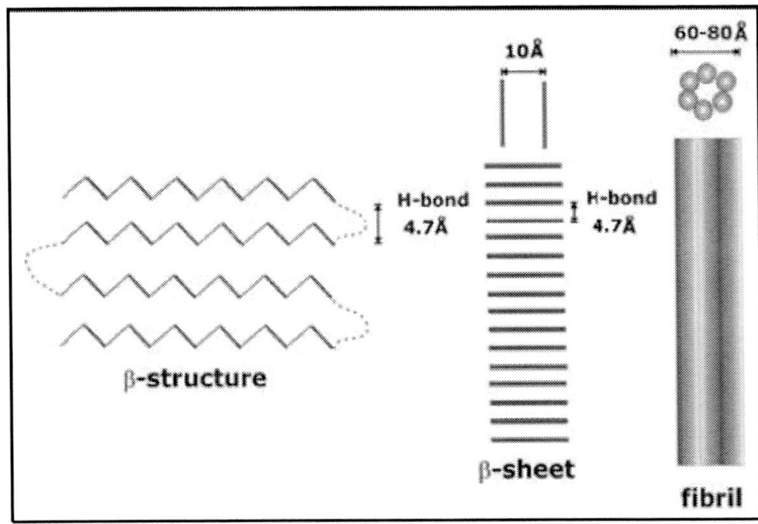

Figure 2B. Distances among β structure elements: β-strands and β-sheet For adaptation of this Figure we thank Saša Jenko Kokalj.

Stefani et al., 2003, proposed that toxicity of the prefibrillar oligomers is a phenomenon in common to pathological and non-pathological proteins. To contribute to the still open questions on the source of toxicity, the present author and co-workers (Anderluh et al. 2005) have studied prefibrillar oligomers toxicity and membrane binding of human stefin B (cystatin B), a protein not involved in any known amyloid pathology. The aggregates bound predominantly to acidic phospholipids such as phoshatidyl glycerol, which also was observed for other amyloidogenic proteins. In line with the "channel hypothesis", correlation between membrane binding and toxicity was demonstrated (Anderluh et al., 2005). It also was shown that the higher oligomers in range of 6 - 16 mers, which were the first toxic entity, bound the strongest in comparison to lower oligomers (Čeru et al., 2008). Finally, it was shown that stefin B indeed perforates artificial lipid membranes (Rabzelj et al., 2008).

Even though amyloid fibril formation is a generic property of proteins (Guijarro et al., 1998), different proteins most likely follow different pathways towards amyloid fibrils, depending on sequential details (Chiti et al, 2000, Lopez de la Paz et al., 2004) and the structural class (Pellarin and Caflish, 2006). Folded and unfolded proteins seem to be able to transform to amyloid-fibrils. A representative of the natively unfolded proteins are: α-synuclein, A-beta and prion fragments, τau and of the folded proteins: β2-microglobulin, transtyretin, cystatins C and B and lysozyme. On the route to the mature fibrils, at least one kind of intermediates accumulate (Walsh & Selkoe, 2007), the so called prefibrillar oligomers, which can line up as short chains, protofibrils (Walsh et al., 2002) or form annular shapes. In is still ongoing debate whereas such oligomeric intermediates are on- or off-pathway to fibrils. There might be parallel pathways leading to amorphous or granular aggregates, protofibrils and fibrils (El Moustaine et al., 2008). It is possible to inhibit the process at several points: at the stage of globular oligomers or from monomers to fibrils, which is important to know when designing aggregation inhibitors as drugs. As well, disaggregation is possible only till a certain stage.

A common underlying mechanism seems nucleation (Plakoutsi et al., 2005). By one scenario (Serio et al., 2000), a conformational change takes place within an oligomeric nucleus, which allows further fibril growth. The newest outlook on the mechanism proposes a 3-stage mechanism: unfolding, nucleation and fibril elongation (Lee et al., 2008). This is very much reminiscent to our model for stefin B fibrillation (Škerget et al., 2008) obtained from the kinetics.

5. Future Perspectives

There is hope to treat "conformational" diseases, among them AD, at their routes by gaining more molecular understanding. New drugs are emerging from the basic science laboratories and moving from testing in cell and animal models towards human trials.

Emerging Therapies for AD

β and γ Secretase Inhibitors

A-beta peptide is produced by proteolytic cleavage in the membrane environment by β and γ secretases. These enzymes are classified as aspartic proteases, the same type of enzymes that are involved in HIV infection and whose inhibitors are used as drugs for treatment AIDS. Inhibitors for β and γ secretases, which are expected to block A-beta formation, are being developed and tested.

Acetylcholinesterase inhibitors, such as donepezil, rivastigmine and galantamine, appear to cause selective muscarinic activation of α-secretase and to induce the translation of APP mRNA; they also seem to diminish amyloid-fibrils formation from A-beta. Activation of N-methyl-D-aspartate receptors is considered a probable cause of chronic neurodegeneration in

AD, and memantine is being used to block glutamate N-methyl-D-aspartate receptors (Zimmerman et al., 2005).

Vaccination

Another approach to combate AD is to stimulate the body's immune system to destroy the amyloid plaques. Scientists have developed a vaccine by injecting A-beta into the blood in the hope of making antibodies to destroy the plaques. The vaccine was successful in transgenic mice, i.e., special mice into which human genes were transferred, that caused them to develop AD-like plaques (Tampellini et al., 2007) But when tested in a human trial, some people showed inflammation of the brain (encephalitis). Further vaccination was stopped, but study participants continue to be followed. Although this particular vaccine may be disappointing, many scientists believe that the strategy of fighting AD by stimulating the immune system still remains an important potential avenue to slow or prevent the disease (Wisniewski & Sigurdsson, 2007).

Metal Complexing Compounds

Beneficial but also somewhat controversial action is claimed for metal complexing compounds. Originally, clioquinol was thought to work by chelating the metal ions, rendering them non-reactive. But it now appears that the drug mediates a process that redistributes the metal ions from their abnormal compartment, outside neurons, back into the neurons themselves. In the new picture, the amyloid plaques fall apart, and the metal ions diffuse back into neurons, restoring their function and causing them to regrow new dendrites.

Recent pre-clinical studies reports by Prana, Co, Bush AI co-funder, are very stimulating. Based on the structural backbone of clioquinol – a new substance PBT2 showed much more potent activity than clioquinol in all pre-clinical tests, including its ability to disaggregate purified amyloid plaque *in vitro*. PBT2 also shows superior solubility to clioquinol, and other characteristics that suggest it should be able to cross the blood-brain barrier in high concentrations. Tested in AD mice, PBT2 proved to significantly reduce amyloid plaque concentrations in the brain. And what is more, the mice recovered very fast.

Researchers at the University of Utrecht in the Netherlands successfully completed phase 1 (safety and tolerability) clinical testing of PBT2 in normal volunteers in March 2006, showing that it was well tolerated at the doses proposed for Alzheimer's therapy. After, http://www.biotechnews.com.au/index.php/id;1827236977

Also the Phase 2a clinical trial of PBT2 was successful. In this double blind multi-centre clinical trial, 78 patients in Sweden and Australia were randomized to receive either a placebo, PBT2 50mg or PBT2 250mg capsule once per day for 12 weeks. [10] http://www.pranabio.com/company_profile/press_releases_item.asp?id=152

[10] **Analysis of the trial data demonstrated that the safety and tolerability profile of PBT2 at both doses was indistinguishable from that of placebo. There were no study withdrawals related to adverse events. There was no serious adverse event (SAE) in any PBT2 treated patient. The study also demonstrated the impact of PBT2 on reducing Abeta 42 in the cerebrospinal fluid (CSF) that surrounds the brain and spinal cord, considered a key biomarker for Alzheimer's Disease. Specifically, PBT2 at the 250mg dose showed a highly significant reduction in CSF Abeta 42 compared to placebo (p=0.006).**

Increase Degradation And Clearance

A-beta peptide itself has been reported as target of a neutral protease neprilysin (Iwata et al., 2004, Farris et al., 2007). The main two degradation routes for mis-folded or unfolded peptides in the cell are ubiquitin proteasome system (UPS) and autophagy (Ding et al., 2007). Even though ubiquinated proteins and UPS connected enzymes are often found in plaques, another clearance route to clear the aggregates operates in the cell: autophagy (Cuervo, 2004). In Huntington's disease models they have found beneficial effects of small molecules, which enhance autophagy (Ravikumar et al., 2006, Sarkar et al., 2007), one of them is a known drug rapamycin. This latter seems to alleviate toxicity of a range of aggregating peptides (Berger et al., 2006). One could imagine that a controlled starvation as a means to induce autophagy could work similarly but ...this may raise some ethical concerns, similarly as giving patients marihuana.

A mini-review paper "Therapeutic approaches for prion and Alzheimer's disease" (Wisniewski & Sigurdsson, 2007) has just appeared at the time of writing this text and can be considered for up-to-date information. Among the discussed treatments are: vaccination trials, inhibition of fibrillation by the so called β-sheet breakers (those with an improved permealization of BBB) and metal chelation in both types of the diseases.

Very exciting are also the new treatments for Huntington's disease (HD) and Parkinson's disease (PD). In HD, gene or protein replacement together with enhanced clearance of the mutant protein might lead to cure and not only to delay of symptoms. In PD, stem cells producing dopamine are being grafted in the brain regions affected with quite some success. Especially, if stem cells from non-faetal tissue will be developed and gene banks of people own fetal cells will be broadly available, the neurodegenerative diseases could indeed become treatable. See, Chapters 4 and 5 of Part A, this volume, on stem cells.

6. Ethical Concerns

Philosophical Questions on Longevity

Prediction goes that within the next 10 years or so, we will at least be able to slow down the disease in people who already have symptoms of AD and do a much better job at identifying people at high risk of getting this disease, who do not yet have symptoms. And once the new treatments come along to slow down the disease, those treatments may be given to people at high risk.

People are expected to live longer in general. Health condition and cell renewal therapies will enable this. Will be AD and other dementia's cured or, will by prolonging life duration, all eventually get AD?

It is disputable if people should correct their appearance and health (artificially). However, juvenilation (trials to prolong youth) are something which was aimed from ancient times. For example, old "church fathers" and oriental religions yogis or simply sheppards who led harsh yet healthy lives normally lived more than 120 years. It is probably nothing wrong with the wish for longevity and even immortality (if only people live the way they enjoy life – healthy and bright-minded).

Concerns About Confidentiality of Genetic Testing

As with all multigene diseases also with late onset AD it is difficult to predict, how much risk one bears if he/she has the risk genes. Not only genes but also environmental and social factors direct the life and lead to or may prevent the disease. Should one know the risk and die of fear or should him vaguely be aware of the danger, similarly to vascular disease, and just stick to "healthy" life style? Some compromise between telling all details and give a first notice by a medical doctor, would be sufficient. Certainly, such information should not be available to anyone interested, such as employers. People should not walk around with genes on a chip…i.e. all recorded on the medical card.

APOE testing, and indeed all genetic testing, raises ethical, legal, and social questions for which there are few answers. Generally, confidentiality laws protect APOE information gathered for research purposes. On the other hand, information obtained in APOE testing may not remain confidential if it becomes part of a person's medical records. Thereafter, employers, insurance companies, and other health care organizations could find out this information, and discrimination could result.

Genetics Counseling for Early Onset Familial AD might be necessary in order to apply a cure when this becomes available. It is important for people who are considering genetic testing for Early Onset Familial AD to have genetic counseling to make sure that they consider all of the possible implications of learning their genetic test results. It is important to think about the impact genetic information could have on a number of factors, such as family relationships, employment and insurance.

References

Anderluh G, Gutierrez-Aguirre I, Rabzelj S, Ceru S, Kopitar-Jerala N, Macek P, Turk V, Zerovnik E. (2005) Interaction of human stefin B in the prefibrillar oligomeric form with membranes. Correlation with cellular toxicity. *FEBS J.* 272, 3042-3051.

Benussi L, Ghidoni R, Steinhoff T, Alberici A, Villa A, Mazzoli F, Nicosia F, Barbiero L, Broglio L, Feudatari E, Signorini S, Finckh U, Nitsch RM, Binetti G. (2003). Alzheimer disease-associated cystatin C variant undergoes impaired secretion. Alzheimer disease-associated cystatin C variant undergoes impaired secretion *Neurobiol. Disease.* 13, 15-21.

Berger Z, Ravikumar B, Menzies FM, Oroz LG, Underwood BR, Pangalos MN, Schmitt I, Wullner U, Evert BO, O'Kane CJ, Rubinsztein DC (2006). Rapamycin alleviates toxicity of different aggregate-prone proteins. *Hum. Mol. Genet.* 15, 433-442.

Bucciantini M, Rigacci S, Berti A, Pieri L, Cecchi C, Nosi D, Formigli L, Chiti F, Stefani M (2005). Patterns of cell death triggered in two different cell lines by HypF-N prefibrillar aggregates. *FASEB J.* 19, 437-439.

Bucciantini M, Giannoni E, Chiti F, Baroni F, Formigli L, Zurdo J, Taddei N, Ramponi G, Dobson CM, Stefani M (2002). Inherent toxicity of aggregates implies a common mechanism for protein misfolding diseases. *Nature.* 416, 507-511.

Bush AI (2001). Therapeutic targets in the biology of Alzheimer's disease. *Curr. Opin. Psychiatry.* 14, 341-348.

Čeru S, Jenko Kokalj S, Rabzelj S, Škarabot M, Gutierrez-Aguirre I, Kopitar-Jerala N, Anderluh G, Turk D, Turk V, Žerovnik E (2008). Size Limits for Toxic Oligomers of Amyloidogenic Proteins: A Case Study of Human Stefin B (Cystatin B) – *Amyloid.* 15, 147-159.

Chiti F, Taddei N, Bucciantini M, White P, Ramponi G, Dobson CM (2000). Mutational analysis of the propensity for amyloid formation by a globular protein. *EMBO J.* 19, 1441-1449.

Cleary JP, Walsh DM, Hofmeister JJ, Shankar GM, Kuskowski MA, Selkoe DJ, Ashe KH (2005). Natural oligomers of the amyloid-beta protein specifically disrupt cognitive function. *Nat. Neurosci.* 8, 79-84.

Crouch PJ, White AR, Bush AI (2007). The modulation of metal bio-availability as a therapeutic strategy for the treatment of Alzheimer's disease. *FEBS J.* 274, 3775–3783.

Cuervo AM. (2004). Autophagy: many paths to the same end. *Mol. Cell. Biochem.* 263, 55-72.

Ding Q, Dimayuga E, Keller JN (2007). Oxidative damage, protein synthesis, and Protein degradation in Alzheimer's disease. *Curr. Alz. Res.* 4, 73–79.

Ding WX, Ni HM, Gao W, Yoshimori T, Stolz DB, Ron D, Yin XM (2007). Linking autophagy to ubiquitin-proteasome system is important for the regulation of endoplasmatic reticulum stress and cell viability. *Am. J. Pathol.* 171, 513-524.

Driscoll I, Resnick SM. (2007). Testosterone and Cognition in Normal Aging and Alzheimer's Disease: an update. *Curr. Alzheimer's Disease.* 4, 33-45.

Ehehalt R, Keller P, Haass C, Thiele C, Simons K. (2003). Amyloidogenic processing of the Alzheimer beta-amyloid precursor protein depends on lipid rafts. *J. Cell. Biol.* 160, 113-123.

Eubanks LM, Rogers CJ, Beuscher IV AE, Koob GF, Olson AJ, Dickerson TJ, and Janda KD (2006). A Molecular Link between the Active Component of Marijuana and Alzheimer's Disease Pathology. *Mol. Pharmaceutics.* 3, 773-777.

Farris W, Schütz SG, Cirrito JR, Shankar GM, Sun X, George A, Leissring MA, Walsh DM, Qiu WQ, Holtzman DM, Selkoe DJ (2007). Loss of neprilysin function promotes amyloid plaque formation and causes cerebral amyloid angiopathy. *Am. J. Pathol.* 171, 241-251.

Finckh U, von der Kammer H, Velden J, Michel T, Andresen B, Deng A, et al., Nitsch RM. (2000). Genetic association of a cystatin C gene polymorphism with late-onset Alzheimer disease. *Arch. Neurol.* 57, 1579-1583.

Ghidoni R, Benussi L, Paterlini A, Missale C, Usardi A, Rossi R, Barbiero L, Spano P, Binetti G. (2007). Presenilin 2 mutations alter cystatin C trafficking in mouse primary neurons. *Neurobiol. Aging.* 28, 371-376.

Goedert M, Spillantini MG, Davies SW (1998). Filamentous nerve cell inclusions in neurodegenerative diseases. *Curr. Opin. Neurobiol.* 8, 619 – 632.

Guijarro JI, Sunde M, Jones JA, Campbell ID, Dobson CM (1998). Amyloid fibril formation by an SH3 domain. *Proc. Natl. Acad. Sci.U. S. A.* 95, 4224-4228.

Hardy JA, Higgins GA (1992). Alzheimer's disease: the amyloid cascade hypothesis. *Science.* 256, 184–185.

Haass C, Selkoe DJ (2007). Soluble protein oligomers in neurodegeneration: lessons from the Alzheimer's amyloid beta-peptide. *Mol. Cell.* 2, 101-112.

Howlett DR (2003). Protein Misfolding in Disease: Cause or Response? *Curr. Med. Chem. – Immunol. Endoc. & Metab. Agents.* 3, 371-383.

Iwata N, Mizukami H, Shirotani K, Takaki Y, Muramatsu S, Lu B, Gerard NP, Gerard C, Ozawa K, Saido TC (2004). Presynaptic localization of neprilysin contributes to efficient clearance of amyloid-beta peptide in mouse brain. *J. Neurosci.* 24, 991-998.

Kagan BL, Hirakura Y, Azimov R, Azimova R and Lin MC (2002). The channel hypothesis of Alzheimer's disease: current status. *Peptides.* 23, 1311 – 1315.

Kawahara M, Kuroda Y, Arispe N, Rojas E. (2000). Alzheimer's beta-amyloid, human islet amylin, and prion protein fragment evoke intracellular free calcium elevations by a common mechanism in a hypothalamic GnRH neuronal cell line. *J. Biol. Chem.* 275, 14077-14083

Kayed R, Head E, Thompson JL, McIntire TM, Milton SC, Cotman CW and Glabe CG (2003). Common structure of soluble amyloid oligomers implies common pathology. *Science.* 300, 487-489.

Kivipelto M, Rovio S, Ngandu T, Kåreholt I, Eskelinen M, Winblad B, Hachinski V, Cedazo-Minguez A, Soininen H, Tuomilehto J, Nissinen A.(2008). Apolipoprotein E epsilon4 Magnifies Lifestyle Risks for Dementia: A Population Based Study. *J. Cell Mol. Med.*, Mar 4; [Epub ahead of print]

Kopito RR (2000). Aggresomes, inclusion bodies and protein aggregation. *Trends Cell Biol.* 10, 524-530. Review.

Kopito RR Sitia R (2000). Aggresomes and Russell bodies. Symptoms of cellular indigestion? *EMBO Rep.* 1, 225-231.

Lane R, Feldman HH, Meyer J, He Y, Ferris SH, Nordberg A, Darreh-Shori T, Soininen H, Pirttilä T, Farlow MR, Sfikas N, Ballard C, Greig NH. (2008). Synergistic effect of apolipoprotein Eε4 and butyrylcholinesterase K-variant on progression from mild cognitive impairment to Alzheimer's disease. *Pharmacogenet Genomics.* 18, 289-298.

Lansbury PT, Lashuel HA (2005). A century-old debate on protein aggregation and neurodegeneration enters the clinic. *Nature.* 443, 774-779.

Lee CC, Nayak A, Sethuraman A, Belfort G, McRae GJ (2007). A three-Stage Kinetic Model of Amyloid fibrillation. *Biophys. J.* 92, 3448 – 3458.

Levy E, Jaskolski M, Grubb A (2006). The Role of Cystatin C in Cerebral Amyloid Angiopathy and Stroke: Cell Biology and Animal Models. *Brain Pathology.* 16 , 60-70.

Lopez de la Paz M, Serrano L (2004). Sequence determinants of amyloid fibril formation. *Proc. Natl. Acad. Sci. USA* 101, 87 – 92.

Maynard CJ, Bush AI, Masters CL, Cappai R, Li QX (2005). Metals and amyloid-β in Alzheimer's disease : Role of amyloid-β in Alzheimer's disease. *Int. J. Exp. Pathol.* 86, 147-159.

Matsuzaki K (2007). Physicochemical interactions of amyloid β-peptide with lipid bilayers. *Biochim. Biophys. Acta.* 1768, 1935–1942.

El Moustaine D, Perrier V, Smeller L, Lange R, Torrent J (2008). Full-length prion protein aggregates to amyloid fibrils and spherical particles by distinct pathways. *FEBS J.* 275, 2021-2031.

Morshedi D, Rezaei-Ghaleh N, Ebrahim-Habibi A, Ahmadian S, Nemat-Gorgani M (2007). Inhibition of amyloid fibrillation of lysozyme by indole derivatives-possible mechanism of action. *FEBS J.* 274, 6415-625.

Pellarin R, Caflish A (2006). Interpreting the Aggregation Kinetics of Amyloid peptides. *J. Mol. Biol.* 360, 882 – 892.

Plakoutsi G, Bemporad F, Calamai M, Taddei N, Dobson CM, Chiti F (2005). Evidence for a mechanism of amyloid formation involving molecular reorganization within native-like precursor aggregates. *J. Mol. Biol.* 351, 910–922.

Rabzelj S, Viero G, Gutiérrez-Aguirre I, Turk V, Dalla Serra M, Anderluh G, Žerovnik E (2008). Interaction with model membranes and pore formation by human stefin B - studying the native and prefibrillar states. *FEBS J.* 275, 2455-2466.

Ravikumar B, Rubinsztein DC (2006). Role of autophagy in the clearance of mutant huntingtin: a step towards therapy? *Mol. Aspects Med.* 27, 520-527.

Risner ME, Saunders AM, Altman JFB, Ormandy GC, Craft S, Foley IM, Zvartau-Hind ME, Hosford DA, Roses AD (2006). Efficacy of rosiglitazone in a genetically defined population with mild-to-moderate Alzheimer's disease. *The Pharmacogenomics J.* 6, 246–254

Ross CA, Poirier MA (2004). Protein aggregation and neurodegenerative disease. *Nature Med.* 10, S10-S17.

Sarkar S, Perlstein EO, Imarisio S, Pineau S, Cordenier A (2007). Small molecules enhance autophagy and reduce toxicity in Huntington's disease models. *Nature Chem. Biol.* 3, 331-338.

Serio TR, Cashikar AG, Kowal AS, Sawicki GJ, Moslehi JJ, Serpell L, Arnsdorf MF, Lindquist SL. (2000). Nucleated conformational conversion and the replication of conformational information by a prion determinant. *Science.* 289, 1317-1321.

Soto C (1999). Alzheimer's and prion disease as disorders of protein conformation: implications for the design of novel therapeutic approaches. *J. Mol. Med.* 77, 412-418. Review.

Soto C, Anderes L, Suardi S, Cardone F, Castilla J, Frossard MJ, Peano S, Saa P, Limido L, Carbonatto M, Ironside J, Torres JM, Pocchiari M, Tagliavini F. (2005). Pre-symptomatic detection of prions by cyclic amplification of protein misfolding. *FEBS Lett.* 579, 638-642.

Stefani M, Dobson CM (2003). Protein aggregation and aggregate toxicity: new insights into protein folding, misfolding diseases and biological evolution. *J. Mol. Med.* 81, 678-699.

Stokin GB, Lillo C, Falzone TL, Brusch RG, Rockenstein E, Mount SL, Raman R, Davies P, Masliah E, Williams DS, Goldstein LS.(2005). Axonopathy and transport deficits early in the pathogenesis of Alzheimer's disease. *Science.* 307, 1282-1288.

Škerget K, Vilfan A, Pompe-Novak M, Turk V, Waltho JP, Turk D, Žerovnik E (2009). The mechanism of amyloid-fibril formation by stefin B; temperature and protein concentration dependence of the rates. *Proteins,* 74, 425-436. (in print, Jul 17. Epub ahead of print)

Tampellini D, Magrané J, Takahashi RH, Li F, Lin MT, Almeida CG, Gouras GK (2007). Internalized antibodies to the Abeta domain of APP reduce neuronal Abeta and protect against synaptic alterations. *J. Biol. Chem.* 282, 18895-18906.

Tun H, Marlow L, Pinnix I, Kinsey R, Sambamurti K. (2002). Lipid rafts play an important role in A beta biogenesis by regulating the beta-secretase pathway. *J. Mol. Neurosci.* 19. 31-35.

Uversky VN. (2008). Amyloidogenesis of Natively Unfolded Proteins. *Curr. Alz. Res.* 5, 260-287.

Walsh DM, Klyubin I, Fadeeva JV, Cullen WK, Anwyl R, Wolfe MS, Rowan MJ, Selkoe DJ (2002). Naturally secreted oligomers of amyloid beta protein potently inhibit hippocampal long-term potentiation in vivo. *Nature.* 416, 535-539.

Walsh DM, Selkoe DJ (2007), Aβ Oligomers – a decade of discovery. *J. Neurochem.* 101, 1172-1184.

Wirths O, Multhaup G, Bayer TA. (2004). A modified beta-amyloid hypothesis: intraneuronal accumulation of the beta-amyloid peptide--the first step of a fatal cascade. *J. Neurochem.* 91, 513-520. Review.

Wisniewski T, Sigurdsson EM (2007). Therapeutic approaches for prion and Alzheimer's diseases. *FEBS J.* 274, 3784 – 3798.

Yang F, Lim GP, Begum AN, Ubeda OJ, Simmons MR, Ambegaokar SS, et al., Cole GM (2005). Curcumin inhibits formation of amyloid beta oligomers and fibrils, binds plaques, and reduces amyloid in vivo. *J. Biol. Chem.* 280, 5892-5901.

Zhang L, Fiala M, Cashman J, et al., 2006. Curcuminoids enhance amyloid-ß uptake by macrophages of Alzheimer's disease patients. *J. Alzheimers Dis.* 10, 1-7.

Zimmermann, M., F. Gardoni, et al. (2005). Molecular rationale for the pharmacological treatment of Alzheimer's disease. *Drugs Aging* 22 Suppl 1, 27-37.

Žerovnik E (2002). Amyloid-fibril formation; Proposed mechanisms and relevance to conformational disease. *Eur. J. Biochem.* 269, 3362-3371.

Source Web Pages

2003 news: in http://www.fda.gov/fdac/features/2003/403_alz.html

2007: new trials with metal chelators: http://www.biotechnews.com.au/index.php/id; 1827236977

2007 news: http://www.mentalhelp.net/poc/view_doc.php? type=weblog&id=287& wlid= 6&cn=231

2008 alter news: http://www.healthandage.com/public/health-center/11/article/3187/Can-Curry-Protect-Against-Alzheimers.html

In: Philosophical Insights about Modern Science
Eds: Eva Žerovnik, Olga Markič and Andrej Ule
ISBN: 978-1-60741-373-8

Chapter 4

ADVANCES IN STEM CELL RESEARCH

Anthony Atala*
Wake Forest Institute for Regenerative Medicine;
Wake Forest University School of Medicine; Medical Center Boulevard;
Winston-Salem, North Carolina 27157, USA

SYNOPSIS

The current shortage of organ donors and the increasing need for organ transplantation has stimulated research on stem cells as a potential resource for cell-based therapies for organ failure. However, established methods for generating human embryonic stem cells are highly regulated or even banned because of ethical concerns. Recently, the development of innovative methods to generate pluripotent stem cells from other sources suggests that there may be new alternatives for cell-based therapies. Here, we provide an overview of pluripotent cell types that can be derived from techniques such as somatic cell nuclear transfer, single cell embryo biopsy, use of arrested embryos, altered nuclear transfer, and somatic cell reprogramming. We also discuss the potential use of the recently discovered amniotic fluid and placental derived stem cells for patient-specific therapies.

INTRODUCTION

Patients with end stage organ failure can be treated with various techniques, but the only method that can fully restore all functions of a diseased organ is transplantation. However, the number of patients in need of new organs far exceeds the organ supply, and this organ shortage is expected to continue to worsen as the aging population increases.

Some primary cells, whether autologous or allogenic, cannot be expanded from particular organs, such as the pancreas. In these situations, pluripotent stem cells are envisioned as an alternative source of cells from which the desired tissue can be derived. Pluripotent stem cells

* Tel 336-716-5701; Fax 336-716-0656; Email: aatala@wfubmc.edu

represent an endless source of versatile cells that could lead to novel sources of replacement organs.

Adult Stem Cells and Tissue Progenitor Cells

Adult stem cells, especially hematopoetic stem cells, are better understood than any other aspect of stem cell biology (Ballas et al. 2002). In fact, cell based therapy using adult stem cells dates back to the first bone marrow transplant in 1956 (Thomas et al. 1957), though the first evidence that a specific, "master" cell type might be responsible for reconstituting the bone marrow was gained from experience with persons exposed to lethal doses of radiation in 1945. Till and McCulloch analyzed the bone marrow in the early 1960s to determine which cells were responsible for marrow reconstitution, and discovered cells with the ability to renew themselves as well as the ability to differentiate into various cell types (McCulloch et al. 1964; Till et al. 1964). These two characteristics are still used to define stem cells today. Later, gastrointestinal stem cells (crypt cells) were discovered as scientists studied the regeneration of the intestinal mucosa after sublethal radiation doses, and this discovery ignited the search for stem cells in other tissues.

Today, that search is still an intense area of study, as progenitor cell therapy is applicable to a myriad of degenerative conditions. Within the last decade, it has been found that progenitor cell populations are present in many adult tissues other than the bone marrow and the gastrointestinal tract. These include the brain, skin, muscle, and amniotic fluid. The discovery of such tissue specific progenitors has opened up new avenues for research, especially because the processes used to obtain these cells can bypass the ethical concerns associated with the use of embryonic stem cells.

Adult stem cells tend to be tissue specific, self-renewing populations of cells which can differentiate into cell types associated with the organ system in which they reside (Presnell et al. 2002; Spradling et al. 2001). They are quite rare, on the order of 1 in 10,000 cells within the tissue of interest (Marshak DR 2001). Currently, it is known that niches of stem cells exist in bone marrow, brain, liver, skin, skeletal muscle, the gastrointestinal tract, the pancreas, the eye, blood, and dental pulp (Al-Rubeai 1999; Jiang et al. 2002b; Presnell et al. 2002; Spradling et al. 2001). Of these, the most studied are CD34+ hematopoietic stem cells isolated from bone marrow. These cells are capable of producing cells of the lymphoid and myeloid lineages in blood. CD34+ cells are the only currently available therapeutic application of stem cells and are used for a variety of purposes, most often the reestablishment of the immune system after a disease or toxic therapy has damaged it.

A notable exception to the tissue specificity of adult stem cells is the mesenchymal stem cell, or what is more recently called the multipotent adult progenitor cell. This cell type is derived from bone marrow stroma (Devine 2002; Jiang et al. 2002a; Spradling et al. 2001). Such a cell has been shown to differentiate *in vitro* into numerous tissue types and to also differentiate developmentally if injected into a blastocyst. Multipotent adult progenitor cells will develop into multiple tissues including neuronal, adipose, muscle, liver, lungs, spleen, and gut, but notably, not bone marrow or gonads (Jiang et al. 2002a).

However, research on adult stem cells has been slow, largely because great difficulty has been encountered in maintaining adult stem cells in culture. To date, there is no known efficient means of maintaining and expanding a long term culture of any adult stem cell in large numbers. Isolation has also proven to be quite problematic as these cells are present in extremely low numbers in the adult tissue. Such cells are often selected utilizing Fluorescent Activated Cell Sorting (FACS) or Magnetic Activated Cell Sorting (MACS) against surface markers specific to the stem cell of interest (Quesenberry PJ 1998), but sometimes there is no known marker specific for a type of stem cell, so these methods cannot be used. However, new markers are being described at a rapid pace.

While current use of adult stem cells is quite limited, there is great potential in future utilization of such cells for the use of tissue specific regenerative therapies. The advantage of adult stem cells is that they can be used in autologous therapies, thus avoiding any immune rejection complications.

Embryonic Stem Cells

According to data from the Centers for Disease Control, as many as 1 million Americans will die every year from disease that, in the future, may be treatable with tissues derived from stem cells (Minino 2004). Diseases that might benefit from embryonic stem cell-based therapies included diabetes, heart disease, cerebrovascular disease, liver and renal failure, spinal cord injuries and Parkinson's disease.

In 1981, pluripotent cells were found in the inner cell mass of the human embryo, and the term "human embryonic stem cell" was coined (Martin 1981). These cells are able to differentiate into all cells of the human body, excluding placental cells (only cells from the morula are totipotent; that is, able to develop into all cells of the human body). These cells have great therapeutic potential, but their use is limited by several factors, both biological and ethical.

The political controversy surrounding stem cells began in 1998 with the creation of human embryonic stem cells (hES) cells derived from discarded human embryos. hES were isolated from the inner cell mass of a blastocyst (an embryo 5 days post-fertilization) using an immunosurgical technique. Using this technique, the blastocyst was incubated with antibodies specific to the trophectoderm. Complement proteins then resulted in lysis of the trophectoderm so that the only surviving cells were the inner cell mass (Solter et al. 1975). Given that some cells cannot be expanded *ex vivo*, ES cells could be the ideal resource for tissue engineering because of their fundamental properties: the ability to self-renew indefinitely and the ability to differentiate into cells from all three embryonic germ layers. However, the clinical application of hES cells is limited because they represent an allogenic resource and thus have the potential to evoke an immune response. In addition, the derivation of hES cells requires the destruction of embryos, and as a result, they are ineligible for federal funding in the United States at this time. New stem cell technologies (such as somatic cell nuclear transfer and reprogramming) promise to overcome these limitations.

Somatic Cell Nuclear Transfer

Somatic cell nuclear transfer (SCNT) entails the removal of an oocyte nucleus in culture, followed by its replacement with a nucleus derived from a somatic cell obtained from a patient. Activation with chemicals or electricity stimulates cell division up to the blastocyst stage, at which time the inner cell mass is isolated and cultured, resulting in ES cells that are genetically identical to the patient. It has been shown that nuclear transferred ES cells derived from fibroblasts, lymphocytes, and olfactory neurons are pluripotetent and generate live pups after tetraploid blastocyst complementation, showing the same developmental potential as fertilized blastocysts (Brambrink et al. 2006; Eggan et al. 2004; Hochedlinger et al. 2002; Rideout et al. 2002). The resulting ES cells are perfectly matched to the patient's immune system and no immunosuppressants would be required to prevent rejection.

Although ES cells derived from SCNT contain the nuclear genome of the donor cells, mitochondrial DNA inherited by the oocyte could lead to immunogenicity after transplantation. To assess the histocompatibilty of nuclear transfer-generated tissue, the nucleus of a bovine skin fibroblast was microinjected into an enucleated oocyte (Lanza et al. 2002). Although the blastocyst was implanted (reproductive cloning), the purpose was to generate renal, cardiac and skeletal muscle cells, which were then harvested, expanded *in vitro*, and seeded onto biodegradable scaffolds. These scaffolds were then implanted into the donor from whom the cells were cloned to determine if cells were histocompatible. Analysis revealed that cloned renal cells showed no evidence of T-cell response, suggesting that rejection will not necessarily occur in the presence of oocyte-derived mtDNA. This finding represents a step forward in overcoming the histocompatibility problem of stem cell therapy.

Although promising, SCNT has certain limitations that require further improvement before its clinical application, in addition to ethical considerations regarding the potential of the resulting embryos to develop into cloned embryos if implanted into a uterus. Many animal studies have shown that blastocysts generated from SCNT can give rise to a liveborn infant that is a clone of the donor when implanted into a uterus. In 1997, for example, a sheep named Dolly was derived from an adult somatic cell using nuclear transfer (Wilmut et al. 1997). This is known a reproductive cloning, which is banned in most countries for human applications. In contrast, therapeutic cloning is used to generate only ES cell lines whose genetic material is identical to that of their source. In this case blastocysts are allowed to grow until a 100 cell-stage where ES cells can be obtained, and thus, the blastocysts are never implanted into a uterus. In addition, this technique has not been shown to work in humans. The initial failures and fraudulent reports of nuclear transfer in humans reduces enthusiasm for human applications (Hwang et al. 2004; Hwang et al. 2005; Simerly et al. 2003). However, it was recently reported that non-human primate ES cell lines were generated by SCNT of nuclei from adult skin fibroblasts (Byrne et al. 2007; Mitalipov 2007). This group used a modified SCNT approach that avoids the use of Hoechst 33342 and ultraviolet light, which is detrimental to the quality of the cytoplasts, to completely remove spindles containing nuclear DNA (Mitalipov 2007). They used an Oosight spindle imaging system to efficiently enucleate primate occytes. A total of 304 oocytes yielded 35 blastocysts, from which two ES cell lines were derived. Both lines demonstrated typical ES cell morphology. They also demonstrate self-renewal and express *OCT4, SSEA4, LEFTYA, TDGF, TRA1-60* and *TRA1-80.* To test their differentiation potential, the cells were exposed to cardiomyocyte differentiation conditions, and this produced contracting aggregates that expressed markers of cardiac muscle tissue. Neural differentiation resulted in the expression of microtubule-associated protein 2 (*MAP2*), B-tubulin, and tyrosine hydroxylase. When injected into SCID mice, SCNT-derived ES cells induced teratomas which contained differentiated cell types representing all three embryonic germ layers. More importantly, microstatellite single nucleotide polymorphism and mtDNA analysis confirmed that nuclear DNA was identical to the donor somatic cells and that the mtDNA originated from the oocyte. This DNA analysis was also confirmed by David Cram of Monash University in Clayton, Australia before publishing the paper.

Before SCNT-derived ES cells can be used as clinical therapy, careful assessment of quality of the lines must be determined. Unfortunately, one of the lines generated by SCNT revealed a translocation consisting of an isochromosome comprised of two copies of the long arm of the Y chromosome. It is not known whether chromosomal abnormalities in SCNT-derived ES cells originate from aneuploid embryos or occurred during ES cell isolation and culture.

The low efficiency of SNCT (0.7%) and the inadequate supply of human oocytes further hinder the therapeutic potential of this technique. Although promising, SCNT has limitations that require further improvement before clinical application, including ethical considerations regarding the potential of the generated blastocysts to develop into cloned embryos if implanted into a uterus. Furthermore, the destruction of embryos is not an ethically acceptable means to generate pluripotent stem cells. On the other hand, this study renews the hope that ES cell lines could be generated by SCNT in humans to generate patient-specific stem cells with the potential to cure/treat many human diseases that are nowadays untreatable.

Single Cell Embryo Biopsy

Since the major objection to human embryonic stem cell research is the destruction of embryos, it would be advantageous to develop a method of isolating these cells from and embryo without destroying it. In 2006, Chung et al. were the first to report the generation of mouse embryonic stem cell lines without destroying the embryo (Chung et al. 2006). This alternative method of generating ES cell lines is based on a technique used to obtain a single cell embryo biopsy for preimplantation genetic diagnosis (PGD) of genetic defects (Handyside et al. 1990). Blastomere-derived ES cells differentiated into derivatives of all three germ layers *in vitro* and as well as into teratomas *in vivo*. In addition, the biopsied mouse embryos developed to term without a reduction in their developmental potential. Experiments are being carried out to determine whether human ES cells can be derived from single blastomeres (Klimanskaya et al. 2006). Additional studies will be needed to determine whether blastomere-derived hES cell lines are different from conventional hES cell lines in their ability to form functional differentiated cell types. It has been recently reported that both blastomere-derived and conventional hES cells can form functional hemangioblasts, which are capable of forming both hematopoeitic and endothelial cell types, suggesting that blastomere-derived hES cells can be used as a cell therapy (Lu et al. 2007a).

Although this technique does not involve the destruction of human embryos, concerns have been raised as to whether individual eight-cell-stage blastomeres could potentially generate a human being. Eight to sixteen cell-stage blastomeres have not been shown to have the intrinsic capacity to generate a complete organism in most mammalian species (Willadsen 1981). Another technical concern is that hES derived from blastomeres require co-culture with a previously established hES cell line. With further studies, it is likely that this technical limitation will be overcome, but the ethical concerns regarding the developmental potential of an individual blastomere will probably remain an issue.

Arrested Embryos

With the current restrictions surrounding hES work, researchers are investigating methods to derive hES cells lines without destroying human embryos. It has been shown that human ES lines can be derived from arrested embryos (Zhang et al. 2006). During *in vitro* fertilization, only a small number of all zygotes produced will develop successfully to the morula and blastocyst stages, and well over half of the embryos stop dividing (Geber et al. 1999; Hardy 1993) and are therefore considered dead embryos (Landry et al. 2004). Such embryos have unequal or fragmented cells and blastomeres and are usually discarded. However, not all the cells within these arrested embryos are abnormal (Martinez et al. 2002; Zhang et al. 2006). In one study, one hundred and sixty one embryos were donated from IVF clinics. Out of these, 119 embryos arrested early at day 3-5 (4-10 cells stage) and 13 embryos arrested late at day 6-7 (16-24 cell stage). One stable and fully characterized hES cell line was derived from the 13 late arrested embryos. This hES cell line expressed *OCT4, NANOG,* REX1, TRA-1-61, TRA-1-81, which are associated with stem cells, and showed normal karyotype. These cells were injected into immunodeficient mice and formed teratomas, just as hES cells would. This data demonstrates that arrested embryos represent a novel source of pluripotent stem cells.

One of the major concerns with this technique is the ambiguity over what constitutes a dead embryo. Arrested embryos were used in this study when no cell or blastomere from the embryo had undergone any cleavage during the last 24 to 48 hours. However, the identification of molecules associated with the loss of embryonic viability will be helpful in defining dead or terminally arrested embryos (Landry et al. 2006). Another concern at this time is the quality of the cell lines derived from arrested embryos. More studies are needed to characterize the full proliferation and differentiation potential of ES cells derived from arrested embyos.

Altered Nuclear Transfer

Altered nuclear transfer is a variation of SCNT in which a genetically modified nucleus from a somatic cell is transferred into a human oocyte. This embryo, which contains a deliberate genetic defect, is capable of developing into a blastocyst but the induced defect prevents the blastocyst from implanting in the uterus. This process has the potential to generate customized ES cells from the blastocyst stage (Hurlbut 2005). It is hypothesized that human embryos with this genetic defect would lack to capacity to develop into a human beings, due to their inability to implant, thus providing an alternative source of stem cells. This concept was proven in mice by Meissener et al., in 2006 (Meissner et al. 2006). Here, scientists inactivated *Cdx2* gene, which is crucial for trophectoderm development. The trophoectoderm is required for the fetal-maternal interface within the placenta (Strumpf et al. 2005). The experiment used a lentiviral vector to insert a floxed Cdx2 short hairpin RNA sequence into the cells. Embryos resulting from nuclear transfers containing the Cdx2 deficiency showed no delay in developing into the early blastocyst stage, but the blastocysts were morphologically abnormal and could not implant into the uteri of pseudopregnant females. In culture, however, even though the Cdx2 knockdown blastocysts did not produce a trophoectoderm, they did produce an inner cell mass structure from which ES cells could be obtained. These ES cell lines were injected into diploid blastocysts and formed chimeras with extensive contributions to most tissues, except for the intestines, consistent with reports that Cdx2 is required for the normal development of the gastrointestinal tract (Chawengsaksophak et al. 2004). Interestingly, restoration of *Cdx2* expression by Cre-mediated deletion of the shRNA vector created ES cell lines that could generate all somatic tissues including normal intestinal cells.

Despite the fact that these altered cells and embryos have little to no potential to form a functional organism, the ethics of the situation is still being debated. In addition, it is not clear whether human *CDX2*-deficent embryos die at the same stage as those from mice,f or whether this mutation restricts their developmental potential into certain lineages. While much research must be done before such a therapy could ever enter the clinic, at this time hES derived from altered nuclear transfer can provide opportunities to study pluripotentiality in hES.

Reprogramming

Reprogramming is a technique that involves de-differentiation of adult somatic cells to produce patient-specific pluripotent stem cells without the use of embryos. Cells generated by reprogramming would be genetically identical to the somatic cells and would not be rejected by the donor. This method also avoids the technical limitations of nuclear transfer into oocytes. Yamanaka was the first to discover that mouse embryonic fibroblasts (MEFs) and adult mouse fibroblasts can be reprogrammed into an induced pluripotent state (IPS) (Takahashi et al. 2006). They examined 24 genes that were thought to be important for embryonic stem cells and identified 4 key genes that were required to bestow embryonic stem cell-like properties on fibroblasts. Mouse embryonic fibroblasts and adult fibroblasts were co-transduced with retroviral vectors, each carrying Oct3/4, Sox2, c-Myc, and Klf4. Reprogrammed cells were selected via drug resistance. In this case, a downstream gene of Oct4, Fbx15, was replaced with a drug resistance gene via homologous recombination. The resultant IPS cells possessed the immortal growth characteristics of self-renewing ES cells, expressed genes specific for ES cells, and generated embryoid bodies *in vitro* and teratomas *in vivo*. When the IPS cells were injected into mouse blastocysts they contributed to a variety of diverse cell types, demonstrating their developmental potential. Although IPS cells selected by *Fbx15* were pluripotent they were not identical to ES cells. Unlike ES cells, chimeras of IPS cells did not result in full-term pregnancies. Gene expression profiles of the IPS cells showed that they possessed a distinct gene expression signature compared to ES cells. The epigenetic state of the IPS cells was somewhere between their somatic origins and fully reprogrammed ES cells, suggesting that the reprogramming was incomplete.

These results were improved significantly by Wernig and Jaenisch in July 2007 (Wernig et al. 2007). Fibroblasts were infected with retroviral vectors and selected for the activation of endogenous *Oct4* or *Nanog* genes. Results from this study showed that DNA methylation, gene expression profiles, and chromatic state of the reprogrammed cells were similar to those of ES cells. Teratomas induced by these cells contained differentiated cell types representing all three embryonic germ layers. Most importantly, the reprogrammed cells from this experiment were able to form viable chimeras and contribute to the germ line like ES cells, suggesting that these IPS cells were completely reprogrammed. This may be due to the fact that Wernig et al. observed that the number of reprogrammed colonies increased when drug selection was initiated later (day 20). This suggests that reprogramming is a slow and gradual process and may explain why the process of using Fbx15 activation on day 3 post-transfection may result in incomplete reprogramming.

It has recently been shown that reprogramming by transduction of four defined factors can be done with human cells (Takahashi et al. 2007; Yu et al. 2007) Yamanaka's group began by optimizing the transduction efficiencies of human dermal fibroblasts (HDF) and determined that the introduction of a mouse receptor for retroviruses into HDF cells using a lentivirus improved the transduction efficiency from 20% to 60%. Yamanaka then showed that retrovirus-mediated transfection of *OCT3/4, SOX2, KLF4,* and *c-MYC* generates human IPS cells that are similar to hES cells in terms of morphology, proliferation, gene expression, surface markers, and teratoma formation. In contrast, Thompson's group showed that retroviral transduction of *OCT4, SOX2, NANOG, and LIN28* could generate pluripotent stem cells without introducing any oncogenes (c-MYC). Both studies showed that human IPS were similar but not identical to hES cells. Another concern is that these IPS cells contain three to six retroviral integrations (one for each factor) which may increase the risk of tumorigenesis.

These studies used retroviral transduction to induce reprogramming of somatic cells into a pluripotent state. Yamanaka et al. studied the tumor formation in chimeric mice generated from Nanog-IPS cells and found 20% of the offspring developed tumors due to the retroviral expression of c-myc (Okita et al. 2007). An alternative approach would be to use a transient expression method, such as adenovirus-mediated system, since both Jaenisch and Yamanaka showed strong silencing of the viral-controlled transcripts in IPS cells (Meissner et al. 2007; Okita et al. 2007). This indicates that they are only required for the induction, not the maintenance, of pluripotency. Another concern is the use of transgenic donor cells for reprogrammed cells in the mouse studies. In both mouse studies, iPS cells were isolated by selecting for the activation of a drug-resistant gene inserted into endogenous *Fbx15, Oct3/4,* or *Nanog*. The use of genetically modified donors hinders its clinical applicability for humans.

To assess whether iPS cells can be derived from genetically unmodified donor cells, MEF and adult skin cells were retrovirally transduced with Oct3/4, Sox2, c-Myc, and Klf4 and ES-like colonies were isolated by morphology, without the use of drug selection for *Oct4* or *Nanog* (Meissner et al. 2007). IPS cells from unmodified donor cells formed teratomas and generated live chimeras. This study suggests that genetically modified donor cells are not necessary to generate IPS cells.

Although this is an exciting phenomenon, it is unclear why reprogramming adult fibroblasts and mesenchymal stromal cells have similar efficiencies (Takahashi; Yamanaka 2006). It would seem that cells that are already multipotent could be reprogrammed with greater efficiency, since the more undifferentiated donor nucleus the better SCNT performs (Blelloch et al. 2006). This further emphasizes our limited understanding of the mechanism of reprogramming using these four factors.

Stem Cells Derived from the Amniotic Fluid and Placenta

The amniotic fluid is known to contain a heterogeneous population of cell types derived from the developing fetus (Polgar et al. 1989; Priest et al. 1978). Cells from this population have been shown to differentiate into adipocytes, osteocytes, neurogenic and endothelial cells (In 't Anker et al. 2003; Prusa et al. 2004; Schmidt et al. 2007; Tsai et al. 2004). Importantly, a subpopulation of OCT4-positive cells have been detected in the amniotic fluid (Prusa et al. 2003; Tsai et al. 2004).

Recently, our group has reported the isolation of human and mouse amniotic fluid-derived stem (AFS) cells. These cells are capable of extensive self-renewal and give rise to adipogenic, osteogenic, myogenic, endothelial, neurogenic and hepatogenic lineages (De Coppi et al. 2007). In this respect, they meet a commonly accepted criterion for pluripotent stem cells.

AFS cells represent approximately 1% of the cells found in the amniotic fluid. The same cells can also be found in the placenta. These cells are immuno-selected for cells that express the surface antigen c-*kit* (CD117), the receptor for stem cell factor (Zsebo et al. 1990). AFS cells express embryonic markers such as OCT4 and SSEA4, but not other markers of ES cells. They also express mesenchymal and/or neuronal markers such as CD29, CD44, CD73, CD90, and CD105. More importantly, clonal analyses using retrovirally-marked human lines confirmed that differentiated cells of various types can be derived from a single cell (De Coppi et al. 2007).

In addition to demonstrating that differentiated cells express lineage-specific markers, we have shown that such cells give rise to specialized functions. Cells differentiated down a neuronal pathway secreted glutamine or expressed G-protein-gated inwardly rectifying potassium channels. Cells of the hepatic lineage secreted urea and α-fetoprotein, while osteogenic cells produced mineralized calcium.

Also, when AFS cells were cultured in neuronal differentiation medium for a time and then grafted into the lateral cerebral ventricles of control mice and the ventricles of the *twitcher* mouse model, in which a progressive loss of oligodendrocytes leads to massive demyelination and neuronal loss. AFS cells integrated into the brains of both strains seamlessly, appeared morphologically indistinguishable from surrounding mouse cells, and survived efficiently for at least 2 months. Interestingly, more of the AFS cells integrated into the injured twitcher brains (70%) than into the normal brains (30%), hinting at the potential for CNS therapies. From a tissue engineering perspective, osteogenic AFS cells were embedded in an alginate/collagen scaffold and implanted subcutaneously into immunodeficient mice. By 18 weeks after implantation, highly mineralized tissues and blocks of bone-like material were observed in the recipient mice using micro CT. These blocks displayed a density somewhat greater than that of mouse femoral bone.

The recent discovery of a stem cell population in the amniotic fluid and placenta offers a very promising alternative source of stem cells for cellular therapy. The full range of adult somatic cells that AFS cells can produce remains to be determined, but their ability to differentiate into cells of all three embryonic germ layers and their high proliferation rate are two advantages over most adult stem cell sources. AFS cells represent a new class of stem cells with properties somewhere between embryonic and adult stem cell types. However, unlike ES cells, AFS cells do not form teratomas and are easily obtained without destruction of embryos. AFS could be use for both autologous and allogenic therapy through matching of histocompatible donor cells with recipients.

The Future of Stem Cells

Before stem cells can be used as any type of clinical therapy, strict guidelines must be established to ensure the quality of the cells, the specificity of differentiation, and the assessment of mixed phenotypes. While lineage-specific gene expression and cell surface markers are commonly used to describe a differentiated phenotype, it is difficult to determine, for example, whether cells are bona fide neurons or merely neuronal-like cells. To address this question, high throughput methodologies using microarrays are being developed to evaluate new stem cell derivatives. In these experiments, hES were differentiated into retinal pigemented epithelial cells (RPE), the site of the major lesions in macular degeneration. Microarrays were used to demonstrate similarities between hES-derived RPE and freshly isolated RPE (Klimanskaya et al. 2004).

Another concern about the clinical potential of stem cells is their tendency to form mixed phenotypes in most differentiation protocols. Approaches to assess heterogeneous populations, also using microarray technologies, are being developed (Lu et al. 2007b). The therapeutic potential of stem cells largely relies on efficient and controlled differentiation towards a specific cell type and the generation of homogeneous cell populations. Many differentiation protocols utilize the formation of progenitors through a stepwise approach. Thus, characterizing and understanding mixed populations of progenitor stages will be of increasing importance in stem cell research. Methods to assess and identify tissue-specific genetic signatures within a heterogeneous population with microarrays through biologically relevant *in silico* comparisons of data sets are being investigated.

Even once these issues are resolved, stem cell therapy still has many hurdles to overcome before it will become a viable and widely used clinical option. For example, the tendency of some stem cell types to produce teratomas *in vivo* is undesirable in clinical practice and must be addressed. Additionally, the ethical issues of stem cell use are still very much in the forefront of public opinion. While overcoming this hurdle is very difficult, a combination of excellent science aimed at obtaining stem cells while avoiding the destruction of embryos and the dissemination of good, scientifically sound information into the public sector via education will certainly be useful in this regard.

Conclusion and Future Directions

At this time, it is unclear which type of stem cell will provide the best approach for cell therapy. Depending on the clinical scenario, one cell may be better than another. For example, tissue progenitor cells are excellent candidates for tissue engineering organs. Even though tissue progenitor cells have restricted growth and differentiation potential, this is advantageous because they can be used without rejection. On the other hand, ES cells have the ability to grow indefinitely and differentiate into cells of all three germ layers. However, their potential to form teratomas, as well as evoking an immune response currently dampen enthusiasm in their clinical potential.

Methods for generating alternative sources of pluripotent stem cells have their advantages and disadvantages (table 1). Somatic cells can be reprogrammed by nuclear transfer into enucleated oocytes and generate pluripotent and patient-specific stem cells. The disadvantages of this method include the use of oocytes, which are in limited supply, and the destruction of a cloned blastocyst. In addition, it is unclear whether SCNT can be successfully performed in humans. Single cell embryo biopsy is a promising approach because it is based on a common, well-established technique in IVF and has been successful in thousands of children.

Table 1. Summary of methods for generating pluripotent stem cells

Methods	Advantages	Limitations
Somatic cell nuclear transfer	-customized stem cells -been shown to work in non-human primates	-requires oocytes -not been shown to work in humans
Single cell embryo biopsy	-patient-specific to embryo -does not destroy or create embryos -has been done in humans	-allogenic cell types -not known if single cells are totipotent -requires coculturing with a previous established hES line
Arrested embryos	-from discarded embryos -has been done in humans	-allogenic cell types -quality of lines maybe questionable
Altered nuclear transfer	-customized stem cells	-ethical concerns regard embryos with no potential -modified genome -not been done with human cells
Reprogrammed	-customized stem cells -no embryos or oocytes needed -has been down with human cells	-retroviral transduction -oncogenes (3 of the 4 studies used oncogenes)
Amniotic fluid-derived stem cells	-express embryonic and adult stem cell markers -does not form teratomas -easily accessible -non-invasive approach -has been done in humans	-full potential not known -patient-specific to embryo -allogenic and autologous if banked

The ability to reprogram an adult somatic cell into a pluripotent stem cell is a scientific breakthrough, but the use of retroviral transduction currently hinders their clinical potential. Amniotic fluid stem cells are easily accessible and do not require technical manipulations, but they may not be as nimble as human embryonic stem cells. In order to determine the best source of stem cell for a given patient, it is important to consider all types of pluripotent stem cells, and the best choice will likely be dictated by the specific clinical application.

Acknowledgements

The authors wish to thank Dr. Jennifer L. Olson for editorial assistance with this manuscript.

References

Al-Rubeai, M. (1999). *Cell Engineering.* Kluwer Academic Publishers.

Ballas, C. B., S. P. Zielske, and S. L. Gerson (2002). Adult bone marrow stem cells for cell and gene therapies: implications for greater use. *Journal of Cellular Biochemistry - Supplement*, 38, 20-28.

Blelloch, R., Z. Wang, A. Meissner, S. Pollard, A. Smith, and R. Jaenisch (2006). Reprogramming efficiency following somatic cell nuclear transfer is influenced by the differentiation and methylation state of the donor nucleus 1. *Stem Cells*, 24, 2007-2013.

Brambrink, T., K. Hochedlinger, G. Bell, and R. Jaenisch (2006). ES cells derived from cloned and fertilized blastocysts are transcriptionally and functionally indistinguishable 1. *Proc. Natl. Acad. Sci. U.S.A*, 103, 933-938.

Byrne, J., D. Pedersen, L. Clepper, M. Nelson, W. Sanger, S. Gokhale, et al. (2007). Producing primate embryonic stem cells by somatic cell nuclear transfer 1. *Nature*.

Chawengsaksophak, K., W. de Graaff, J. Rossant, J. Deschamps, and F. Beck (2004). Cdx2 is essential for axial elongation in mouse development 1. *Proc. Natl. Acad. Sci. U.S.A*, 101, 7641-7645.

Chung, Y., I. Klimanskaya, S. Becker, J. Marh, S. J. Lu, J. Johnson, et al. (2006). Embryonic and extraembryonic stem cell lines derived from single mouse blastomeres 3. *Nature*, 439, 216-219.

De Coppi, P., G. Bartsch, Jr., M. M. Siddiqui, T. Xu, C. C. Santos, L. Perin, et al. (2007). Isolation of amniotic stem cell lines with potential for therapy.[see comment]. *Nat. Biotechnol.*, 25, 100-106.

Devine, S. M. (2002). Mesenchymal stem cells: will they have a role in the clinic? *Journal of Cellular Biochemistry - Supplement*, 38, 73-79.

Eggan, K., K. Baldwin, M. Tackett, J. Osborne, J. Gogos, A. Chess, et al. (2004). Mice cloned from olfactory sensory neurons 1. *Nature*, 428, 44-49.

Geber, S., and M. Sampaio (1999). Blastomere development after embryo biopsy: a new model to predict embryo development and to select for transfer. *Hum. Reprod.*, 14, 782-786.

Handyside, A. H., E. H. Kontogianni, K. Hardy, and R. M. Winston (1990). Pregnancies from biopsied human preimplantation embryos sexed by Y-specific DNA amplification 1. *Nature*, 344, 768-770.

Hardy, K. (1993), Preimplantation Embryo Development. In, B. Bavister (Ed., Springer, New York, 184-199.

Hochedlinger, K., and R. Jaenisch (2002). Monoclonal mice generated by nuclear transfer from mature B and T donor cells. *Nature*, 415, 1035-1038.

Hurlbut, W. B. (2005). Altered nuclear transfer as a morally acceptable means for the procurement of human embryonic stem cells. *Perspect. Biol. Med.*, 48, 211-228.

Hwang, W. S., Y. J. Ryu, J. H. Park, E. S. Park, E. G. Lee, J. M. Koo, et al. (2004). Evidence of a pluripotent human embryonic stem cell line derived from a cloned blastocyst 1. *Science*, 303, 1669-1674.

Hwang, W. S., S. I. Roh, B. C. Lee, S. K. Kang, D. K. Kwon, S. Kim, et al. (2005). Patient-specific embryonic stem cells derived from human SCNT blastocysts 1. *Science*, 308, 1777-1783.

In 't Anker, P. S., S. A. Scherjon, C. Kleijburg-van der Keur, W. A. Noort, F. H. Claas, R. Willemze, et al. (2003). Amniotic fluid as a novel source of mesenchymal stem cells for therapeutic transplantation. *Blood*, 102, 1548-1549.

Jiang, Y., B. N. Jahagirdar, R. L. Reinhardt, R. E. Schwartz, C. D. Keene, X. R. Ortiz-Gonzalez, et al. (2002a). Pluripotency of mesenchymal stem cells derived from adult marrow.[see comment][erratum appears in Nature. 2007 Jun 14;447(7146):879-80]. *Nature*, 418, 41-49.

Jiang, Y., B. Vaessen, T. Lenvik, M. Blackstad, M. Reyes, and C. M. Verfaillie (2002b). Multipotent progenitor cells can be isolated from postnatal murine bone marrow, muscle, and brain.[erratum appears in Exp Hematol. 2006 Jun;34(6):809]. *Exp. Hematol.*, 30, 896-904.

Klimanskaya, I., J. Hipp, K. A. Rezai, M. West, A. Atala, and R. Lanza (2004). Derivation and comparative assessment of retinal pigment epithelium from human embryonic stem cells using transcriptomics. *Cloning & Stem Cells*, 6, 217-245.

Klimanskaya, I., Y. Chung, S. Becker, S. J. Lu, and R. Lanza (2006). Human embryonic stem cell lines derived from single blastomeres 2. *Nature*, 444, 481-485.

Landry, D. W., and H. A. Zucker (2004). Embryonic death and the creation of human embryonic stem cells 2. *J. Clin. Invest*, 114, 1184-1186.

Landry, D. W., H. A. Zucker, M. V. Sauer, M. Reznik, and L. Wiebe (2006). Hypocellularity and absence of compaction as criteria for embryonic death 1. *Regen. Med.*, 1, 367-371.

Lanza, R. P., H. Y. Chung, J. J. Yoo, P. J. Wettstein, C. Blackwell, N. Borson, et al. (2002). Generation of histocompatible tissues using nuclear transplantation 1. *Nat. Biotechnol.*, 20, 689-696.

Lu, S. J., Q. Feng, S. Caballero, Y. Chen, M. A. Moore, M. B. Grant, et al. (2007a). Generation of functional hemangioblasts from human embryonic stem cells 1. *Nat. Methods*, 4, 501-509.

Lu, S. J., J. A. Hipp, Q. Feng, J. D. Hipp, R. Lanza, and A. Atala (2007b). Genechip analysis of human embryonic stem cell differentiation into hemangioblasts: An in silico dissection of mixed phenotype 1. *Genome Biol.*, 8, R240.

Marshak DR, G. R., Gottlieb D (2001). *Stem Cell Biology.* Cold Spring Harbor Laboratory Press, New York.

Martin, G. R. (1981). Isolation of a pluripotent cell line from early mouse embryos cultured in medium conditioned by teratocarcinoma stem cells. *Proc. Natl. Acad. Sci. U. S. A.*, 78, 7634-7638.

Martinez, F., L. Rienzi, M. Iacobelli, F. Ubaldi, C. Mendoza, E. Greco, et al. (2002). Caspase activity in preimplantation human embryos is not associated with apoptosis. *Hum. Reprod.*, 17, 1584-1590.

McCulloch, E. A., and J. E. Till (1964). Proliferation of Hemopoietic Colony-Forming Cells Transplanted into Irradiated Mice. *Radiat. Res.*, 22, 383-397.

Meissner, A., and R. Jaenisch (2006). Generation of nuclear transfer-derived pluripotent ES cells from cloned Cdx2-deficient blastocysts 10. *Nature*, 439, 212-215.

Meissner, A., M. Wernig, and R. Jaenisch (2007). Direct reprogramming of genetically unmodified fibroblasts into pluripotent stem cells 1. *Nat. Biotechnol.*, 25, 1177-1181.

Minino, A., 2004: Deaths: Final Data for 2004.

Mitalipov, S. (2007). Reprogramming following somatic cell nuclear transfer in primates is dependent upon nuclear remodeling. *Hum. Reprod.*, 22, 2232-2242.

Okita, K., T. Ichisaka, and S. Yamanaka (2007). Generation of germline-competent induced pluripotent stem cells 1. *Nature*, 448, 313-317.

Polgar, K., R. Adany, G. Abel, J. Kappelmayer, L. Muszbek, and Z. Papp (1989). Characterization of rapidly adhering amniotic fluid cells by combined immunofluorescence and phagocytosis assays. *Am. J. Hum. Genet.*, 45, 786-792.

Presnell, S. C., B. Petersen, and M. Heidaran (2002). Stem cells in adult tissues. *Semin. Cell Dev. Biol.*, 13, 369-376.

Priest, R. E., K. M. Marimuthu, and J. H. Priest (1978). Origin of cells in human amniotic fluid cultures: ultrastructural features. *Lab. Invest.*, 39, 106-109.

Prusa, A. R., E. Marton, M. Rosner, G. Bernaschek, and M. Hengstschlager (2003). Oct-4-expressing cells in human amniotic fluid: a new source for stem cell research? *Hum. Reprod.*, 18, 1489-1493.

Prusa, A. R., E. Marton, M. Rosner, D. Bettelheim, G. Lubec, A. Pollack, et al. (2004). Neurogenic cells in human amniotic fluid. *Am. J. Obstet. Gynecol.*, 191, 309-314.

Quesenberry PJ, S. G., Forget B, Weissman S. (1998). *Stem Cell Biology and Gene Therapy.* Wiley-Liss, New York.

Rideout, W. M., III, K. Hochedlinger, M. Kyba, G. Q. Daley, and R. Jaenisch (2002). Correction of a genetic defect by nuclear transplantation and combined cell and gene therapy. *Cell*, 109, 17-27.

Schmidt, D., J. Achermann, B. Odermatt, C. Breymann, A. Mol, M. Genoni, et al. (2007). Prenatally fabricated autologous human living heart valves based on amniotic fluid derived progenitor cells as single cell source 1. *Circulation*, 116, I64-I70.

Simerly, C., T. Dominko, C. Navara, C. Payne, S. Capuano, G. Gosman, et al. (2003). Molecular correlates of primate nuclear transfer failures 2. *Science*, 300, 297.

Solter, D., and B. B. Knowles (1975). Immunosurgery of mouse blastocyst. *Proc. Natl. Acad. Sci. U.S.A*, 72, 5099-5102.

Spradling, A., D. Drummond-Barbosa, and T. Kai (2001). Stem cells find their niche. *Nature*, 414, 98-104.

Strumpf, D., C. A. Mao, Y. Yamanaka, A. Ralston, K. Chawengsaksophak, F. Beck, et al. (2005). Cdx2 is required for correct cell fate specification and differentiation of trophectoderm in the mouse blastocyst 1. *Development*, 132, 2093-2102.

Takahashi, K., and S. Yamanaka (2006). Induction of pluripotent stem cells from mouse embryonic and adult fibroblast cultures by defined factors 2. *Cell*, 126, 663-676.

Takahashi, K., K. Tanabe, M. Ohnuki, M. Narita, T. Ichisaka, K. Tomoda, et al. (2007). Induction of Pluripotent Stem Cells from Adult Human Fibroblasts by Defined Factors. *Cell*, 131, 861-872.

Thomas, E. D., H. L. Lochte, Jr., W. C. Lu, and J. W. Ferrebee (1957). Intravenous infusion of bone marrow in patients receiving radiation and chemotherapy. *N. Engl. J. Med.*, 257, 491-496.

Till, J. E., E. A. McCulloch, and L. Siminovitch (1964). A Stochastic Model of Stem Cell Proliferation, Based on the Growth of Spleen Colony-Forming Cells. *Proc. Natl. Acad. Sci. U. S. A.*, 51, 29-36.

Tsai, M. S., J. L. Lee, Y. J. Chang, and S. M. Hwang (2004). Isolation of human multipotent mesenchymal stem cells from second-trimester amniotic fluid using a novel two-stage culture protocol 2. *Hum. Reprod.*, 19, 1450-1456.

Wernig, M., A. Meissner, R. Foreman, T. Brambrink, M. Ku, K. Hochedlinger, et al. (2007). In vitro reprogramming of fibroblasts into a pluripotent ES-cell-like state 1. *Nature*, 448, 318-324.

Willadsen, S. M. (1981). The development capacity of blastomeres from 4- and 8-cell sheep embryos. *J. Embryol. Exp. Morphol.*, 65, 165-172.

Wilmut, I., A. E. Schnieke, J. McWhir, A. J. Kind, and K. H. Campbell (1997). Viable offspring derived from fetal and adult mammalian cells.[see comment][erratum appears in Nature 1997 Mar 13;386(6621):200]. *Nature*, 385, 810-813.

Yu, J., M. A. Vodyanik, K. Smuga-Otto, J. Antosiewicz-Bourget, J. L. Frane, S. Tian, et al. (2007). Induced Pluripotent Stem Cell Lines Derived from Human Somatic Cells. *Science*, 1151526.

Zhang, X., P. Stojkovic, S. Przyborski, M. Cooke, L. Armstrong, M. Lako, et al. (2006). Derivation of human embryonic stem cells from developing and arrested embryos 1. *Stem Cells*, 24, 2669-2676.

Zsebo, K. M., D. A. Williams, E. N. Geissler, V. C. Broudy, F. H. Martin, H. L. Atkins, et al. (1990). Stem cell factor is encoded at the Sl locus of the mouse and is the ligand for the c-kit tyrosine kinase receptor 1. *Cell*, 63, 213-224.

In: Philosophical Insights about Modern Science
Eds: Eva Žerovnik, Olga Markič and Andrej Ule
ISBN: 978-1-60741-373-8

Chapter 5

Reflections on the Use of Stem Cells for Restoring Neurodegenerative Damage

Franz-Josef Müller[1,2], Jeanne F. Loring[1] and Paul Christian Baier[2]

[1] Center for Regenerative Medicine, The Scripps Research Institute, 10550 North Torrey Pines Road, La Jolla, California 92037, USA.
[2.] Center for Psychiatry and Psychotherapy, University Hospital Schleswig Holstein, Niemannsweg 147, D-24105 Kiel, Germany

Here I sit, I form humans
After my own image;
Prometheus, J.W. Goethe, 1827

Who wants to live forever ...?
Queen, 1985/1986

Synopsis

Stem cells are commonly characterized as cell types which have the ability both to proliferate indefinitely without changing their properties (self-renewal") and to mature into different specialized cell types that are the building blocks of functional organ systems ("differentiation potential").

This potential has been further subdivided into pluripotency and multipotency, indicating, respectively, the ability to form many cell types or just a few.

In general, defining cells as stem cells requires experimental proof of functional properties after the cells have undergone differentiation. In the absence of an unequivocal biological definition for the stem cell state we will use the term "stem cell" in this manuscript in an inclusive way, while keeping in mind that the stem cell concept is rigorously defined only in a few model systems.[1]

In this discussion we will use Parkinson's disease (PD) as an example for which stem cell therapies are currently being investigated. Neural stem cells are believed to the first prototypical multipotent solid organ stem cell to have been identified and characterized.[2] Transplantation of neural stem cells has been proposed as a treatment of last resort for several incurable neurological disorders that involve functional loss and death of nerve cells. The rationale for cell replacement therapy for PD is based on the observation that during embryological development, cells with stem cell properties appear to be the source of neuronal populations that later become dysfunctional. Hence, successful repair may be achieved by introducing "replacement" progenitor cells into the pathologically altered brain. Based on this rationale, patients with PD were treated in the 1980's and1990s with fetal tissue preparations assumed to contain stem cells. In spite of very inconsistent results from those clinical studies, optimism for the idea of a stem cell therapy for PD remains high.

The concept of stem cells as a powerful therapeutic tool has gained much attention beyond the scientific community, from the public as well as from policy makers. There is currently a surge of funding for translational stem cell research worldwide, which will most likely result in the first clinical stem cell trials within a few years. One reason for the heightened attention may be that the idea of stem cell therapy seems simple and in tune with many of our current sociocultural memes.

Having memorized,
What to say and what to do,
With my powers of will I can
Do some witching, too!

The Sorcerers Apprentice, J. W. Goethe

INTRODUCTION

The concept of stem cells was first developed in the second half of the nineteenth century in the field of embryology and research on the blood forming organ system.[3] Hematopoietic stem cell (HSC) transplantation was the first (and so far only) clinical application in which stem cells have proved to be routinely successful for the reconstitution of lost organ function.[4]

The success of HSC transplantation lies in the fact that the umbilical cord and the bone marrow contain cells types that are able to fully reconstitute a dysfunctional blood-forming organ system.[5] The discovery of the bone marrow as the cellular substrate of blood formation has been ascribed to Ernst Neumann's work in the 1860s.[6] About 70 years later, almost immediately after the bone marrow biopsy methodology was developed, the first attempts at treating anemia were undertaken.[7,8] These cell preparations can be harvested relatively easily from living mammals and can be studied at the single cell level more readily than cell preparations from other organs.[9] Importantly, hematopoietic cells do not need to be cultured *in vitro* for subsequent *in vivo* experimentation in order to demonstrate their self-renewal and differentiation potential.[10]

With the development of more and more advanced cell culture techniques in the early 1980s, embryonic stem cells were derived for the first time from mouse embryos:.[11,12] These cells have the ability to give rise to all cell types when they are mixed with host mouse embryos.[13] Embryonic stem cells alone (without the aid of a host embryo) can differentiate into all three embryonic germ layers (ectoderm, mesoderm and endoderm) but not into extraembryonic cell types, which defines embryonic stem cells as being pluripotent ("many powers") and not totipotent ("all powers").

During embryonic development the three germ layers give rise to all organ systems, and embryonic tissues retain multipotent (but not pluripotent) stem cells. Because of their extensive developmental potential, pluripotent stem cells (PSCs, such as embryonic stem cells) are currently considered to be most important cell type for a broad range of regenerative medicine applications.

In the early 1990s, genetic immortalization strategies[14] and improved culture methods led to the generation of many neural stem cell preparations from the embryonic and later from the adult rodent nervous system.[15-17] These *in vitro* preparations appeared to integrate well into developing brains.[18] Soon after, these cells were introduced into pathologically altered brains to repair functional impairments.[19]

The concept of cell transplantations as a strategy for nervous system repair, though, predates the stem cell field by several decades. Beginning in the 1950s, researchers started to transplant pituitary preparations into various heterotopous regions, among them the central nervous system, in order to study hormonal regulatory circuitry in rodents.[20] Yet 40 years earlier the embryologist Hans Spemann had pioneered the use of transplanting developmental organizers from and to embryonic amphibians, molluscs and sea urchins.[21] Conceived as an experimental tool for studying the effects of certain cell types with specific functions on an experimental organism, this concept turned out to be extremely versatile and powerful in many biological disciplines.

Transplantation of cells to mitigate the loss of neurons in PD came from two lines of thought. First, the identification of a highly localized population of neurons[22] with a unique function (the production of dopamine) and "natural stain"[23] (neuromelanin, a byproduct of the dopamine synthesis[24]) that die in the course of the disease suggested a "focal" problem that could be solved with a targeted intervention.

Second, the discovery of dopamine's function as a neurotransmitter[25,26] and the positive effects of the reconstitution of diminished dopamine levels in PD patients by pharmacologic means[27] suggested a simply "replaceable" single neuronal function.

This has led to the million-dollar question to which thousands of scientists try to find an answer today: if cells can replace lost endocrine functions in certain instances, can we reconstitute dopamine deficiency in PD patients with dopamine-secreting cells?

Dopamine is also secreted by the adrenal glands.[28] It is relatively easy to generate cell grafts from the adrenal glands of patients.[28,29] The accessibility and the lack of immune rejection of autologous cell grafts enabled several clinical trials in the 1980s.[29] At the same time, with more and more knowledge of brain development, fetal tissue grafts from aborted fetuses appeared to be a better alternative, since the clinical results of adrenal grafts were disappointing.[30] The next logical step was to work on "pure" neural stem cell preparations engineered and optimized for transplantation in PD.[19]

> "In adult centres the nerve paths are something fixed, ended, immutable. Everything may die, nothing may be regenerated. It is fort he science of the future to change, if possible, this harsh decree"
>
> Santiago Ramon Y Cajal, Degeneration and Regeneration of the Nervous System 1928

"But although, at present, uninformed as to the precise nature of the disease, still it ought not to be considered as one against which there exists no countervailing remedy. On the contrary, there appears to be sufficient reason for hoping that some remedial process may ere long be discovered, by which, at least, the progress of the disease may be stopped."

James Parkinson, Essay on the shaking palsy, 1817

Parkinson's Disease as a Prototype for Clinical Stem Cell Therapies

PD is a common progressive neurodegenerative disorder. It is characterized by the clinical triad rigidity, tremor and akinesia, as well as by vegetative disturbances, and is frequently associated with depression and dementia. In its early stages pharmacological treatment with dopamine-agonistic drugs is usually successful. However, drug treatment does not halt the disease's progression and in the later stages of PD it becomes limited by "wearing-off" and the occurrence of unwanted side effects (e.g., dyskinesias and drug-induced psychosis). Hence, there is an urgent need for alternative therapeutic approaches.

There is no widely accepted definition of PD. Instead, depending on the context, PD is defined based on clinical, neuropathological or combined criteria. Although the causative factors contributing to PD remain largely unknown, traditionally it has been considered a well-defined clinicopathologic entity due to a relatively selective and circumscript loss of dopaminergic neurons in the substantia nigra pars compacta. Under this assumption, cell replacement therapy appears to be a straightforward option and PD has become the prototypical neurodegenerative disorder that is believed to be amenable to stem cell therapy.

However, growing knowledge in the past years, in particular of genetic forms of PD, has raised the question of whether we can continue to believe that the pathology of PD has a focal nature. Concepts of a more widespread neurodegeneration also involving noradrenergic, serotonergic, cholinergic and other central and peripheral circuits have been proposed as alternatives.[31] Secondly, the scientific concept of PD has recently undergone a significant change: PD is most likely not a single nosologic entity but most appropriately represents the clinical manifestation of several different patho-mechanisms.[31]

The wide spectrum of signs, symptoms and features that are unassociated or merely loosely associated with the demise of dopaminergic projections emanating from the substantia nigra pars compacta demonstrates clearly that correction of a striatal dopaminergic deficit is necessary but by no means sufficient to mitigate the suffering of patients with PD.

"A few treatments, preventions and cures in mice:

- Heart Attack, damage reversible 1996
- Cancer, cured 1997
- Baldness, cured 1998
- Creutzfeldt-Jakob Disease, incubation prolonged indefinitely 1999
- Sickle Cell Disease, cured 2001
- Blindness from Leber's Congenital Amaurosis, symptoms reversed 2002
- Type 1 Diabetes, cured 2003

- Parkinson's Disease, cured 2003
- Multiple Sclerosis, symptoms reversed 2003
- Early-Stage Alzheimer's Disease, progression halted 2004
- Phenylketonuria (PKU), cured 2005
- Hemophilia Type B, symptoms reversed 2005
- West Nile Virus, cured 2005
- Severe Acute Respiratory Syndrome (SARS), cured 2005
- Diabetic Blindness, prevented 2005"

Josh Braun, Seed Magazine 1, Oct/Nov 2005

Preclinical Disease Models

Several animal models for PD have been developed, of which the most widely used is the 6-hydroxy-dopamine (6-OHDA) model in rodents.[32] It models a striatal deficiency of dopaminergic innervation from the substantia nigra (see figure 1).

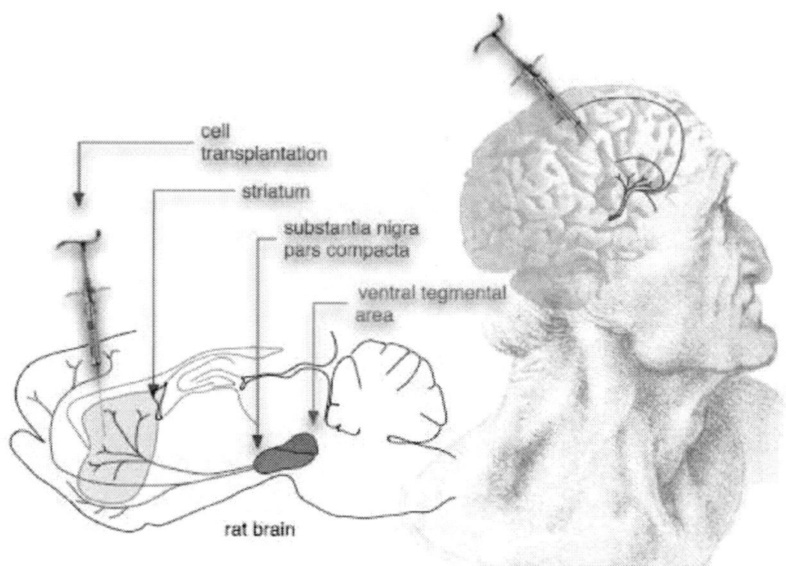

Figure 1. Heterotopic transplantation of dopaminergic grafts in animal PD models and patients with PD. Dopaminergic nerve cells degenerate in the substantia nigra and lead to a dopamine deficit in the parts of the brain which they innervate, mainly the striatum. Grafts are transplanted to the denervated regions in order to mitigate symptoms due to the dopamine deficit.

Briefly rats or, more technically challenging, mice are being treated with an intracerebral injection of 6-OHDA. The agent is injected into either the substantia nigra pars compacta, the striatum or the medial forebrain bundle representing the main connection between them. Typically, unilateral lesions are performed, as complete lesions of both sides are commonly associated with severe impairments in drinking and feeding that result in a high rate of mortality of animals. Also, the functional imbalance between sides provides the most specific readout for the researcher: an asymmetric lack of dopamine leads to spontaneous and amphetamine reinforced turning-behavior towards the lesioned side (ipsilateral) and to contralateral turning behavior after the application of drugs directly stimulating the dopamine receptor (levodopa, apomorphine). This turning behavior correlates well with the extent of dopaminergic loss and can be restored by several treatments.

To illustrate the significance of this "animal model of Parkinson's disease" one need look only as far as the most significant recent scientific discovery in the stem cell field: Yamanaka and colleagues recently demonstrated the induction of pluripotence in somatic cells by defined factors.[33-35] These findings were confirmed, replicated and extended by a group of prominent researchers worldwide.[36-38] The first preclinical model of a neurological disorder that was treated with neural cells differentiated from this novel source was, not by chance, the 6-hydroxy-dopamine rat model of PD.[39]

Neural stem cells (NSCs) and PSCs have been differentiated into cell types that secrete dopamine, and hence have been proposed as a suitable treatment for the striatal dopamine deficiency in PD.[40,41] We have shown that transplanted elements do not need to be integrated into the neuronal circuitry in order to mitigate a striatal dopamine deficiency and improve behavioral deficits.[42] Other equally effective strategies in preclinical models are the chronic infusion of dopamine by cannulas[30,43], encapsulated PC12-cells (a tumor line derived from a rat pheochromocytoma)[44,45] or the pharmacological delivery of dopamine by means of oral administration of its biological precursor 3,4-dihydroxy-L-phenylaalanine (L-DOPA)[46], which is also among the most important drugs in the treatment of patients suffering from PD.[27,47]

All the animal models for PD that have been proposed, including the 6-OHDA model, have in common that they incompletely mimic a very limited number of features of the complex and not yet understood patho-physiological sequence of a multi-system degeneration that eventually leads to PD.

Yet all attempts at neural transplantation, from adrenal grafts to fetal mesencephalic grafts to, possibly, stem cells in the near future, have been essentially based on a rationale developed from data gathered with the very same preclinical animal models.

> "Your eyes have seen my unformed limbs"
> The Bible, Psalm 139, Verse 16

> "And He breathed into his nostrils the breath of life; and man became a living soul."
> The Bible, Genesis, 2 Verse 7

Stem Cell Sources

Cell types that have been termed "stem cells" or "progenitors" have been isolated from a plethora of tissue sources using even more methods and protocols (see figure 2). It has been proposed that these preparations have the potential to differentiate into multiple cell types ("potency") and, under suitable conditions, to remain undifferentiated in spite of proliferation ("self-renewal"). The stringency and consistency in the assessment of these postulated stem cell features varies strongly between studies and laboratories.[1]

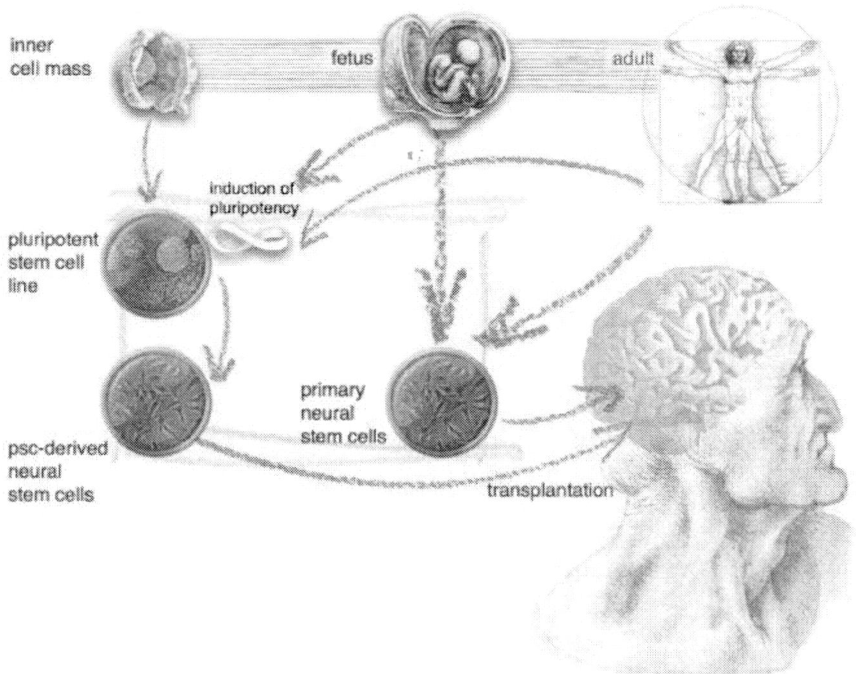

Figure 2. Sources of stem cells. Stem cells with varying developmental potential can be generated from embryos, fetuses and adult humans. Pluripotent stem cells (PSCs), which theoretically have the ability to contribute to every organ system, can either be cultured from preimplantation embryos or induced from somatic cells from fetuses or adult humans. For the transplantation into the dopamine-deficient striatum, neural stem cells are believed to be the most preferable cell type. These somatic stem cells may be differentiated from PSCs or derived from fetuses or adult individuals.

A recent systematic survey of whole-genome transcriptional phenotypes has suggested that a pluripotency-associated self-renewal phenotype exists in embryonic, tumor-derived and induced PSCs. "Differentiated" stem cell phenotypes appeared to be highly heterogeneous.[48] Our analysis argues against a "stemness" program that different stem cells have in common, as it has been proposed following a much more limited data analysis in 2002 by two independent groups in high profile studies.[49,50]

While it is too early to draw a definitive conclusion about the true identity of stem cells and the biological basis of their properties, it is possible to generally summarize the technical means by which stem cells have been engineered.

PSC lines can be derived from early embryos that have not yet implanted in utero.[11,12,51] These stem cell lines have pluripotent features and a much higher proliferation potential than any other stem cell type that has been isolated so far. At the same time, these cell lines are surrounded by much controversy since in most cases their generation requires the destruction of a human embryo.[52,53]

Every embryo starts as a fertilized egg. Procedures that involve somatic nuclear transfer ("cloning") do start by replacing the nucleus of an unfertilized egg with the nucleus of a differentiated cell from another individual (donor). This manipulated cell is then brought to the pre-implantation embryonic stage *in vitro.[54]* In the case of "therapeutic cloning" this embryo is then used to generate an embryonic PSC line.[55] Hence, therapeutic cloning is an extended procedure for the generation of embryonic PSCs.

"True cloning" stands in this context for the generation of a fully developed individual that is genetically identical to another living individual.[56] "True cloning" is banned worldwide for humans but has been demonstrated for many sub-primate animals.[56] In this case, the embryo is implanted into a pseudo-pregnant uterus and grows to become a fully developed living organism. In 2006, the first induction of pluripotency in differentiated cells (induced PSCs, iPSCs) by a genetic strategy was reported and offers theoretical independence from problematic cell sources like embryos and human oocytes.[33] iPSCs are not reliant on sources other than the potential cell transplantation recipient. Thus, induced pluripotency is a potential means for the industrial-scale generation of individualized stem cells.

Different from this "raw material" type of concept are tissue-specific stem cells, which have acquired an identity and function within their biological compartment. From the conceptual perspective of PSCs, which are imagined to be able to differentiate into every cell type (including all somatic stem cells), somatic stem cells appear to be of limited use. The opposite is true when the actual clinical relevance of somatic stem cells versus that of PSCs is considered: so far no patient has been treated with preparations derived from PSCs, but many thousands of patients have been cured with HSCs in more than 500 clinical centers worldwide.

In contrast to HSC transplantations, fetal neural stem cell transplantations are in their infancy and there are ethical problems surrounding the use of biological material from aborted fetuses. There is currently one clinical trial where six children with a neurologically devastating metabolic disorder (Batten's disease) have received NSC grafts.[57] The rationale behind this approach is to deliver missing enzymatic activity by means of NSCs but not primarily to replace lost cellular elements.[57]

The key requirement for any stem cell therapy, proposed or real, is that the outcomes have to be predictable: predictability in the case of hematopoeitic stem cells is due to their bioequivalence. This means that cells that are part of an organ system in a donor organism are used to replace the very same cellular elements in a host organism. It has been very difficult to demonstrate bioequivalence of PSC-derived cells, even in preclinical models. One recent example is the study by Roy et al. (2007), where the authors transplanted NSCs derived from human PSCs into rats with a 6-OHDA-induced lesion.[58,59] The lab animals experienced a symptomatic recovery, possibly resulting from increased dopamine production.[58,59] At the same time, uncontrolled growth of tumor-like NSC-cytomas originating from the stem cell grafts has raised serious safety issues. This and many other reports add a cautionary note to early claims of bioequivalence for PSC-derived cell types.[60]

> "Everything gives way and nothing stays fixed."
>
> Heracleitus

> "By midnight, an enourmous mound of clay lay before the rabbi. Praying softly, he plunged his hands into the vast lump, shaping it. Hours later, he arose and stood back. A crude clay giant lay lifeless on the riverbank."
>
> Golem, David Wisniewsky, 1996

DIFFERENTIATION

A comprehensive description of the current technologies and protocols being used to derive NSCs and eventually functional neural elements from pluripotent stem cells is beyond the scope of this manuscript (for more detailed accounts see[1] and[61]). We will highlight instead certain important issues that appear to be relevant to future developments in the field.

Initially, all cells of the nervous system start off as neural stem cells during embryonic development.[62] As brain development progresses they start to differentiate into neural cells, which then in turn begin to build connections as they mature. In later stages of development, NSCs also begin to differentiate into glial cells, which were long believed to be just "glue cells": elements holding neurons together.[63] In recent years, glial cells have gained more and more attention from neuroscientists and are likely to be as important as neurons for higher brain functions.[64,65] NSCs decrease in number as more differentiated cell types emerge and retain progenitor functions in only a few niches in the adult brain.[62,66]

The preceding paragraph is an almost impossible simplification of probably one of the most complex processes to have evolved on this planet, summarizing a developmental process that produces several tens of thousands of distinct cell types distributed in an intricate architecture among several billion cells. These elements interact and react in a highly coordinated and orchestrated fashion which enables the reader of this text to perceive, understand and reflect on the lines of text printed above and below.[67]

NSCs were first isolated from embryonic rodent brains. Subsequently, cell preparations which were claimed to have similar properties have been derived from human embryonic, fetal and adult tissues.[68] Currently, isolation and growth in culture is the instrumental first step in manipulating a specific cell type in order to generate grafts in transplantation studies.

Initially, knowledge of NSCs was gathered mostly from *in vitro* neural cell preparations from embryonic and fetal brains.[69,70] Because of this, successive NSC-like cell types from other sources have been "defined" on the basis of observations made with this particular fetal NSC phenotype as the reference point.[71] Certain genes termed "stem cell markers" that initially appeared to be specific for NSCs became important for the "characterization" of NSC preparations.[72] This assumption, however, appears to be an oversimplification: "whole genome" technologies are becoming more and more important in the biomedical field and gather data from, for example, all known genes in the human genome (~ 35,000 genes) at the same time in the same cell preparation. With this methodology it has been shown that even stem cells with the same "marker profiles" can be extremely different with regard to functional biological processes inside the cell.[48] These differences appear to extend to actual functional properties of specific stem cell preparations

This adds another underappreciated level of complexity to the envisioned stem cell therapies: the stem cell state is not the stage in which a cell type has been considered to be useful for mitigating impairment by neurodegenerative disorders.

Mature, differentiated progeny derived from NSCs are believed to be necessary to functionally replace the loss of neural elements in the brain. One study has demonstrated that when injected into developing mammalian brains, human PSC-derived NSCs can give rise to seemingly bioequivalent neurons.[73]

An unresolved conceptual problem emerges from this study. The authors demonstrate their ability to culture cells which upon exposure to the right cues and signals in a developing neural system assume neuronal functions that are similar to normal. However, this does not automatically indicate that grafts from NSCs will function normally after terminal differentiation *in vitro* or in the context of a more mature, possibly pathologically altered brain.

Regeneration by stem cells does not occur at functionally relevant levels in the adult brain or after pathological insults.[74] Therefore, no convincing concept of the "normal" therapeutic function of NSCs in neurodegenerative disorders exists.

Most scientists believe that NSCs induced to recapitulate developmental stages *in vitro* will work as a replacement-type "motor" after transplantation. Today, *in vitro* differentiation protocols are highly artificial systems which try to mimic sequential developmental cues and signals.[61] These signal sequences are in most cases factors that are believed to play a role in the location-specific maturation of specific neural cell types.

For example, when the goal is to derive dopaminergic neurons, NSCs are exposed to different proteins that have been shown to play an important role in the maturation of dopaminergic neurons during normal fetal development.[75] Eventually, the production of dopamine by the cellular end-products of this process is assessed as an endpoint for the successful conversion to dopaminergic neurons, simply because we are not exactly sure what other relevant physiological properties dopaminergic neurons might possess.

> "In order to ascertain whether the boy, after feeling so slight an affection of the system from the cow-pox virus, was secure from the contagion of the smallpox, he was inoculated the 1st of July following with variolous matter, immediately taken from a pustule."
>
> Edward Jenner, An Inquiry Into the Causes and Effects of the Variolæ Vaccinæ, Or Cow-Pox. 1798

> "I have already told you with what care they look after their sick, so that nothing is left undone that can contribute either to their case or health; and for those who are taken with fixed and incurable diseases, they use all possible ways to cherish them and to make their lives as comfortable as possible."
>
> Thomas Moore, Utopia, 1516

From Bench to Bedside

The outcomes of preclinical transplantation studies with 6-OHDA-treated animals were broadly considered as successful in the late 1970s and early 1980s.[28,76-78] This led medical scientists in the US and Europe to conduct several open-label transplantation series with fetal mesencephalic grafts in patients with Parkinson's disease.[30, 79-81]

The principal transplantation strategy from the 6-OHDA model was not changed: grafts were transplanted heterotopically into the striatum. These open-label trials demonstrated promising results[30] and built the foundation for today's perception of PD as a reasonable target for stem cell therapy.

At the end of the 1990s, the lead researchers in the US then turned to the most rigorous clinical study type in the arsenal of experimental medicine, a placebo controlled, double-blinded study design,[82,83] and should be applauded for this choice. Then, it was highly controversial whether sham neuro-surgery was ethical, since the treatment was considered to be highly effective.

The results of these trials came as somewhat of a surprise: the data of one study demonstrated clearly that only a subgroup of patients responded to the experimental treatment and that older patients in particular (>60 years of age) were less likely to benefit from the procedure.[84] In a second study, patients did not improve as a group, but some of them showed prominent functional recovery, which was correlated with the preoperative L-dopa response.[85]

Currently, the first long-term results for graft survival are being published. Cells transplanted to the striatum do seem to survive for up to 16 years.[86,87] Possibly the most significant novel finding was the detection of Lewi bodies in heterotopically transplanted TH-positive grafts:[86,88] apparently, the pathologic mechanisms leading to neuro-degeneration in PD are also active in brain regions distant from the substantia nigra.

What remains is that fetal grafts in the dopamine-deficient striatum do offer symptomatic relief on a par with other treatment options for Parkinson's disease that lead to the tonic dopaminergic signalling in the striatum.[89]

Still, it appears that the common perception of Parkinson's disease as "low hanging fruit" for clinical stem cell transplantation trials has not fundamentally changed.

A more realistic outlook might show give what biotech companies, as one of the driving forces for the commercial application of stem cell treatments, are currently pursuing. For example, Geron, a company that has licensed patents on the application of human embryonic stem cells to treat insulin-dependent diabetes, cardio-vascular disorders and disorders of the central nervous system from the WiCell Inc.[90], is currently planning trials of treating spinal cord injury with human ePSC-derived oligodendrocytes.[90]

> "The man said, 'This is now bone of my bones, And flesh of my flesh; She shall be called Woman, Because she was taken out of Man.' "
>
> Genesis 2:23
>
> In the kingdom of ends everything has either a price or a dignity.
>
> Immanuel Kant, Groundwork of the Metaphysics of Morals

The Scientific Method, the Public and Stem Cell Policies

In 2004, Proposition 71, a referendum on state-specific financial support for stem cell research with a focus on human PSCs, was supported by 59.1% of the electorate in California.[91] This has led to the commitment of US$3 billion over ten years to this field of research. This amount of money equals the Manhattan project (which has brought us the nuclear bomb) or the Apollo program (which has brought man to the moon) with regard to the amount of money spent per single resident.[92]

Similar measures have been proposed and legislated in other states of the US; efforts of comparable scope have been initiated in several countries in Asia and in Europe. These efforts dwarf the amounts committed to follow through on tested treatments or find novel approaches to fight e.g., tuberculosis or child malnourishment in developing countries. Yet, it appears that no other scientific field is surrounded by more controversy than stem cell research.

One important factor contributing to this surge in public and fiscal attention may be a unique situation created inadvertently by stem cell scientists.

In stem cell research, borrowing most methodologies from developmental biology, tissues, cells and other materials from all stages of development can be taken from one individual multi-cellular organism and reintroduced into another individual organism.

The way experimental mice husbandry is maintained may serve as an example: several mouse strains have been inbred for many generations with the result of uniformity in their genetic set-up.[93,94] The main advantage of such an effort becomes obvious in transplantation studies: within one mouse strain, theoretically, every individual animal can be used as a tissue/organ donor as well as a recipient, since they will not reject tissue with an identical genetic and thus identical immunological set-up.

This is in stark contrast to our current humanistic and individualized medical approach, which requires first and foremost not to cause harm to any human individuals.

This contrast turns into conflict, once basic biological studies are being conducted with the intention to translate the findings into clinical practice.

For lab animals, it is generally not regarded as problematic to harvest tissues from adult, fetal or embryonic mammals and extract e.g., oocytes from mice, not unlike animals that are produced and exploited by the food industry.

The same procedures, e.g., harvesting oocytes for somatic nuclear transfer in humans, causes major logistical, medical, legal and ethical problems and even more so in the generation of human embryos from such an engineered cell in order to generate stem cell lines (also called therapeutic cloning).[54]

The scientific method clashes in this instance with a conceptual framework represented by the term "empathy". Certain primate species (chimpanzees and humans) have been proposed to be capable of empathy, i.e., an altruistic impulse in response to the perception of other individuals in distress, need or pain.[95-97]

Stem cell research following the scientific method, in which empathy is not a major determinant or motivation, creates a situation where two subjects, the patient or the stem cell source (which may be an embryo, fetus or adult donor), compete for the empathy of human individuals outside the scientific process.

Because the postulated potential of stem cell research for not yet realized "rescue missions" has been made a commonplace public topic by research institutions in order to generate new funding opportunities, public spending has increased dramatically as a perceived altruistic act in response to empathy for either patient or the stem cell donor.

Yet, this conflict between stem cell donor and recipient cannot be resolved by biological experimentation, since it does not exist in the conceptual framework of the scientific method.

> "I've seen things you people wouldn't believe. [...] All those ... moments will be lost in time, like tears...in rain. Time to die."
>
> Blade Runner, Hampton Fancher, David Peoples, 1982

"The event on which this fiction is founded has been supposed by Dr. Darwin, and some of the physiological writers of Germany as not of impossible occurrence."

Frankenstein, Mary Wollstonecraft-Shelly, 1823

PERSPECTIVE

In the mid-18th century, French physician Charles Le Roy began experimenting with electricity to cure blindness and other malaises.[98] Luigi Galvani discovered bio-electromagnetic phenomena in the 1780s by making a dead frog's leg move by touching the innervating nerves with electrically charged instruments.[99] This research, as well as his fellow scientist Allesandro Volta's work, extended and also popularized the concept of "animal electricity".[100] The results and even more so their interpretation were discussed controversially among scientists. Volta also discovered that when he placed metal rods into his own ears and connected them to a 50 volt circuit, acoustic sensations could be experienced. "Galvanic operations" were imagined in the ensuing years, during which electromagnetism was considered to be a potential cure for all types of paralysis, blindness and other functional impairments by medical experts of the time.[101] When Mary Wollstonecraft Shelley's Frankenstein was published in 1823, the word "galvanism" also implied the release, through electricity, of mysterious life forces.[102] Galvani's nephew, Giovanni Aldini, among others proposed that "electrical therapy" could revive cadavers.[103]

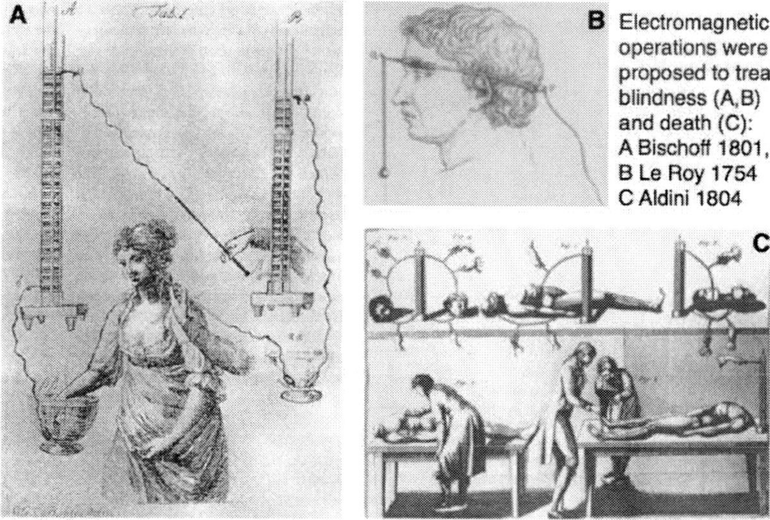

Figure 3. Two hundred years ago, electromagnetic operations were imagined to cure blindness and death.

Today, 200 years later, about 100 000 patients with impaired hearing worldwide have at least gained some auditory capabilities with cochlear implants, about 100 000 patients with heart disorders receive live-saving pacemakers annually in Germany alone and 50 000 patients with Parkinson's disease have received deep brain stimulation implants in the US alone.

The electromagnetic principles discovered by Galvani and Volta have so far not led to a cure for disease and death. These treatments are all symptomatic and do not change the underlying patho-physiological processes, yet can be hailed for substantially decreasing mortality or at least improving quality of life.

This might apply in an analog way to translational stem cell therapies: stem cell research most certainly will lead to scientific insights and eventually to novel therapeutical strategies.

Yet, today we probably have no clue what these therapies might be.

REFERENCES

[1] Muller, F.J., Snyder, E.Y. & Loring, J.F. Gene therapy: can neural stem cells deliver? *Nat. Rev. Neurosci.* 7, 75-84 (2006).

[2] Imitola, J., et al. Stem cells: cross-talk and developmental programs. *Philos. Trans. R. Soc. Lond. B. Biol. Sci.* 359, 823-837 (2004).

[3] Ramalho-Santos, M. & Willenbring, H. On the origin of the term "stem cell". *Cell Stem. Cell.* 1, 35-38 (2007).

[4] Thomas, E.D., Lochte, H.L., Jr., Lu, W.C. & Ferrebee, J.W. Intravenous infusion of bone marrow in patients receiving radiation and chemotherapy. *N. Engl. J. Med.* 257, 491-496 (1957).

[5] Barnes, D.W. & Loutit, J.F. Protective effects of implants of splenic tissue. *Proc. R. Soc. Med.* 46, 251-252 (1953).

[6] S. Hakim, N.a.d.e.y. & E. Papalois, V.a.s.s.i.l.i.o.s. *History of Organ and Cell Transplantation.* 464 (2003).

[7] Schretzenmayr, A. Aämiebehandlung mit Knochenmarksinjektionen. *Klin. Wissenschaftschreiben.* 16, 1010 (1937).

[8] Osgood, E.E., Riddle, M.C. & Mathews, T.J. Aplastic anaemia treated with daily transfusions and intravenous marrow. *Ann. Intern. Med.* 13, 357 (1939).

[9] Siminovitch, L., McCulloch, E.A. & Till, J.E. The Distribution of Colony-Forming Cells among Spleen Colonies. *J. Cell Physiol.* 62, 327-336 (1963).

[10] Lorenz, E., Uphoff, D., Reid, T.R. & Shelton, E. Modification of irradiation injury in mice and guinea pigs by bone marrow injections. *J. Natl. Cancer Inst.* 12, 197-201 (1951).

[11] Evans, M.J. & Kaufman, M.H. Establishment in culture of pluripotential cells from mouse embryos. *Nature.* 292, 154-156 (1981).

[12] Martin, G.R. Isolation of a pluripotent cell line from early mouse embryos cultured in medium conditioned by teratocarcinoma stem cells. *Proc. Natl. Acad. Sci. U. S. A.* 78, 7634-7638 (1981).

[13] Bradley, A., Evans, M., Kaufman, M.H. & Robertson, E. Formation of germ-line chimaeras from embryo-derived teratocarcinoma cell lines. *Nature.* 309, 255-256 (1984).

[14] Ryder, E.F., Snyder, E.Y. & Cepko, C.L. Establishment and characterization of multipotent neural cell lines using retrovirus vector-mediated oncogene transfer. *J. Neurobiol.* 21, 356-375 (1990).

[15] Cattaneo, E. & McKay, R. Proliferation and differentiation of neuronal stem cells regulated by nerve growth factor. *Nature.* 347, 762-765 (1990).
[16] Reynolds, B.A. & Weiss, S. Generation of neurons and astrocytes from isolated cells of the adult mammalian central nervous system. *Science.* 255, 1707-1710 (1992).
[17] Morshead, C.M., et al. Neural stem cells in the adult mammalian forebrain: a relatively quiescent subpopulation of subependymal cells. *Neuron.* 13, 1071-1082 (1994).
[18] Renfranz, P.J., Cunningham, M.G. & McKay, R.D. Region-specific differentiation of the hippocampal stem cell line HiB5 upon implantation into the developing mammalian brain. *Cell.* 66, 713-729 (1991).
[19] Nikkhah, G., Cunningham, M.G., McKay, R. & Bjorklund, A. Dopaminergic microtransplants into the substantia nigra of neonatal rats with bilateral 6-OHDA lesions. II. Transplant-induced behavioral recovery. *J. Neurosci.* 15, 3562-3570 (1995).
[20] Martinovitch, P.N. & Radivojevitch, D.V. Long-term transplants of infantile rat pituitaries cultivated in vitro and grafted into the anterior eye chamber of young cats. *Nature.* 176, 699-700 (1955).
[21] Spemann, H.a.n.s. & Mangold, H. Induction of embryonic primordia by implantation of organizers from a different species. 1923. *Int. J. Dev. Biol.* 45, 13-38 (2001).
[22] Foix, C. & Nicolesco, J. *Anatomie cérébrale. Les noyaux gris centraux et la région Mésencéphalo-sous-optique. Suivi d'un appendice sur l'anatomie pathologique de la maladie de Parkinson.*, (Masson et Cie, Paris, 1925).
[23] Hassler, R. Zur Pathologie der Paralysis agitains und des postenzephalitischen Parkinsonismus. *J. Psychol. Neurol.* 48, 387–476 (1938).
[24] Graham, D.G. On the origin and significance of neuromelanin. *Arch. Pathol. Lab. Med.* 103, 359-362 (1979).
[25] Carlsson, A. & Waldeck, B. Release of 3H-metaraminol by different mechanisms. *Acta Physiol. Scand.* 67, 471-480 (1966).
[26] Ehringer, H. & Hornykiewicz, O. [Distribution of noradrenaline and dopamine (3-hydroxytyramine) in the human brain and their behavior in diseases of the extrapyramidal system.]. *Klin. Wochenschr.* 38, 1236-1239 (1960).
[27] Birkmayer, W. & Hornykiewicz, O. [The L-3,4-dioxyphenylalanine (DOPA)-effect in Parkinson-akinesia.]. *Wien. Klin. Wochenschr.* 73, 787-788 (1961).
[28] Kolata, G. Grafts correct brain damage. *Science* 217, 342-344 (1982).
[29] Backlund, E.O., et al. Transplantation of adrenal medullary tissue to striatum in parkinsonism. First clinical trials. *J. Neurosurg.* 62, 169-173 (1985).
[30] Madrazo, I., et al. Transplantation of fetal substantia nigra and adrenal medulla to the caudate nucleus in two patients with Parkinson's disease. *N. Engl. J. Med.* 318, 51 (1988).
[31] Marras, C. & Lang, A. Invited article: changing concepts in Parkinson disease: moving beyond the decade of the brain. *Neurology.* 70, 1996-2003 (2008).
[32] Simola, N., Morelli, M. & Carta, A.R. The 6-hydroxydopamine model of Parkinson's disease. *Neurotox. Res.* 11, 151-167 (2007).
[33] Takahashi, K. & Yamanaka, S. Induction of pluripotent stem cells from mouse embryonic and adult fibroblast cultures by defined factors. *Cell.* 126, 663-676 (2006).
[34] Takahashi, K., et al. Induction of pluripotent stem cells from adult human fibroblasts by defined factors. *Cell.* 131, 861-872 (2007).

[35] Okita, K., Ichisaka, T. & Yamanaka, S. Generation of germline-competent induced pluripotent stem cells. *Nature.* (2007).

[36] Meissner, A., Wernig, M. & Jaenisch, R. Direct reprogramming of genetically unmodified fibroblasts into pluripotent stem cells. *Nat. Biotechnol.* (2007).

[37] Park, I.H., et al. Reprogramming of human somatic cells to pluripotency with defined factors. *Nature.* 451, 141-146 (2008).

[38] Yu, J., et al. Induced pluripotent stem cell lines derived from human somatic cells. *Science.* 318, 1917-1920 (2007).

[39] Wernig, M., et al. Neurons derived from reprogrammed fibroblasts functionally integrate into the fetal brain and improve symptoms of rats with Parkinson's disease. *Proc. Natl. Acad. Sci. U. S. A.* 105, 5856-5861 (2008).

[40] Lee, S.H., Lumelsky, N., Studer, L., Auerbach, J.M. & McKay, R.D. Efficient generation of midbrain and hindbrain neurons from mouse embryonic stem cells. *Nat. Biotechnol.* 18, 675-679 (2000).

[41] Kawasaki, H., et al. Induction of midbrain dopaminergic neurons from ES cells by stromal cell-derived inducing activity. *Neuron.* 28, 31-40 (2000).

[42] Baier, P.C., et al. Behavioral changes in unilaterally 6-hydroxy-dopamine lesioned rats after transplantation of differentiated mouse embryonic stem cells without morphological integration. *Stem Cells.* 22, 396-404 (2004).

[43] Horne, M.K., et al. Intraventricular infusion of Dopamine in Parkinson's disease. *Ann. Neurol.* 26, 792-794 (1989).

[44] Jaeger, C.B., Greene, L.A., Tresco, P.A., Winn, S.R. & Aebischer, P. Polymer encapsulated dopaminergic cell lines as "alternative neural grafts". *Prog. Brain Res.* 82, 41-46 (1990).

[45] Aebischer, P., Winn, S.R., Tresco, P.A., Jaeger, C.B. & Greene, L.A. Transplantation of polymer encapsulated neurotransmitter secreting cells: effect of the encapsulation technique. *J. Biomech. Eng.* 113, 178-183 (1991).

[46] Ng, K.Y., Chase, T.N., Colburn, R.W. & Kopin, I.J. Dopamine: stimulation-induced release from central neurons. *Science.* 172, 487-489 (1971).

[47] Schapira, A.H. Future directions in the treatment of Parkinson's disease. *Mov. Disord.* 22 *Suppl.* 17, S385-391 (2007).

[48] Mueller, F.J., et al. Regulatory networks define phenotypic classes of human stem cell lines. *Nature.* (2008 (in press)).

[49] Ivanova, N.B., et al. A stem cell molecular signature. *Science.* 298, 601-604 (2002).

[50] Ramalho-Santos, M., Yoon, S., Matsuzaki, Y., Mulligan, R.C. & Melton, D.A. "Stemness": transcriptional profiling of embryonic and adult stem cells. *Science.* 298, 597-600 (2002).

[51] Thomson, J.A., et al. Embryonic stem cell lines derived from human blastocysts. *Science.* 282, 1145-1147 (1998).

[52] Chung, Y., et al. Human embryonic stem cell lines generated without embryo destruction. *Cell Stem. Cell.* 2, 113-117 (2008).

[53] Klimanskaya, I., Chung, Y., Becker, S., Lu, S.J. & Lanza, R. Derivation of human embryonic stem cells from single blastomeres. *Nat. Protoc.* 2, 1963-1972 (2007).

[54] French, A.J., et al. Development of human cloned blastocysts following somatic cell nuclear transfer with adult fibroblasts. *Stem Cells.* 26, 485-493 (2008).

[55] Cervera, R.P. & Stojkovic, M. Commentary: somatic cell nuclear transfer--progress and promise. *Stem Cells.* 26, 494-495 (2008).

[56] Wilmut, I., Schnieke, A.E., McWhir, J., Kind, A.J. & Campbell, K.H. Viable offspring derived from fetal and adult mammalian cells. *Nature.* 385, 810-813 (1997).

[57] Taupin, P. HuCNS-SC (StemCells). *Curr. Opin. Mol. Ther.* 8, 156-163 (2006).

[58] Goldman, S.A., Roy, N.S., Beal, M.F. & Cleren, C. Large stem cell grafts could lead to erroneous interpretations of behavioral results? *Nat. Med.* 13, 118-119 (2007).

[59] Roy, N.S., et al. Functional engraftment of human ES cell-derived dopaminergic neurons enriched by coculture with telomerase-immortalized midbrain astrocytes. *Nat. Med.* 12, 1259-1268 (2006).

[60] Thinyane, K., et al. Fate of pre-differentiated mouse embryonic stem cells transplanted in unilaterally 6-hydroxydopamine lesioned rats: histological characterization of the grafted cells. *Brain Res.* 1045, 80-87 (2005).

[61] Schwartz, P.H., Brick, D.J., Stover, A.E., Loring, J.F. & Mueller, F.J. Differentiation of Neural Lineage Cells from Human Pluripotent Stem Cell. *Methods.* in print(2008).

[62] Temple, S. The development of neural stem cells. *Nature.* 414, 112-117 (2001).

[63] Volterra, A. & Meldolesi, J. Astrocytes, from brain glue to communication elements: the revolution continues. *Nat. Rev. Neurosci.* 6, 626-640 (2005).

[64] Horner, P.J. & Palmer, T.D. New roles for astrocytes: the nightlife of an 'astrocyte'. La vida loca! *Trends Neurosci.* 26, 597-603 (2003).

[65] Karadottir, R., Hamilton, N.B., Bakiri, Y. & Attwell, D. Spiking and nonspiking classes of oligodendrocyte precursor glia in CNS white matter. *Nat. Neurosci.* 11, 450-456 (2008).

[66] Suh, H., et al. In Vivo Fate Analysis Reveals the Multipotent and Self-Renewal Capacities of Sox2(+) Neural Stem Cells in the Adult Hippocampus. *Cell Stem. Cell.* 1, 515-528 (2007).

[67] Hawkins, J.e.f.f. & Blakeslee, S.a.n.d.r.a. *On Intelligence.* 272 (2005).

[68] Pevny, L. & Rao, M.S. The stem-cell menagerie. *Trends Neurosci.* 26, 351-359 (2003).

[69] Gage, F.H., Ray, J. & Fisher, L.J. Isolation, characterization, and use of stem cells from the CNS. *Annu. Rev. Neurosci.* 18, 159-192 (1995).

[70] Gage, F.H. Mammalian neural stem cells. *Science.* 287, 1433-1438 (2000).

[71] Shin, S., et al. Whole genome analysis of human neural stem cells derived from embryonic stem cells and stem and progenitor cells isolated from fetal tissue. *Stem Cells.* 25, 1298-1306 (2007).

[72] Kornblum, H.I. & Geschwind, D.H. Molecular markers in CNS stem cell research: hitting a moving target. *Nat. Rev. Neurosci.* 2, 843-846 (2001).

[73] Muotri, A.R., Nakashima, K., Toni, N., Sandler, V.M. & Gage, F.H. Development of functional human embryonic stem cell-derived neurons in mouse brain. *Proc. Natl. Acad. Sci. U. S. A.* 102, 18644-18648 (2005).

[74] Androutsellis-Theotokis, A., et al. Notch signalling regulates stem cell numbers in vitro and in vivo. *Nature.* 442, 823-826 (2006).

[75] Prakash, N., et al. A Wnt1-regulated genetic network controls the identity and fate of midbrain-dopaminergic progenitors in vivo. *Development.* 133, 89-98 (2006).

[76] Perlow, M.J., et al. Brain grafts reduce motor abnormalities produced by destruction of nigrostriatal dopamine system. *Science.* 204, 643-647 (1979).

[77] Brundin, P., et al. Behavioural effects of human fetal dopamine neurons grafted in a rat model of Parkinson's disease. *Exp. Brain Res.* 65, 235-240 (1986).

[78] Redmond, D.E., et al. Fetal neuronal grafts in monkeys given methylphenyltetrahydropyridine. *Lancet.* 1, 1125-1127 (1986).

[79] Lindvall, O., et al. Human fetal dopamine neurons grafted into the striatum in two patients with severe Parkinson's disease. A detailed account of methodology and a 6-month follow-up. *Arch. Neurol.* 46, 615-631 (1989).

[80] Lindvall, O., et al. Grafts of fetal dopamine neurons survive and improve motor function in Parkinson's disease. *Science.* 247, 574-577 (1990).

[81] Marx, J. Fetal nerve grafts show promise in Parkinson's. *Science.* 247, 529 (1990).

[82] Macklin, R. The ethical problems with sham surgery in clinical research. *N. Engl. J. Med.* 341, 992-996 (1999).

[83] Freeman, T.B., et al. Use of placebo surgery in controlled trials of a cellular-based therapy for Parkinson's disease. *N. Engl. J. Med.* 341, 988-992 (1999).

[84] Freed, C.R., et al. Transplantation of embryonic dopamine neurons for severe Parkinson's disease. *N. Engl. J. Med.* 344, 710-719 (2001).

[85] Olanow, C.W., et al. A double-blind controlled trial of bilateral fetal nigral transplantation in Parkinson's disease. *Ann. Neurol.* 54, 403-414 (2003).

[86] Li, J.Y., et al. Lewy bodies in grafted neurons in subjects with Parkinson's disease suggest host-to-graft disease propagation. *Nat. Med.* 14, 501-503 (2008).

[87] Mendez, I., et al. Dopamine neurons implanted into people with Parkinson's disease survive without pathology for 14 years. *Nat. Med.* 14, 507-509 (2008).

[88] Kordower, J.H., Chu, Y., Hauser, R.A., Freeman, T.B. & Olanow, C.W. Lewy body-like pathology in long-term embryonic nigral transplants in Parkinson's disease. *Nat. Med.* 14, 504-506 (2008).

[89] Nilsson, D., Nyholm, D. & Aquilonius, S.M. Duodenal levodopa infusion in Parkinson's disease--long-term experience. *Acta Neurol. Scand.* 104, 343-348 (2001).

[90] Vogel, G. Cell biology. Ready or not? Human ES cells head toward the clinic. *Science.* 308, 1534-1538 (2005).

[91] Trounson, A., Klein, R. & Murphy, R. Stem cell research in California: the game is on. *Cell.* 132, 522-524 (2008).

[92] Longaker, M.T., Baker, L.C. & Greely, H.T. Proposition 71 and CIRM--assessing the return on investment. *Nat. Biotechnol.* 25, 513-521 (2007).

[93] Deacon, R.M. Housing, husbandry and handling of rodents for behavioral experiments. *Nat. Protoc.* 1, 936-946 (2006).

[94] Linder, C.C. Mouse nomenclature and maintenance of genetically engineered mice. *Comp. Med.* 53, 119-125 (2003).

[95] Warneken, F. & Tomasello, M. Altruistic helping in human infants and young chimpanzees. *Science.* 311, 1301-1303 (2006).

[96] Warneken, F., Hare, B., Melis, A.P., Hanus, D. & Tomasello, M. Spontaneous Altruism by Chimpanzees and Young Children. *PLoS Biol.* 5, e184 (2007).

[97] de Waal, F.B. Putting the altruism back into altruism: the evolution of empathy. *Annu. Rev. Psychol.* 59, 279-300 (2008).

[98] Wagner, T., Valero-Cabre, A. & Pascual-Leone, A. Noninvasive human brain stimulation. *Annu. Rev. Biomed. Eng.* 9, 527-565 (2007).

[99] Galvani, L. Aloysii Galvani de viribus electricitatis in motu musculari commentarius. (1791).

[100] G.A. Volta, A.l.e.s.s.a.n.d.r.o. Schriften Uber Die Thierische Elektrizitat. (1793).

[101] Bischoff, C.H.E. *Commentatio de usu Galvanismi in arte medica speciatem ver in morbis nervorum paralyticis*, (Jena, 1801).

[102] Lederer, S.E., Fee, F. & Tuohy, P. *Frankenstein: Penetrating the Secrets of Nature.*, (National Library of Medicine (U.S.), , Washington, DC, 1998).

[103] Aldini, G. *Essai théori...e galvanisme.*, (Fournier Fils, Paris, 1804).

In: Philosophical Insights about Modern Science
Eds: Eva Žerovnik, Olga Markič and Andrej Ule
ISBN: 978-1-60741-373-8

Chapter 6

How Intelligent Can Robots Become: Implications and Concerns

Matjaž Gams*

Institut Jožef Stefan, Department of Intelligent Systems
Jamova 39, 1000 Ljubljana, Slovenia

Synopsis

In this overview, basic dilemmas about intelligent robots are reexamined. The question "Can robots become intelligent?" – is very similar to the question – "Can computers become intelligent?" The major hypothesis of this paper is that the human mind, the brains and body are different than those of computers and robots; that the human mind is a supermind compared to digital computing powers. The principle reason for this belief is the way human minds perform thinking – according to the multiple-world theory in many worlds/dimensions. However, there are indications that it is only a matter of time till something comparable to humans will emerge – and here are some speculations, how it will look.

Introduction: Basic Dilemmas About Robots

When dealing with robotics, there are some basic concepts worth mentioning. One is "The Three Laws of Robotics". They concern the relation between humans and robots, i.e., how to protect humans from (un)intentional harm from these mechanical beings. One of the best known science fiction writers at his time, Isaac Asimov addressed this problem in 1942 when he published a story called "Runaround", in which he stated the Three Laws of Robotics:

* Tel: +386 1 477 3644; fax: +386 1 425 1038; matjaz.gams@ijs.si

- First Law -A robot should not injure a human being, or, through inaction, allow a human being to be harmed.
- Second Law -A robot must obey the orders given to it by human beings except where such orders would conflict with the First Law.
- Third Law -A robot must protect its own existence as long as such protection does not conflict the First and the Second Law.

To natural/technical scientists and engineers, the three laws had no major impact, quite the contrary to the public opinion and social scientists. The engineering problem is how to design advanced robots and make them cost-effective. Robots, like computers, currently and in the near future will not going to have free will to decide, rather they do what they are designed for. It is true that robots occasionally kill humans (see figure 1), some purposely, like the Predator military airplane, or by accident as cars do. But many less people get killed or injured by robots than by cars. In fact, the harm done by robots is irrelevant compared to the harm done by car accidents, and we still use cars, don't we?

Common Popular Questions On Robots

Before going into more scientific details, let us now analyze some popular questions about robots, widely discussed over the internet. However, the viewpoints presented by the present author are clearly based on scientific knowledge, presented later and mostly similar to (Gams, 2001; 2004).

Figure 1. Some robots are designed to kill humans, but all robots currently do what humans made them for.

Will Robots Be Slaves/Machines?

Today's robots are mainly dumb machines, at least those living in the physical world (not totally so in the virtual world). But imagine a truly intelligent robot (Flynn 2007) – such a robot will clearly have its own will, which will allow it to decide about itself. Is a dog a slave? If yes, then robots will be slaves as well. If no, then robots will not be slaves. But be sure that dogs and robots will take a long long time before they free themselves from the reign of humans – if ever.

Will Robots Be Mentally or Physically Superior to Humans?

Some authors in machine intelligence or science fiction see robots as all knowing, purely logical and physically superior to humans. This seems unrealistic. Certainly robots and computers will be superior and already currently are in several tasks, and will be so more and more in the future. For example, a car robot is already much faster than a human on a flat surface. But there will always be certain things humans will know better and perform better. Just consider a simple task of opening a door – at which best robots are currently several times slower – if successful at all? On the other hand computers already play chess better than any human and even more so for most of the computer games. And intelligent autonomous vehicle like cars of choppers successfully move in an unknown area for hours without a crash. However, the progress in certain areas is slower than anybody could anticipate – just imagine how it is possible that today humans unmask a computer program faking a human faster than we did 10 years ago, while at the same time the computers got 30 times more powerful?

Will Robots Have Legal Rights Like Humans?

Until robots by-pass dogs, humans will create laws protecting robots – something close to treating them as machines. But one could imagine that truly intelligent robots will sooner or later start taking care of themselves with the help of some dedicated humans. The exact relation between humans and robots at that point is not clear to this author. Some authors insist that intelligent robots will have the same rights as humans, and some disagree.

It seems reasonable to assume that robots will not have inbuilt just logical intelligence but also an emotional one, e.g., their senses (inputs) will simulate human touch and /or pain. Yes, then there will likely be "robots rights" as humans will not accept hurting such an entity who would suffer. Alike there are "animal rights".

Which Is the Coolest Current Commercial Robot?

Some of the coolest robots in the world were and are being developed at the MIT Artificial Intelligence Lab. Commercially the most successful intelligent robot at the moment is iRoomba, a vacuum-cleaning robot, sold in millions all over the world (figure 2). These robots are modeled on insects. Along with being the iRobot Chairman and CTO, Rod Brooks is also the Director of the MIT Computer Science and Artificial Intelligence Lab, where his goal is to understand and eventually have the robots display human intelligence (http://www.amazon.com/b/?node=10287641).

Figure 2. iRoomba, the most useful and commercially successful robot is far from being fully intelligent, yet its intelligence can match those of primitive bugs.

Will There Be Varieties of Varieties of Robots?

Currently, there is a variety of robots from those working in factories to those autonomous vehicles like choppers or submarines. In future, hardware solutions will certainly be more specialized. But the question is different – will each robot of the same type be unique at least in the mental sense, as humans or animals are? Again, when speaking about intelligent robots, no robot will likely be the same in the "mental" sense. So there will be varieties of varieties of intelligent robots, when they will emerge. As every human being, each intelligent robot will most likely be "unique".

How Will Robots Be Created?

Currently, robots are produced like cars and computers. But robots of the future – those truly intelligent robots – will have to learn a bit like children, following some rules of evolution (Kononenko, Kukar, 2007). Even if those robots would be self-born or self-created, meaning robots will produce new robots, the task of creating a mind has its own rules of design and quite probably there is no way of circumstancing that. How will that be done, is unclear. One idea is that instead of merging father's and mother's DNA such as occurs in humans, the basic knowledge of one (father) robot, at least in the symbolic form, will be merged with the knowledge of another robot (mother) into a new "robot child" that will also learn a lot from the environment. In this way, a kind of evolutionary genetic game will be introduced.

Will There Be Male and Female Robots?

Ursula le Guin in her book "The Left Hand of Darkness" considers the possibility of a race with three genders. In Star Trek, Mr. Data is a male robot, although totally logical. Regarding male and female robots, or robot sex itself, there might be different laws than in our real life. There will be laws regulating relations like long-term relations, maybe also love, sex and partnerships between robots and humans and between robots themselves.

Will There Be Robotic Sex?

If we have already accepted the idea that robot minds will have to be created through some evolutionary process, then it will be beneficial that robots will have some parents – one, two, or many. After all, sex can be regarded as a machine to produce new, more advanced offspring, and why should the principle be much different for robots – of course if we neglect technical details. In genetic algorithms, one typically deals with a population of subjects, competing with the environment and each other, where best parents produce offspring with a mixed DNA from two parents (each). If it were beneficial to have say 10 parents, these software algorithms would surely be applied, but the experiments with more parents than two did not show any advantage (Eiben et al, 1995). It might sound a bit confusing, because when discussing robotic sex, this might be even more platonic than in humans: producing an offspring might be similar to merging "data bases" of parents, while courting might be more similar to humans since two or several partners will have to get sure about each other qualities and prosperity.

Will Robots Marry, Fall in Love and Have Children? Will Robots Have Feelings?

To have true social life is probably not inside a totally logical (calculating) mind. Until various kinds of intelligence and senses will be inbuilt into robots (Guid, Strnad, 2007; Flynn, 2007), they could not be imagined to feel good or to feel lonely. But some very shallow feelings can be simulated already even on current computers. In summary, it is not clear to this author whether robots will marry, or if they will even want to marry. Some robots might prefer to live on their own, others might prefer to live in communities of robots, or in mixed communities with humans or with animals (figure 3). It is not even clear if marriage between humans will be a social norm in a couple of decades, and since other forms of marriage like same-sex is becoming more and more popular, some still other forms of marriage – e.g., with an animal or a plant or a thing is also possible. So, why should not some human marry a robot, in particular one specialized for marriage tasks? However, it is likely (by the authors opinion) that some mental connections and partnerships will prevail – short or long term. All social being have some relations with other actors, and robots will certainly be actors as well. When robots will achieve an interesting mental level, say of a dog, partnerships between humans and robots will probably emerge, be it formal or informal. The author of this paper wrote a SF story "Romeo and Juliet", describing a mental relation between a man and a war robot, resulting in a new kind of offsprings. Robot parents like all other parents, or their owners, might desire that some of their own design characteristics were incorporated within the new robot child. And there is an evolutionary advantage in that. Will robots be "loving" (caring) parents; i.e., will such an emotion evolve by itself or will it have to be inbuilt by humans?

Will Robots Have Different Lives to Humans?

Sure. Robots will certainly have some capabilities humans will not and vice versa. First of all, robots are currently machines consisting of rods, sensors, wires etc. being produced for the last hundred years or so, while humans are biological cognitive beings, the top design of the billion-year evolution (Kordes, Markic, 2007). Even e.g., fully intelligent specialized marriage robots will not very likely be fully android or biological. They will be functional and cost-effective. Some SF authors see that specialized robots will serve all the human needs and at the same time sometimes alienating humans from each other, while other authors see robots as a self-evolving independent species.

Figure 3. Will robot form special partnerships with humans?.

Will Robots Think Like Humans?

Robots will think with computers instead of the human brain, so the question is related to the human-computer question. Computers/robots currently already think much faster than humans, and have approximately the same amount of information as humans. In technical terms, robot-robot communication will be much faster than human-human communication. However, intelligent computers are not inside sight and so are not intelligent robots (Gams et al, 1997). It should be noted, however, that we are not dealing with typical IQ tests (Murdoch 2007), but with something recognizable by humans as resembling the integrated human-level intelligence. We should also not mix the above thinking with ideological reasoning that machines will not ever achieve human-type intelligence. In scientific community, there is no question that intelligent robots and intelligent computers will arise eventually, most probably inside future decades and not centuries. But that should not worry us in terms of cataclysmic wars between humans and machines. In reality, the shortage of oil is much nearer the horizon and that will certainly affect our civilization in an unpredictable negative way, while better computers and robots will no doubt enhance human progress.

Will Robots Lie or Cheat?

Asimov thought that robots will be totally logical creatures, totally incorruptible in theory if not in practice. However, even today there are some attempts of creating robots coursing, lying or just being unpredictable. In extreme, some of the robots are already designed to kill people or to entangle in war situations of various kinds. But all robots do what they were designed for in the sense of a dedicated slave. There is nothing similar to disobedience or cheating on robot's own. On the other hand, according to this author's Principle of multiple knowledge, when true intelligence will be incorporated into robots/computers, robots will have their own will and will by definition become unpredictable and will do things like lying, stealing etc. (figure 4). Like all beings do.

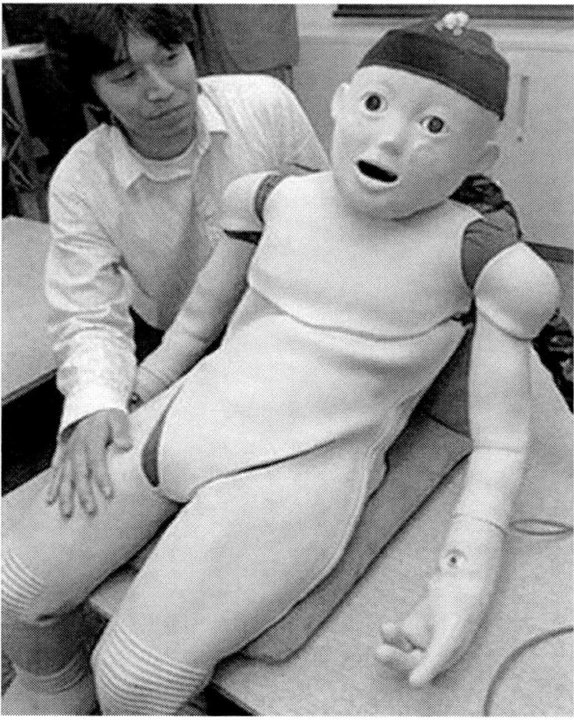

Figure 4. Will robots be very similar to humans - androids? Will they die? Maybe, in the far future.

Will Robots Die / Get Dumped?

In reality, all machines and beings die or get dumped. There is no reason to believe that with robots there will be anything different. However, in principle, it is possible to transform a car to become something like immortal. When a part becomes absolute or nonfunctional, it can get replaced with the same new or a newer version. This is of course impractical and stalls the progress; therefore it is meaningless as is cloning of beloved dogs at the moment. Even when looking at the knowledge that robot develops over time, so at the mental level, things seem similar. Given enough time, say a few ten or hundred years, an old robot would inevitably be out of date compared to younger models. Very likely, it is not only newer hardware, it is that the robot's previous knowledge inevitably shapes the way it understands the world. The first few years experience of the robot will probably shape the rest of its life. As a result of this changing nature of the robot as it gets older, it will get more and more out of touch with the current modern world and will become more and more alienated and as such not as interesting as the newer models. Somehow, robots will have to be "cut of power" in this way or another (figure 3). Until they become recognized as free-will owners, it will be much easier, since they will be treated as machines. Later on, probably a robotic clinic will not repair too damaged robots without enough money being paid by their owners or those not willing to be repaired. Perhaps a kind of euthanasia will be applied as it is now to very ill dogs.

In the next chapter we will analyze advanced computing mechanisms, those that can potentially achieve advanced computing compared to the universal Turing machine. Namely, robots basically combine a HW body and a computer, and all the intelligence is in the computer - besides being embodied in the body as well.

SUPERCOMPUTING MECHANISMS

The debate about artificial intelligence in comparison to humans is as old as computers themselves. Alan Turing (1912-1954), founder of computing science, introduced the universal Turing machine (TM) for simulating procedural/mechanic thinking of a human while accessing that this might not be sufficient for creative thinking (Turing, 1947; Teuscher, 2002). At the same time he found no formal reason why computers in some future should not outperform humans in thinking and even in feelings. His estimate of this turning point was around year 2000. Now that we are in 2008, it seems that he might be wrong, not because of a couple of years have passed from 2000, but because the today's computers do not seem intelligent at all – they remain very fast computing machines, not yet resembling human thinking or reasoning and falling light-years behind true human intelligence. In general, there are several viewpoints regarding true intelligence (Penrose, 1989; 1991; 1994):

1. Computer intelligence will sooner or later achieve human levels and then even bypass them.
2. Human intelligence is so different than that of computers that even though they are in principle of the same powers, there will be tasks where the human brain will always significantly outperform computers.

3. Human intelligence is in principle stronger than that of current computers, however fast or with whatever memory, yet new-type computers will sooner or later become intelligent.
4. We humans are unique and no machine can become close to our computing powers. No machine can become fully intelligent, or possess consciousness and other human-like qualities such as love.

Needless to say, the last viewpoint seems suspiciously ideological and non-scientific. It is much harder to distinguish between the first three viewpoints; however, the author finds no. 3 as most plausible.

In theory, there are several stronger-than-TM computing mechanisms with interesting properties (Copeland, 1997; 2002). Terms like "hypercomputation" and "superminds" (Bringsjord, Zenzen, 2003) are introduced. Turing also proposed a formally totally correct stronger computing mechanism – the Turing machine with an oracle, capable of answering any question with always correct Yes/No (Turing, 1948). This computing mechanism can easily solve several problems like the halting problem, i.e. whether a TM performing a program will stop or not under any condition. The only problem is that there is no known physical implementation of an oracle, while digital computers are very good implementations of the universal Turing machine, performing tasks as predicted in theory also in reality. Even if there were a TM with an oracle in real life, it would certainly not perform like humans.

There are several other supercomputing mechanisms. For example, Scarpellini (1963) suggested that nonrecursive functions, i.e. those demanding stronger mechanisms than the universal Turing machine, are abundant in real life. This distinction is important, since obviously most of simple processes are computable, and several simple mental processes are computable as well. Komar (1964) proposed that an appropriate quantum system might be hypercomputational. This is unlike Penrose who proposed that only the undefined transition between quantum and macroscopic is nonrecursive. Putnam (1965) described a trial-and-error Turing machine, which can compute also the Turing-incomputable functions like the halting problem. Abramson (1971; 1994) introduced the Extended Turing machine, capable of storing real numbers on its tape. Since not all numbers are Turing-computable, Turing machines cannot compute with those numbers, and are there inferior in principle. Boolos and Jeffrey (1974) introduced the Zeus machine, a Turing machine capable of surveying its own indefinitely long computations. The Zeus machine is another version of the stronger-than-UTM. It is also proposed as an appropriate computing mechanism by Bringsjord (Bringsjord, Zenzen, 2003). Karp and Lipton (1980) introduced McCulloch-Pitts neurons, which can be described by Turing machines, but not if growing at will. Rubel (1985) proposed that brains are analog and cannot be modeled in digital ways. Kononenko (2008) complies with the idea that TM are digital and cannot fully simulate analog events. Gams (2001) proposes multiple Turing machines interacting with open environment as stronger mechanisms than UTMs.

One of the most famous and reputed scientists proposing a new supercomputing mechanism was an Oxford professor Roger Penrose (1989; 1994). According to him, humans are not as constrained as formal systems like UTMs; they are stronger than formal systems (computers). Humans either use nonrecursive mechanisms and are computationally stronger or are not stronger in principle, but practically more effective. The supercomputing mechanisms are in the nerve tissue, according to the Penrose-Hameroff theory (Hameroff et al, 1998), based on quantum effects in connections between neurons. One of the stronger-than-UTMs are interacting Turing machines (Wegner, 1997) based on open, truly interactive environment. Such computing mechanisms are achieved already by groups of social intelligent agents on the Internet (Wellman et al, 2007). Quite similar to the interaction TM are coupled Turing machines (Copeland, Sylvan, 1999). The improvement is in the input channel, which enables undefined input and thus makes it impossible for the universal Turing machine to copy its behavior. A similar idea comes from partially random machines (Turing, 1948; Copeland, 2000). These Turing machines get random inputs and therefore cannot be modeled by a Turing-computable function as shown already by Church. It is only fair to notice that Penrose later softened his ideas a bit and that some of his claims were shown to be theoretically wrong. But Penrose and Hameroff remain self-confident that sooner or later their ideas will get confirmed by scientific methods. Indeed, are so many of their critics actually making mistakes or is it just so hard to formally prove the difference that is obvious to a normal human?

An example of the differences between computers and humans by the author of this paper opinion is simply the lack of intelligence and consciousness in computers that all humans have (Gams 2002). There is a clear distinction between the physical world and the mental world. The Turing machine might well be sufficient to perform practically all meaningful practical tasks in real life. But the mental world is a challenge that is beyond the universal Turing machine. Every moment in our heads a new computing mechanism rumbles on. In analogy to physics, existing computer single-models correspond to Newtonian models of the world. Intelligent computer models have additional properties corresponding to quantum models of the world valid in the atomic universe. The Heisenberg's principle of uncertainty in the quantum world is strikingly similar to multiple interaction computing introduced later.

One of the most interesting recent books is The Supermind book, subtitled "People harness hypercomputation and more", authored by Selmer Bringsjord and Michael Zenzen. They aggressively attack the strong viewpoint that human thinking processes are computationally as strong as those of computers. Indeed, the author of this paper agrees with several viewpoints presented in the book. For example, consider the following argument (Zlatev, 2001; Bringsjord, 2004): suppose you live with somebody for years and that person seems human in all respects, but after death a mechanical device is found in its/his head (reminding the Terminator trilogy). Would your first thought be that you have found a true robot or that it was some kind of a trick? And more, if it were some kind of mechanical or digital device, very strong evidence would be needed that such a thing is possible, because all the empirical data and theoretical investigations in weak AI strongly indicate that such a thing is not possible in practice and in principle. Bringsjord and Zenzen base their superminds theory on the Zeus machine. This machine computes each further step faster and can thus easily compute problems demanding infinite time to solve. In this way, the halting problem of the UTM is solved – if it stops in infinite time, then it stops, otherwise not. Besides the Zeus machine, several other machines are considered as stronger than UTM, such as trial-and-error machines or analog chaotic NNs. However, the book was not accepted as a trustful scientific material and after some discussions it seems that this is not a generally accepted new computing mechanism, stronger than UTM.

The author of this paper published a book introducing the principle of multiple knowledge (Gams 2001). The "technical" version of the Principle of multiple knowledge claims that on average it is reasonable to expect improvements over the best single model when single models are sensibly combined. The "creative" version of the Principle of multiple knowledge states that multiple models are an integral and necessary part of any creative process. Creative processes are top-level processes demanding top performance, and top performance can not be achieved by any "single" process. In other words: a sequential single model executable on the universal Turing machine will not achieve as good performance as combined models in majority of real-life domains. Therefore, no Turing machine executing a single model (e.g., no computer model constructed as a single model) will be able to achieve creative performance. Mind, "multiple" should not be confused with "parallel".

The Principle does not directly imply that digital computers cannot achieve creative behavior. Rather, it implies that current computers need substantial improvements to become creative. The Principle is based on comparing performance of two cooperative actors and one of them alone, in particular the best one of them. The probability of successful performance of combined interaction actors is obtained as a sum over all possible situations; for two independent actors/models it is:

$$p_M = \sum_{s \in S} p_s q_s = p_2(1-p_1)\,q_{FT} + p_1(1-p_2)\,q_{TF} + p_1 p_2$$

and for two dependent models:

$$\begin{aligned} p_M^* &= \sum_{s \in S} p_s q_s \\ &= (1-d)(\,p_2(1-p_1)\,q_{FT} + p_1(1-p_2)\,q_{TF} + p_1 p_2) \;+\; dp_1 . \end{aligned}$$

without any loss of generality we can assume that $p_1 \geq p_2$. Now, p_M^* can be expressed in relation to p_M :

$$p_M^* = (1-d)\,p_M + d\,p_1 .$$

For $0 < d < 1$, the accuracy p_M* lies between p_1 and p_M. Moreover, whenever two combined independent models indicate better overall performance than the best model alone, the model for dependent models will also indicate it, and the improvement of accuracy p_M* - p_1 will be directly proportional to p_M - p_1 of the independent model with a factor of 1 - *d*. The last two conclusions imply that the model of two independent models can reveal basic conditions in the 4-dimensional $(p_1, p_2, q_{TF}, q_{FT})$ space, under which the combined dependent models are also more successful than the best single model alone. For first analyses we shrink the analyzed 4-dimensional space into a 3-dimensional space by predefining one variable. The idea is to determine conditions under which the use of combined classification pays off, i.e. when the classification accuracy p_M is greater than the one of the best single model.

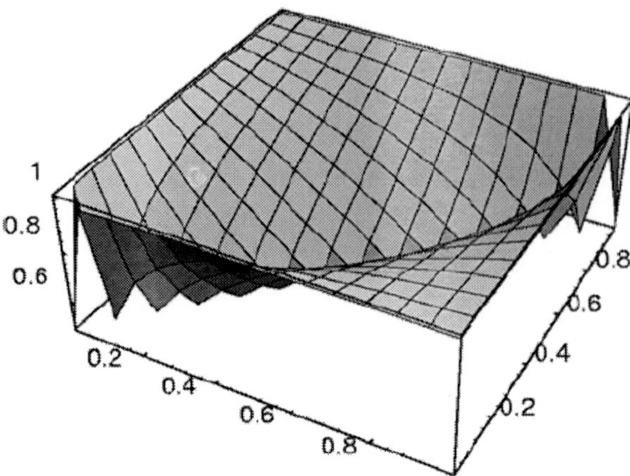

Figure 5. Two actors perform better than the best one alone under certain conditions which seem reasonable in real life.

Further research analyses more complex relations, yet conclusions basically remain the same – under reasonable conditions, that can be expected in real life, a reasonable improvement will emerge. In lay terms, the conditions are:

- The two actors/models perform with similar accuracy meaning that one is not superior to the other in most of the cases.
- The two actors/models should be as different as possible since this indicates a potential amount of improvement.
- The combining mechanism should guess in as many cases as possible, which of the two actors/models is correct and which not. Quite often, 50% accuracy is already enough for a small improvement.

Although the conclusions seem trivial, they basically confirm old saying that more heads know more. In reality, the benefit is not based only on knowing more. For example, the processes in the human brain/mind seem to comply with the Principle – several processes are created during solving a task and these processes constantly interact with each other. Finally, not the best process is chosen, it is the best combination of potential solutions that produces the best results. The real mystery is how such an algorithm actually performs in the brain/mind.

Form of Robotic Knowledge

Principles of multiple knowledge offer several conclusions regarding the properties of knowledge including answers to the following questions:

- Can knowledge be stored?
- Is knowledge predictable?
- Is knowledge understandable?
- What is the form of knowledge?

Can Knowledge Be Stored in a Robot/Computer?

There is a thesis that knowledge stored in a computer can not capture the meaning because it consists of strings of meaningless symbols (Searle, 1992). For example, so far we have been able to store descriptions of only relatively simple sequential domains while for complex real-life domains storing the knowledge in existing knowledge representations is beyond our existing capabilities and knowledge. Says Clancey (1989; 1993): ``Machine learning will never progress beyond its current state until people realize that knowledge is not a substance that can be stored." Today, it is more or less accepted that anything describable can be in principle stored in a string of symbols. One possibility left is that thinking is not describable through symbols at all. On the other hand, even theoretical confirmation may not mean much for practical feasibility. It may be that although theoretically possible, the process of transforming the knowledge of one human into the fastest possible digital computer would take, say, thousands of years. Therefore, there are many options opened for further study.

The principle of multiple knowledge can enhance our understanding - problems could emerge due to the inappropriateness of existing ways of describing knowledge. Currently, formal science and computers can handle well simple sequential (or parallel) implementations of knowledge. But, the principle of multiple knowledge implies that direct storing of (human) knowledge in a simple sequential way is not feasible. If storing of knowledge is feasible, it must be multiple, i.e., more complex than previously thought. This is not a minor difference since we are talking about several orders of magnitude more difficult problems and more capable solving mechanisms which, on the other hand, can not solve better formal problems appropriate for formal systems. Multiple systems enable only a couple of percents better performance in real-life problems at the price of more powerful computing mechanisms - intelligence and consciousness and several other properties like feelings. Intelligence is seen by some authors as redundant or as a by-product of human evolution. But according to the principle, this is not so. Essential human processing properties like intelligence are part of the superior computing mechanism. Automatic transformation from multiple systems into sequential ones seem feasible, given sufficient time and space; however, the future performance is lost. In analogy with the human brain, one has to store several types, forms and representations of knowledge. At the top, there is a clear distinction between two types of thinking that are performed by the brain hemispheres. It is quite possible that any computer system emulating human intelligence would have to have similar top-level architecture consisting of modules with different computing approaches.

Will Intelligent Robotic Knowledge Be Predictable?

Can we predict the behavior of stored knowledge, i.e., a computer model of a real-world domain? In most artificial domains a positive answer is a common case. For example, theoretically we can predict the behavior of a computer program on the bases of its listing. On the other hand, it is commonly known that the Turing halting problem is unsolvable on the Turing machine. The halting problem can be solved by other, more powerful computing mechanisms; however, each so far known computing mechanism has its own unsolvable questions. Therefore, it is theoretically impossible to predict all future behavior even of digital computers, although in reality it is so in 99.99%. Even simple random generators based on system clock and the contents of a random page on Internet produce unpredictable numbers. Introducing multiple knowledge makes things orders of magnitude worse. Complex real-world models consist of several levels of submodels, each potentially similar to the Turing halting problem. Realistic predictability of details seems unfeasible, in general. But for most formal and real-life domains, much can be predicted in normal circumstances because only rational performance is evolutionally rewarded. In addition, it is often possible to execute, i.e., to simulate knowledge models, and observe the performance.

Will Robotic Knowledge Be Understandable?

The question is whether humans will be able to understand knowledge of truly intelligent robots. A related matter was the famous ``Feigenbaum's Bottleneck" (Feigenbaum 1985), based on observations in the process of acquisition of expert's knowledge. Experts were able to achieve very good performances and at the same time they were able to explain specific solutions in an understandable and meaningful way to humans. But when trying to formalize or even only to describe their knowledge in the form of rules or trees they found it very difficult and even frustrating. As if the expert knowledge would not be appropriate for transparent single knowledge models. Accepting the principle we can see why it is so difficult formulate the knowledge on paper. Not only that submodules are changing over time, the possible combinatorial explosion in the combining mechanism makes knowledge very difficult to understand at the top integrated level. But after each specific conclusion is made, it can be easily explained. This makes the combining mechanism that utilizes knowledge one of the core secrets of human thinking. Quite probably, it is at least partially hardware-brain coded. Maybe, it is not executable with formal mechanisms after all.

Then, how can it be expected to explain human knowledge with say one (huge) decision tree? Constructing a set of multiple models seems better although not the ultimate idea and in our experiments they enabled much better possibilities to analyze the laws of the domain. Unfortunately, it is more difficult to explain solutions with multiple models than with one model alone. At this stage it seems that expert's knowledge can not be simply transformed into one model (i.e. knowledge base) through knowledge acquisition or learning without loosing performance in the sense of top performance and consistency with expert's knowledge. In another thought, it is only a rather common myth that knowledge stored in computers is fully understandable. And, that a formal computer program cannot be intelligent because it can be fully explained (which we argue as unfeasible).

What Will Be the Form of Robotic Knowledge?

Parts of knowledge can be stored, can be consistent, and can be stored in a string of symbols. On the other hand, it is argued that complex real-life knowledge can not be easily stored in a consistent, compact and sequential form. Real-life knowledge has different properties than knowledge describing formal domains. As declared by the principle of multiple knowledge, the form of knowledge in real-life domains must be truly multiple.

DISCUSSION

There is a slight difference between robots and computers - robots have bodies and are made with an aim to live (help humans) in real worlds. However, basic questions about robots' intelligence seem similar to those of the quest for intelligent computers. And here, scientific knowledge can predict a lot about the future progress, and can help to limit discussion to reasonable variations.

The real relations between computers, robots and humans are much more complicated that realized in every-day life. In this paper, some of the analyses both popular, near to SF views and scientific research were presented, shading some new light on what is to come based on scientific backgrounds. What some find as shocking is nearly sure to come, e.g., intelligent computers and intelligent robots – but in different forms as predicted today; and what some view as oblivious, is far from certain and not likely at all – e.g., that robots will clash with humans. Intelligent robots and computers already are, and will be even more so in the future, essential for the progress of human civilization. But the actual future in detail is not fully predictable – it is for all of us involved, humans and robots alike, still to be seen.

REFERENCES

Abramson, F.G. (1971). Effective computation over the real numbers, Twelfth annual symposium on switching and automata theory, *Nortridge,* CA.

Abrahamson, J.R. (1994). Mind, Evolution, and Computers, AI magazine, pp. 19-22.

Boolos, G.S., Jeffrey, R.C. (1974). Computability and logic, Cambridge Computability Press.

Bringsjord, S., Zenzen, M. J. (2003). *Superminds, Kluwer.*

Bringsjord, S. (2004). On building robot persons: response to Zlatev, *Mind and Machines,* pp. 381-385.

Clancey, W.J. (1989). The Knowledge Level Reinterpreted: *Modeling How Systems Interact, Machine Learning,* 4, pp. 255-257.

Clancey, W.J. (1993). The biology of consciousness, Artificial Intelligence, 60, pp. 313-356.

Copeland, B.J. (1997). The Church-Turing thesis, in E. Zalta (ed.), *Stanford Encyclopedia of Philosophy.*

Copeland B.J. (2000). Narrow versus wide mechanisms, *Journal of philosophy.* 96, pp. 5-32.

Copeland, B.J. (2002). *Hypercomputation, Minds and Machines,* Vol. 12, No. 4, pp. 461-502.

Copeland, B.J., Sylvan, R. (1999). *Beyond the Universal Turing Machine, Australian journal of philosophy.* 77, pp. 46-66.

Eiben, A.E., van Kamenade, C.H.M., Kok, J.N. (1995). Orgy in the Computer: Multi-Parent Reproduction in Genetic Algorithms, *Lecture Notes In Computer Science;* Vol. 929, *Proc. of the Third European Conference on Advances in Artificial Life,* pp. 934 – 945, Springer-Verlag.

Feigenbaum, E.A. (1985). Themes and Case Studies of Knowledge Engineering, Expert Systems in the Microelectronic Age, ed. Michie D., Edinburgh University Press.

Flynn, J.R. (2007). What is Intelligence?: Beyond the Flynn Effect, Cambridge University Press.

Gams, M., Paprzycki M., Wu X. (eds.) (1997). Mind Versus Computer: Were Dreyfus and Winograd Right?, IOS Press.

Gams, M. (2001). Weak intelligence: Through the principle and paradox of multiple knowledge, Advances in computation: Theory and practice, Volume 6, *Nova science publishers*, inc., NY, ISBN 1-56072-898-1, pp. 245.

Gams, M. (2002). The Turing Machine May Not Be The Universal Machine. *Minds and machines,* 2002, vol. 12, pp. 137-142.

Gams, M. (2004). Computational analysis of human thinking processes. International journal of computational cognition. [Online], Vol. 2, str. 1-19. http://www.yangsky.com/ijcc23.htm.

Guid, N., Strnad, D. (2007). *Artificial Intelligence, FERI Maribor.*

Hameroff, Kaszniak, Scott, Lukes (eds.), (1998). Consciousness Research Abstracts, Towards a science of consciousness, *Tucson,* USA.

Karp, R.M., Lipton, R.J. (1982). Turing machines that take an advice, in *Logic and algorithmic, Enseighement Mathematique.*

Komar, A. (1964). Undicedability of macroscopically distinguishable states in quantum field theory, *Physical Review,* 133B, pp. 542-544.

Kononenko, I. (2008). Some Viewpoints on Machine Learning, Artificial Intelligence and Consciousness, http://lkm.fri.uni-lj.si/xaigor/slo/Nekateri%20vidiki.doc.

Kononenko, I., Kukar, M. (2007). Machine Learning and Data Mining, Introduction to Principles and Algorithms, Horwood Publishing, UK.

Kordes, U., Markic, O. (eds.) (2007). *Cognitive Sciences in Ljubljana,* PF, Uni LJ.

Murdoch, S. (2007). *IQ: A Smart History of a Failed Idea, Wiley.*

Penrose, R. (1989; 1991). The Emperor's New Mind: Concerning computers, minds, and the laws of physics, Oxford University Press.

Penrose, R. (1994). Shadows of the Mind, A Search for the Missing Science of Consciousness, Oxford University Press.

Putnam, H. (1965). Trial and error predicates and the solution of a problem of Mostowski, *Journal of Symbolic Logic.* 30, pp. 49-57.

Searle, J.R. (1992). The Rediscovery of the Mind, MIT Press.

Scarpellini, B. (1963). Zwei Unentscheitbare Probleme der Analysis, Zeitschrift fuer Mathematische Logik und Grundlagen der Mathematik, pp. 265-354.

Teuscher, C. (ed.) (2002). Alan Turing: Life and Legacy if a Great Thinker, Springer.

Turing, A.M. (1947). Lecture to the London Mathematical Society on 20 February 1947, in Carpenter, Doran (eds.) A.M. Turing's ACE Report of 1946 and other papers, MIT Press.

Turing, A.M. (1948). Intelligent Machinery, in B. Meltzer, D. Michie (Eds.), *Machine Intelligence* 5, Edinburgh University Press.

Wegner, P. (1997). Why Interaction is More Powerful than Computing, Communications of the ACM, Vol. 40, No. 5, pp. 81-91.

Wellman, M.P., Greenwald, A., Stone, P. (eds.) (2007). Autonomous Bidding Agents: Strategies and Lessons from the Trading Agent Competition (Intelligent Robotics and Autonomous Agents) , MIT Press.

Zlatev, J. (2001). The epigenesis of meaning in human beings, and possibly in robots, Mind and Machines 11, pp. 155-195.

In: Philosophical Insights about Modern Science
Eds: Eva Žerovnik, Olga Markič and Andrej Ule
ISBN: 978-1-60741-373-8

Chapter 7

THE CONCEPT OF FREE WILL ENTERING THE FIELD OF NEUROLOGICAL SCIENCES

Zvezdan Pirtošek*
Department of Neurology, Medical Faculty and Faculty of Arts,
Ljubljana University, Ljubljana, Slovenia

SYNOPSIS

The chapter reviews the emerging concept of free will in the areas of neurological sciences.

For a very long time the concept of "freedom of the will" dwelled in the domains of philosophy, theology and law. With detailed and astute clinical observations in neurology and with the the rise of new cognitive sciences (quantum physics, neuroscience) neurology and neurophysiology too could not ignore the concept of free will. These sciences - the easier part of the problem being the study of phenomenology, anatomy and mechanism of what is referred to as "free will" – often expressed scepticism about the existence of "free will" and pleaded for different concept and terminology.

In clinicopathological settings, neurologists and psychiatrists have for long time observed and described syndromes of a "sick will", characterized by inactivity, poverty of movement and thought: various movement disorders, schizophrenia, depression, autism, ADHD syndrome, dementia, parkinsonism, hysteria, apraxia. The concept of volition has been studied particularly well in the domain of movement. From the point of volition movements can be categorized into four groups: voluntary, semivoluntary or unvoluntary, involuntary and automatic movements.

Main methods used in the study of volition and free will are functional imaging and electrophysiological methods, The findings of functional brain studies suggest that willed acts are formulated in the prefrontal cortex along the structures involving so called cortico-subcortico-frontal loops and imply a relationship between the volition and the dorsolateral prefrontal cortex, particularly on the left side. Electrophysiological techniques provide characteristics of brain activity in temporal terms and they can be particularly suited to elucidate timing between conscious intention to act and the act

* Zvezdan Pirtošek, M.D., Ph.D. is Head of Department of Neurology and Professor of Neurology at the Medical Faculty and Faculty of Arts at the Ljubljana University. His main areas of clinical and scientific research include neurodegenerative disorders and cognitive neuroscience.

itself. Libet's famous but controversial study implies that conscious awareness actually follows the intention to act rather than preceding (and thus causing) it. Libet, trying to "save" the concept of free will, proposed that voluntary acts are unconsciously initiated but are subject to conscious control which can either promote or veto the unconsciously initiated process shortly before the execution. Therefore, free will and volition are associated more with the vetoing of action than its instigation implying the existence of free won't rather than free will. Clinical case reports, electrophysiological studies and new brain imaging techniques confirm the assumption that free will is "localizable" to a certain extent. Several brain regions contribute to the execution of consciously chosen, volitional action. Most important structures are located in the prefrontal cortex and its related cortical and subcortical structures, along the cortico-subcortico-frontal and cortico-frontal circuits.

1. The Idea of Free Will

What does it mean to be human? Half a century ago biologists, psychologists and anthropologists would propose consciousness, language, sense of self awareness, the ability to reason, the feeling of right and wrong as the features delineating human being from the rest of the Universe. As cognitive sciences advanced, it became clear that higher animals too may have a form consciousness, language, they too may be aware of their own self and they too are able to solve complex problems. The difference between *Homo sapiens sapiens* and other beings seems to be more a matter of degree than of the kind. One of the last bastions of the cognitive properties uniquely human may thus be "free will", an ability to generate conscious decisions and subsequent actions in a self determined way.

For a very long time the concept of "freedom of the will" dwelled in the domains of philosophy, theology and law. Philosophers took various, often opposing views (from pure determinism to pure libertarianism), and Kant (1788) declared it as one of those (three) metaphysical problems which lie beyond the powers of human intellect. Science remained closed and hostile to the concept of free will, mostly insisting on the universe as a thoroughly deterministic even if not totally predictable system.

Libertarianism defined the concept of free will as "the power of agents to be the ultimate creators (or originators) and sustainers of their own ends and purposes" (Kane 1996). It understood free choices as absolutely uncaused choices: a free choice is made when, without any antecedent cause the decision is made that results in an action. An example of a freely chosen action would be my decision to start walking now. Indeterminacy or the noncausality occurs at the moment of deciding; its nature or origin is presently unknown, but it may be quantum-level indeterminacy. Important questions remain unanswered: for example, what exactly is the agent who chooses? Is it a central processing "self" or a long-distance "operator"?

On the other hand, Hume argued that responsible choice is inconsistent with uncaused choice. Our free choices are in fact internally caused by various antecedent mental and physiological processes – feelings, beliefs, faith, ethical bounds, hunger, thirst, etc. My decision to start walking towards refrigerator now is caused by my antedecent feeling of thirst (even though I may not be aware of it) and not by random unpredictability. However, not all kinds of causes are consistent with free choice; many are consistent with more or less forced choice. As Hume put it in his great work:

> Where (actions) proceed not from some cause in the characters and disposition of the person, who perform'd them, they infix not themselves upon him, and can neither redound to his honor if good, nor infamy, if evil (Hume 1739)

However, clinical observations in neurology and the rise of new sciences (quantum physics, neuroscience) could not ignore the concept of free will. These sciences - apart from trying to elucidate the phenomenon itself and the mechanism – often expressed scepticism about the existence of "free will" and pleaded for different concept and terminology. There is a general agreement that the causal machine of choices and antecedent processes appear to be the brain and neuronal processes and many scientists who believe in "free will", such as Libet, design and perform creative studies of brain mechanisms. There are other, on the opposite side of the spectrum, who deny the concept (Wegner 2002). They criticize the implication that brain functions are governed by what they derisively refer to as homunculus (little man).

Choices are not made by homunculus, they arise from the constant flow of homeostatic and other unconscious neural functions, similar to what is demonstrated in more primitive reflexive or automatic neural processing. Causal chains between stimulus and response in voluntary, "freely determined" acts leading to voluntary behaviour may be longer, more complex and largely unconscious, but they remain causal chains. The difference is a matter of degree not of a kind.

Of course, easier part of the problem is the study of phenomenology, anatomy and mechanism of what is referred as "free will" – regardless to whether the free will exists or not. The phenomenon can be elegantly studied on the model of voluntary vs. involuntary motor action and the important question is: are there systematic neuronal-based differences between voluntary and involuntary actions. Neurology, functional imaging and electrophysiology of the brain can provide some insights by studying healthy people and patients with brain disorders characterized by impaired volition of movement and/or mind.

2. Neurological and Psychiatric Disorders Affecting Decision Making & Free Choice

It is common observation that both, low levels (drowsiness, lack of sleep, boredom) as well as high levels of conscious awareness (anxiety, pain, intense emotion) can physiologically reduce "will power". In clinicopathological settings, neurologists and psychiatrists have for long time observed and described syndromes of a "sick will", characterized by inactivity, poverty of movement and thought:

- Movement disorders (tic, chorea, myoclonus)
- Cognitive disorders (apraxia, alien hand syndrome)
- Schizophrenia
- Depression
- Autism
- ADHD syndrome
- Dementia
- Parkinsonism

- Hysteria

The concept of volition has been studied particularly well in the domain of movement. Movement can be categorized into four groups (Jankovic 1992):

1. voluntary movements
   a. intentional (planned, self-initiated, internally generated)
   b. externally triggered in response to some external stimulus.

2. semivoluntary or unvoluntary movements
   a. triggered by an inner sensory stimulus (the need to scratch, akathisia)
   b. triggered by a compulsive or unwanted feeling (compulsive touching)

3. involuntary movements
   a. nonsupressible (reflexes, seizures, myoclonus)
   b. suppresible (tics, chorea, tremor, dystonia, stereotypy) (Koller and Biary 1989) ;

4. automatic movements (walking, speaking, swinging of arms during walking) are learned motor behaviours performed without conscious effort. Automatic learned behaviours appear to be encoded into the basal ganglia (Jog et al., 1999)

Conversive Disorder ('Hysteria')

It is defined as a mental disorder characterized by emotional excitability and sometimes by amnesia or a physical deficit, such as paralysis, or a sensory deficit (blindness), without an organic cause and with no detectable clinical, MR, EEG or CSF abnormalities.

Abulia

Abulia is a state in which an individual seems to have lost will or motivation to move, speak, act. Its extreme form - akinetic mutism (Fisher 1983) - is the symptom of complete loss of voluntary movement. The affected areas are in the mesial frontal regions (supplemental motor area, cingulate).

Apraxia

Apraxia is a disorder of motor control, characterized by a loss of a motor programme, not explicable by more basic motor, sensory or cognitive impairments. In the most common form of apraxia, ideomotor apraxia, knowledge of tasks is present, but gestures are distorted due to temporal and spatial errors. It is probably due to a disconnection between cortical parietal and premotor areas.

Alien Hand Syndrome

The alien hand phenomenon is characterized by unwanted movements that arise without any sense of their being willed. The hand behaves as though it has a will of its own, not consistent with patient's intentions.. The spectrum of movement disorder ranges from simple quasi-reflex movements (waving, levitation, grasping) to a structured intermanual conflict (Fisher 2000), when hand movements may appear purposeful (grabbing cookies). In such cases not only has the patient no control over them, these movements may exert certain control over him. There appears to be a difficulty in self-initiating movement and excessive ease in the production of involuntary and triggered movements. The lesion is usually in the middle area of the cingulate.

Chorea

The disease is characterized by nonpurposeful, jerky movements due to the neuronal loss in the basal ganglia (striatum). The movements are involuntary, not consistent with patient's intentions. However, early in the course of the disease patients with chorea often do not recognize that there are any involuntary movements. Later they do, but they feel having no control over them. Involuntary movements may be improved with dopamine antagonists, but their involuntary character remains.

Myoclonus

Myoclonus refers to sudden, brief, shock-like movements, resulting in contraction of a muscle or multiple muscles. This nonsuppressible movement often have a characteristic saw-tooth pattern and usually disappears during sleep.

In cortical myoclonus, the brain makes the movement, yet the patient interprets the movement as involuntary.

Obsessive-Compulsive Disorder (OCD)

An overwhelming urge to think about or to perform an act. The patient is fully aware of his urge and intentions as being his own. It is he who performs the act, but the act is out of his control and, if possible, he would choose to get rid of the thought or the act. Symptoms of OCD are due to impaired circuitry connecting the orbito-frontal cortex, anterior cingulate gyrus and the basal ganglia, particularly the caudate nucleus.

Tourette Syndrome

Tourette syndrome (TS) is a neurological disorder characterized by repetitive, stereotyped, involuntary movements and vocalizations called tics. The disorder starts in childhood, with the average onset between the ages of 7 and 10 years. Tics are relatively brief and intermittent movements (motor tics) or sounds (phonic tics) that often look like voluntary movements. However, patients often cannot say whether the movements are voluntary or involuntary. They can supress them or just let them happen and are thus categorized either as semivoluntary or involuntary (suppressible).

Parkinson's Disease

Due to the degeneration of the dopaminergic neurons in the substantia nigra, the disordered basal ganglia lead to excessive inhibition of supplementary motor area (SMA) and dorsolateral prefrontal cortex (DLPFC). Although the patient knows exactly what action he wants to perform, his motor (and sometimes mental) actions are slowed down (bradykinesia, bradyphrenia) require more effort (hypokinesia, abulia) and the action is sometimes impossible to execute (freezing off, mental block). Some aspects of the disordered action can be improved with levodopa.

3. Instrumental Methods

3.1. Functional Imaging

Imaging techniques provide an image of regional brain activity in spatial terms. Overt or imagined willed acts have been studied with various techniques: 2-D extracranial measurements of the regional cortical blood flow (rCBF) (Ingvar and Philipson 1977) as well as with high resolution 3-D positron emission tomography (PET) (Frith et al, 1991). The findings of functional brain studies suggest that willed acts are formulated in the prefrontal cortex and imply a relationship between the volition and the dorsolateral prefrontal cortex, particularly on the left side. Reduction of the prefrontal activity has been observed in the majority of clinical disorders characterized by a "sick will", such as schizophrenia, depression and dementia.

In OCD, neuroimaging studies reveal markedly changed activity in orbito-frontal cortex, in the anterior cingulate gyrus and in the basal ganglia, particularly the caudate nucleus (Rauch et al, 1994; Breiter et al, 1996). There is also imaging evidence that the two interventions - treatment with SSRI as well as psychological treatment - independently cause similar changes in patterns of cerebral glucose metabolism (Baxter et al, 1992).

In Parkinson's disease, PET studies demonstrate hypoactivity of SMA, DLPFC and anterior cingulate cortex and preserved primary motor and premotor cortical activity.

In hysteria, functional imaging studies revealed an interesting pattern: when a healthy person performs a movement (wiggling finger, moving the leg), two brain regions become active – (i) the motor cortex (executing the act) and (ii) the premotor area (preparing the act). In a hysterically paralyzed patients the motor area failed to activate and an activation of the anterior cingulated and the orbito-frontal lobes was observed instead (as if the anterior cingulated and orbito-frontal regions were inhibiting or vetoing the patient's attempts to move the paralyzed limb) (Marshall et al, 1997).

3.2. Electrophysiology

Electrophysiological techniques provide characteristics of brain activity in temporal terms and they can be particularly suited to elucidate timing between conscious intention to act and the act itself.

Grey Walter (1963) studied patients with electrodes implanted in the motor cortex. They were asked to look at a sequence of slides which they projected at their own chosen speed, by pushing a button. The button, however, was a dummy and what actually advanced the slide was a burst of activity in the motor cortex, transmitted directly to the projector via the implanted electrodes. The patients felt like the projector had anticipated their decision, initiating a slide change just as they were about to advance, but before they had decided to press the button.

An interesting tool for studying the role of awareness in motor planning is provided by so called readiness potential (Bereitschaftspotential, BP). This is a back-averaged slow negative electrical cortical brain wave arising 1 – 2 sec prior to an intentional self-paced voluntary movement (e.g., lifting a finger at a time of subject's choosing) (Kornhuber and Deecke, 1965). BP does not appear with other movements, including the externally triggered voluntary movements (Papa et al., 1991). From approximately 2000 to 1000 ms to approximately 650 – 450 ms the BP emerges bilaterally over premotor areas, has a slowly increasing negative slope with the maximum in the vertex. Later – after 650 to 450 before the voluntary movement it becomes enhanced over the primary motor area contralateral to the movement. Earlier components of BP most likely emanate from the supplementary motor area (SMA) and later components from contralateral primary motor and sensory cortices.

A willed act starts with the formulation of a goal. This step takes a certain time. Deecke and Lang (1996) studied preplaned willed movements and found that BP preceding a willed finger movement may take several seconds. Grey Walter reported that BP recorded from RAF bombardiers preceded the conscious decision to drop a simulated bomb (reported by Claxton 1999). Libet wanted to know if voluntary acts are initiated by a conscious decision to act and whether physiological facts are compatible with the belief that free will determines voluntary acts. He (Libet et al, 1983 a, b) studied spontaneous willed movements and tried to determine the timing of the conscious intention to act (from the clock display), in relation to the BP and to the act itself. He found that the readiness potential over premotor areas clearly preceded the subject's conscious intention to act which again occurred about 200 ms before the muscular contraction. This famous but controversial study implies that conscious awareness actually follows the intention to act rather than preceding (and thus causing) it. Libet, trying to "save" the concept of free will, proposed that voluntary acts are unconsciously initiated but are subject to conscious control which can either promote or veto the unconsciously initiated process. Therefore, free will and volition are associated more with the vetoing of action than its instigation: we don't have free will, but we have free won't.

In a study on patients with Parkinson's disease, the BP was found to be reduced in the early part, whilst it was larger than normal in the later part. This may be due to the underactivity of a source in the SMA and compensatory overactivity in lateral motor areas (Dick et al, 1987).

In myoclonus BP is absent. In patients with tics, Obeso et al (1981) reported that there was no BP prior to tics, but it was present and normal for similar movements produced voluntarily by the same patients. Results imply that tics differ from voluntary movements because of their lack of cortical preparatory activity.

4. Functional Anatomy of Volition

Much is known about the anatomy and physiology of the motor system and the movement, but much less about the concept of voluntariness and free will. There are certain movements that appear to be internally triggered and humans have the sense that they have willed the movement; and there are other involuntary or automatic movements clearly devoid of such a sense .

Clinical case reports, electrophysiological studies and new brain imaging techniques confirm the assumption that free will is "localizable" to a certain extent. Several brain regions contribute to the execution of consciously chosen, volitional action. Most important structures are located in the prefrontal cortex and its related cortical and subcortical structures, along the cortico-subcortico-frontal and cortico-frontal circuits. The cortico-subcortico (thalamo)-frontal loops are semiclosed circuits, connecting the basal ganglia, specific thalamic nuclei and relevant regions of the frontal cortex (Alexander et al,1986). Particularly important seem to be (i) a mesial loop closing on the anterior cingulate cortex, the orbito-frontal and the supplementary motor area and (ii) the lateral loop closing on the dorsololateral prefrontal cortex.

The anterior cingulate is a place of convergence for and selection of motor, emotional, homeostatic and cognitive drives. The supplementary motor area plays a role in the initiation, sequencing and programming of motor acts to fit a motor plan. The orbitofrontal area is involved in the modulation of action by reward, punishment and social context. The dorsolateral prefrontal cortex formulates action goals and is implicated in response selection, particularly in the context of novelty, in the internal generation of action, in the switch to alternative actions and plays an important role in the working memory.

Willed acts concern the future and our future oriented goals and intentions. They arise particularly in the prefrontal brain areas which are involved in the serial programming of motor behaviour, speech, and cognition. At the same time, a formulation of the future goal is associated with an active supression of representations which do not contribute to the willed activity. There is evidence that prefrontal cortex and the cingulate gyrus may participate in the supression of the irrelevant representations (Frith e tal,1991; Posner and Raichle 1994).

5. Discussion and Implications

Observational clinical studies of patients with certain neurological disorders, their functional imaging and results of electrophysiological tests to a large degree support the notion that we are recipients rather than architects of our actions. A notion which may not be very popular, but which certainly is concordant with our daily experiences: so many important turning points in our lives are unaccompanied by conscious decision; so many intelligent actions occur without any preceding intention – they simply happen to us or we perform them by "intuition".

And yet – the belief or hope of having free will is firmly anchored in our consciousness of intention. Intention seems to be strongly associated with the sense of "self", which, in contemporary Western culture is comprised of an intricate web of goals, interests, preferences and threats which are summed up in a conscious or in a subconscious prediction or intention. With conscious intentions we probably feel these intimations as the causes and the instigators of the impending act. If intentions remain subconscious, we will predict or intend without knowing why, without being able to rationally explain the cause of our act. The results of clinical observations, of functional imaging and of electrophysiological studies do support certain scepticism about the existence of "free will" and do, at least, plead for different concept and terminology. These results should also evoke certain uneasiness in the field of law. It would probably be wrong to rush into hasty change of legal concepts of responsibility but lawmakers and society should make the door for debate wide open and promote new interdisciplinary research of the field.

References

Alexander GE, DeLong MR,Strick PL 1986. Parallel organization of functionally segregated circuits linking basal ganglia and cortex. *Ann. Rev. Neurosci.* 9; 357-81.

Baxter LR Schwartz JM, e tal. 1992. Caudate glucose metabolic rate changes with both drug and behavior therapy for obsessive-compulsive disorder. *Arch. of General Psychiatry.* 49: 681-9.

Breiter HC, Rauch SL et al. (1996). Functional magnetic resonance imaging of symptoms provocation in obsessive-compulsive disorder. *Archives of General Psychiatry.* 53, 595-606.

Claxton G 1999.Whodunnit? Unpicking the "Seems" of Free Will. *J. of Consciousness Studies.* 6; 8-9: 99-113.

Deecke L. and Lang W. (1996).Generation of movement-related potentials and fields in the supplementary sensorimotor area and the primary motor area. *Advances in Neurology,* 70. *Supplementary Sensorimotor Area.* 127-146.

Dick JP, Cantello R, Buruma O, et al. (1987). The Bereitschaftspotential, l-dopa and Parkinson's disease. *Electroencephalogr. Clin. Neurophysiol..* 66: 263-274.

Fisher CM 1983. Honored guest presentation: Abulia minor vs. Agitated behavior. *Clin. Neurosurg.* 31: 9 – 31.

Fisher CM 2000. Alien hand phenomena: A review with the addition of six personal cases. *An. J. Neurol. Sci.* 27: 192 – 203.

Frith C,D, Friston K, Liddle PE, and Frackowiack RSJ. (1991). Willed action and the prefrontalcortex in man. A study with PET. *Proc. R. Soc. Lond.* (B), 244: 241-6.

GreyWalter W 1963. Presentation to the Osler Society, Oxford University; quoted in Consciousness Explained, D Dennett (Boston, MA: Little Brown, 1991)

Hume D.1739. A Treatise of Human Nature. Edited by L.A. Selby-Bigge, 1888 and 1896. Oxford: Oxford University Press

Ingvar DH and Philipson (1977).Distribution of cerebral blood flow in the dominant hemisphere during motor ideation and motor performance. *Ann. Neurol.* 2 (3): 230-7.

Jankovic J 1992. Diagnosis and classification of tics and Tourette's syndrome. In Chase T, Friedhoff A, Cohen DJ (eds): Tourette's syndrome. *Advances in Neurology.* vol. 58, New York, Raven Press, pp 7 – 14.

Jog MS, Kubota Y, Connoly CI,e tal 1999. Building neural representations of habits. *Science.* 286: 1745-1749.

Kane R. 1996. The significance of free will.(New York: Oxford University Press)

Kant I.1788. Critique of Practical Reason, ed. and tr. LW Beck (Macmillan)

Koller WC, Biary NM 1989. Volitional control of involuntary movements. *Mov. Disord.* 4: 153 – 156.

Kornhuber HH and Deecke L 1965. Hirnpotentialaenderungen bei Wuellkurbewegungen und passiven Bewegungen des Menschen:Bereitschaftspotential and reafferente Potentiale. *Pfuegers Arch. Ges. Physiol.* 284: 1-17.

Libet B, Gleason CA, Wright Jr. EW, Gleason CA, Pearl DK 1.983 a.Time of conscious intention to act in relation to onset of cerebral activity (readiness potential). *Brain.* 106: 623-42.

Libet B,Wright Jr. EW, Gleason CA 1.983 b.Preparation- or intention-to-act, in relation to pre-event potentials recorded at the vertex. *Electroencephalography and Clinical neurophysiology.*56: 367-72

Marshall JC, Halligan PW, Fink GR, et al. (1997). The functional anatomy of a hysterical paralysis. *Cognition.* 64: B1-B8.

Obeso JA, Rothwell JC, Marsden CD 1981. Simple tics in Gilles de la Tourette's syndrome are not prefaced by a normal premovement potential. *J. Neurol. Neurosurg. Psychiatry.* 44: 735-738.

Papa SM, Artieda J, Obeso JA (1991). Cortical activity preceding self-initiated and externally triggered voluntary movement. *Mov. Disord.* 6: 217-224.

Posner MI and Raichle ME (1994). Images of Mind. New York: WH Freeman and Company.

Rauch SL, Jenike MA, et al. (1994). Regional cerebral blood flow measured during symptoms provocation in obsessive-compulsive disorder using 15O-labeled CO2 and positron emission tomography. *Arch. of General Psychiatry.* 51: 62-70.

In: Philosophical Insights about Modern Science
Eds: Eva Žerovnik, Olga Markič and Andrej Ule
ISBN: 978-1-60741-373-8

Chapter 8

NEUROSCIENCE AND THE IMAGE OF THE MIND

Olga Markič
Department of Philosophy, Faculty of Arts, University of Ljubljana,
Aškerčeva 2, 1000 Ljubljana, Slovenia

SYNOPSIS

The chapter deals with the question of the relationship between neuroscience and our image of the mind. Flanagan (2002) has described two competing images of who we are: the humanistic image and the scientific image. The humanistic image has its roots in theology and dualistic philosophy and is also much in accordance with everyday thinking about the mind. The scientific image says that we are animals that evolve according to the principles of natural selection and cannot circumvent the laws of nature. This image takes consciousness, cognition and volition as natural capacities of embodied creatures that live in natural and social environments. It seems that these two approaches are incompatible. Recent developments in neuroscience raise the worry that understanding how brains cause behavior will radically change our understanding of the mind and undermine our views about free will and, consequently, about moral responsibility. The traditional problem of free will and determinism raises the objective question of whether we have free will. The worry many cognitive scientists express is the following: decisions, choices and actions are generally thought of as freely willed; but if they were to be revealed as results of neural mechanisms, they could not be seen as free anymore and would not support moral responsibility. Free will would be best seen as just an illusion. In fact, Daniel Wegner (2002) argues that experiments in neuroscience and psychology have already demonstrated such a conclusion. The author suggests that Wegner is influenced by the dualistic humanistic image and sets a standard for free will so high that only a supernatural being could reach it. She argues that the psychological experiments to which Wegner refers support only the weaker interpretation of illusion, i.e. we do not have direct access to the causal link between thought and action. Understanding illusion in this weaker sense does not preclude our basic intuitions about human mind and moral responsibility. Advances in theoretical and clinical neurosciences open a path to a better understanding of the human mind and to a new neurophilosophical approach.

1. The Humanistic and the Scientific Image

The development of science in recent centuries has brought major changes in our everyday understanding of the world as well as our human selves. It is often said that science has dealt with three major blows to the folk-theoretical picture: first, Copernicus and Galileo removed the earth from the center of the universe, then Darwin's theory of evolution showed that animal species, including humans, are produced by natural processes and not by God(s), and finally Freud contributed to the third revolutionary turn by opening the doors to the unconsciousness. These three shifts changed views on cosmology, biology and psychology but did not completely change what Owen Flanagan (2002) has called the humanistic image. He believes that in the Western tradition we have two grand images of who we are: the humanistic and the scientific. Flanagan describes the humanistic image as a set of beliefs about ourselves based on the assumption that we are spiritual beings with free will and consequently able to lead a moral and meaningful life. It has its roots in religion and in perennial philosophy (Plato, Aristotle, Augustine, Aquinas, Descartes, Hume, Kant) and refers to some supernatural concepts like the soul, God and immortality. In contrast, the scientific image says that we are animals that evolve according to the principles of natural selection and cannot circumvent the laws of nature. The question is whether these two images are compatible. Flanagan suggests that they can coexist if we understand the humanistic image to reveal our spiritual nature and science as unlocking the secrets of the external world and our animal essence. He thinks this coexistence is not possible without the premise that we are only partly animal (Flanagan, 2002, p. xii). But the advances in evolutionary biology, cognitive science and especially cognitive neuroscience make this premise highly problematic. Namely, the scientific image takes consciousness, cognition and volition as natural capacities of embodied creatures that live in natural and social environments. According to this view, humans do not possess any extra ingredients that could do the work as traditionally conceived. Some scientists (e.g., D. Wegner, 2002) and scientifically oriented philosophers (e.g., Paul Churchland, 1988) are even more radical and think that many concepts employed by the humanistic image are just illusions without real reference. According to them, new discoveries in neuroscience herald even more major changes and will lead to the abandonment of the humanistic image. On the other side, the defenders of the traditional humanistic image think that the scientific image leads to an impoverished image of the mind which cannot support the concept of a person able to live morally and meaningfully. It seems that we have two competing and incompatible approaches.

But do we really have to abandon our intuitions of what it means to be a human if we treat the mind as a natural phenomenon? My aim in this paper is to examine recent investigations in neuroscience that pose a major threat to our intuition that we are, at least sometimes, free and morally responsible agents.

2. The Traditional Problem of Determinism and Free Will

Let us first look at common sense intuition about the free will. It is often said that philosophers and cognitive scientists are already biased and thus the best way to check intuition about free will is to ask ordinary people who have not yet been exposed to the problem. Hodgson (2005) calls it "a plain person's free will". It contains a cluster of phenomenological observations, for example:

- We are aware of making choices in the world.
- We have the recurring experience that we initiate our actions.
- We sometimes weigh up the alternatives, sometimes we seem to follow our habits.
- We become aware of the consequences of the things that we have done.
- We are held accountable for our actions.

Hodgson thinks that such intuition supports the strong, libertarian notion of free will, namely that free will exists and is inconsistent with determinism. He also argues that this position is closer to the truth than opposing views (Hodgson, 2005, p. 3). But this is not the only viewpoint. Philosophers have been trying to resolve the problem of free will and determinism for a long time (Kane, 2005; Pareboom, 1997). A schematic mapping of the main philosophical positions on freedom of the will is as follows:

	Determinism	Indeterminism
We have free will	Compatibilist position (Soft determinists)	Incompatibilist position Libertarians
No free will	Incompatibilist position (Hard determinists)	

Doubts about humans having free will have come from two sides – from theological doctrines and from the natural sciences. If there is an omnipotent God who controls our actions, then it is questionable if we are free agents able to do otherwise. Namely, even if God refrains from controlling us, as some theologians argue, he is still able to have foreknowledge of our actions. Therefore our actions are predetermined and we are not free. A similar threat from determinism comes from the natural sciences. A naturalist who takes humans as part of nature and believes that the world is entirely determined by its prior states and the laws of nature has similar problems to preserve freedom as theologians had. The most discussed argument for incompatibilism is the Consequence Argument. Peter van Inwagen has informally stated it as follows:

> "If determinism is true, then our acts are consequences of the laws of nature and events in the remote past. But it is not up to us what went on before we were born; and neither is up to us what the laws of nature are. Therefore the consequences of these things (including our own acts) are not up to us. (van Inwagen, 1983, p. 16)

Because it seems that the problem for naturalists results from the assumption of determinism, they think that if the universe is not deterministic, they can have both, freedom and a scientific world view. The most common move is to appeal to the indeterministic interpretation of quantum mechanics and argue that it saves freedom. However, saving free will with indeterminism is not easy. First, it seems that if our actions are not determined, then they are random. But if our actions are due to chance events, "arbitrary", "uncontrolled", "irrational", "matters of luck and chance", then they are anything but free and responsible actions (Kane, 2005, p.38). And second, even if indeterminism is supporting free will, we need evidence that indeterminism is at the right level (Weber, 2005). So, to defend libertarian free will many libertarians have adopted different extra-factor strategies. The most obvious is dualism of mind and body – a disembodied mind or soul that is outside the physical world and is not governed by the laws of nature (like Descartes' *res cogitans*). The second strategy is Kant's. He distinguishes between practical or moral reasoning, which requires that we believe in libertarian free will, and theoretical or scientific reasoning, which cannot explain freedom. Parallel to this are the phenomenal and noumenal selves. Only noumenal selves can be free because they are not constrained by space and time or the laws of nature. The third strategy is called theory of agent-causation. According to this theory, the causal chain begins with the agent, who is a "prime mover unmoved". One of the defenders of this view, Roderick Chisholm, explained:

> "We may say that the hand was moved by the man, but we may also say that the motion of the hand was caused by the motion of certain muscles; and we may say that the motion of the muscle was caused by certain events that took place within the brain. But some event, and presumably one of these that took place within the brain, was caused by the agent and not by any other events." (Chisholm, "Human Freedom and the Self", quoted from Kane, 2005, p. 46)

The most common objection to extra-factor strategies is that in order to have free will they postulate a mysterious metaphysical entity.

In contrast to libertarians, compatibilists do not see a real conflict between determinism and free will. Classical compatibilists (Hume, Mill, Dennett) see confusions about the nature of freedom and nature of determinism. They start with what they think is the ordinary meaning of free action or choice. They see freedom as (1) a power or ability to do what someone wants or desires to do and (2) absence of constraints. It does seem that it captures the freedom of action and the freedom of choice or decisions that is morally important and so worth wanting. Compatibilists also point out that people think determinism is a threat to freedom because they confuse determinism with (1) constraint and coercion (Hume), (2) control by other agents, (3) fatalism and (4) mechanism (Kane, 2005, p. 18-21).

Philosophers arguing for one of the above-mentioned positions all think that we have a stronger or weaker freedom of will. The third traditional position is hard determinism. Hard determinists affirm determinism but, in contrast to compatibilists, think that there is no free will worth wanting. Free will does not exist in a true sense required for genuine responsibility. The kernel of this position is a rejection of both compatibilism and libertarianism and those who accept it are skeptics about free will. The skeptical position requires living without belief in free will and true moral responsibility. Kane thinks that unqualified endorsement of it is rare and compares it to the principle behind the exclamation of a Victorian lady when she first heard of Darwin's theory of evolution: "Descended from the apes. Let's hope it isn't true. But if it is, let's hope it does not become generally known." (Kane, 2005, p.70)

This short overview of the main traditional answers to the question of free will and its compatibility with determinism will help us better see the neuroscientific challenge.

3. Neuroscientific Challenges to Free Will and Moral Responsibility

Recent developments in neuroscience raise the worry that understanding how brains cause behavior will radically change our understanding of the mind and undermine our views about free will and, consequently, about moral responsibility. It is only recently that neuroscientists have been able to investigate the cognitive phenomena that are the hallmarks of what it is to be human. Such investigations are made possible because of methods like electroencephalography (EEG) and especially new brain imaging techniques like computed tomography (CT), magnetic resonance imaging (MRI), positron emission tomography (PET), and functional magnetic resonance imaging (fMRI), which can display the structure and function of the brain regions that regulate human capacity for impulse control, reasoning, and decision-making. PET and fMRI scans are especially significant because they can display real-time brain function. These techniques can measure activity in the cerebral cortex while subjects are engaged in cognitive tasks and also in subcortical areas associated with emotions (Glannon, 2005).

The worry many cognitive scientist express is the following: decisions, choices and actions are generally thought of as freely willed; but if they were to be revealed as results of neural mechanisms, they could not be seen as free anymore and would not support moral responsibility. Free will would be best seen as just an illusion. In fact, Daniel Wegner (2002) argues that experiments in neuroscience and psychology have already demonstrated such a conclusion.

I will tackle the challenges as follows: I will first discuss Wegner's evidence for the illusion thesis and then go to the more general observations about the possible threats for moral responsibility.

3.1. Wegner's Illusion of Conscious Will

In his book "The Illusion of Conscious Will" (2002) Wegner argues that our conscious will, which means both the experience of willing and the perception of the causation by conscious thought, is an illusion. He supports his thesis by numerous examples and experiments from neuroscience and psychology, including Libet's (1985) famous studies on the unconscious cerebral initiative and the role of conscious will in voluntary action (see also Pirtošek this volume), experiments with transcranial magnetic simulation and examples of absence of the experience of will in the case of motor automatisms (table-turning, Ouija-board spelling, pendulum diving). He sees conscious will as an illusion in a sense that "the experience of consciously willing an action is not a direct indication that the conscious thought has caused the action" (Wegner, 2002, p.2) or in other words:

> "Although our thoughts may have deep, important, and unconscious causal connections to our actions, the experience of conscious will arises from a process that interprets these connections, not from the connections themselves. Believing that our conscious thoughts cause our actions is an error based on the illusory experience of will – much like believing that a rabbit has indeed popped out of an empty hat." (Wegner & Wheatley, 1999, p. 490)

He suggests that the experience of consciously willing our actions arises primarily when we believe our thoughts have caused our actions. This happens when the following three principles are satisfied (Wegner & Wheatley, 1999, p. 483-486):

> Priority: The thought should precede the action at a proper interval.
> Consistency: The thought should be compatible with the action.
> Exclusivity: The thought should be the only apparent cause of action.

Wegner argues that the interpretative process that creates the experience of conscious will works according to the theory of apparent mental causation (Wegner, 2002; Wegner & Wheatley, 1999). The theory tells us that the actual causal paths are not present in the person's consciousness. It is the principles of priority, consistency and exclusivity that govern the inferences people make about the causal influence of their thoughts on their actions, and thus underlie the experience that their actions are willed.

So, do the data gathered by Wegner really support the interpretation that conscious will does not play a causal role? Wegner discusses experiments that show that conscious willing of an action *can* be separated from the action and that sometimes people have a conscious feeling of not owning an action and not being responsible for it, but they in fact are, and vice versa. But I think it is a hasty generalization to conclude from these specific examples that conscious willing is never causally relevant, as would be wrong to conclude that perceptual illusions show us that our perception is always misleading. I agree with Nahmias that Wegner's examples show "that there are various exceptions to the rule that our conscious experiences of our actions correspond with those actions. But the fact that there are these exceptions does not show that, in the *normal* cases of correspondence, conscious will is causally irrelevant." (Nahmias, 2002, p. 533). Several other philosophers (Bayne, 2006; Mele, 2008) similarly criticize Wegner's epiphenomenalism of conscious intentions.

It is not easy to find out what exactly Wegner means by the thesis of illusion of conscious will (Nahmias, 2002). One suggestion is that he is denying the prevalent common sense view that Flanagan (2002) considers as a part of the humanistic image of the mind, namely the Cartesian soul or what Ryle called "a ghost in the machine" (Ryle, 1949). Denying it causes troubles for the libertarian "who sets the bar for free will ridiculously high" (Mele, 2008, p. 124). But it does not threaten compatibilists like Dennett:

> "If you are one of those who think that free will is only *really* free will if it springs from an immaterial soul that hovers happily in your brain, shooting arrows of decision into your motor cortex, then, given what *you* mean by free will, my view is that there is no free will at all. If, on the other hand, you think free will might be morally important without being supernatural, then my view is that free will is indeed real, but just not quite what you probably thought it was." (Dennett, 2003, p. 223)

We can also find support for this suggestion in his article "Self Is Magic":

> "We are enchanted by the operation of our minds and bodies into believing that we are "uncaused causes", the origins of our own behavior. Each self is magic in its own mind. Unfortunately, the magic self stands squarely in the way of the scientific understanding of the psychological, neural and social origins of our behavior and thought." (Wegner, 2008, p. 226)

And two pages further he writes: "seeing one's own causal influence as supernatural is part of being human" (Wegner, 2008, p. 228).

I think that the psychological experiments to which Wegner refers support only the weaker interpretation of illusion, i.e. we do not have direct access to the causal link between thought and action. Wegner is right to take our folk-psychological belief in direct access to the causes of our action as false. But in contrast to him I believe that this is not enough to infer that conscious intentions could not cause actions. This weaker interpretation does not say that we could not have causally relevant conscious intentions, it just indicates a false understanding of mental processes and of our own agency. Recent research in neuroscience suggests that taking a process through which a person makes her volitions as a purely rational decision is false, and that we have to look at the emotional aspects and subjective feelings in constructing more sophisticated model of agency (Damasio, 1994, 2003).

3.2. Threats to Our Common Sense Notion of Responsibility

We have seen that a naïve dualistic picture and the concept of free will as an uncaused supernatural entity on the one side, and the belief that deliberating, choosing and decision-making are a deterministic brain mechanism on the other side, lead to the conclusion that there is no free will and moral responsibility. Although scientists would perhaps not overtly admit the dualistic standpoint, this is probably behind the jump to the claim of the illusion. Martha Farah's passage illustrates this reasoning:

> "How do these scientific advances affect our understanding of moral and legal responsibility? We do not blame people for acts committed reflexively (e.g., as the result of a literal knee-jerk), in states of diminished awareness or control (e.g., while sleep-walking or under hypnosis) or under duress (e.g., with a gun held to the head), because in these cases we perceive the acts as not resulting from the exercise of free will (Danno, 2003). The problem with neuroscience accounts of behavior is that everything we do is like a knee-jerk in the following important way: it results from a chain of purely physical events that are as impossible to resist as the laws of physics." (Farah, 2005, p. 37-38).

We have briefly discussed the traditional problem of free will and determinism and we have seen that the objective question of whether we have free will can be raised regardless of what we know about the brain. If neuroscientific theories reject libertarian extra-factor strategies, there are still ways to refrain from the hard determinist position. First, the question whether processes at the right level are determinist or indeterminist is not yet finally resolved. Second, if the processes are deterministic, one can argue for compatibilism. As Adina Roskies said:

> "A view of ourselves as biological mechanisms should not undermine our notion of ourselves as free and responsible agents. After all, some causal notion is needed for attributions of moral responsibility to make sense. The predictive power of our high-level psychological generalizations grounds our views of agency, so further evidence that we behave in a law-like fashion should not undermine our notions of freedom." Roskies, A. (2006, p. 421)

The third point concerns everyday moral judgments of responsibility. P. F. Strawson (1982) argued that to regard people as responsible agents we have to treat them in certain ways and adopt various attitudes toward them, such as resentment, admiration, gratitude, guilt, blame, and forgiveness. People must be part of a moral community in order to appropriately take such reactive attitudes to one another and thus hold each other responsible. The judgments about moral responsibility are almost always produced in concrete, emotionally-charged situations.

Philosophers often appeal to intuition when they argue for their own solution, incompatibilist or compatibilist. The claims were not empirically tested until the recent rise of experimental philosophy with studies designed to investigate intuitions about particular cases. Nichols and Knobe (2007) investigated the role of affect in connection with compatibilist/incompatibilist intuition of moral responsibility. They design a study in which they describe a deterministic and indeterministic world. In both worlds they depicted one high-affect and one low-affect scenario. The results showed that in a deterministic world approximately twice the percentage of subjects gave a compatibilist answer in the high-affect case compared to the low-affect case. Nichols and Knobe think that people's conflicting intuitions about incompatibilism and compatibilism are due to the operation of different subsystems governing reasoning about moral responsibility. In emotionally neutral scenarios when guided by abstract reasoning, people tend to produce incompatibilist intuitions, whereas in emotionally charged situations judgments are more in line with compatibilist intuitions. Roskies thinks that if this is correct, then "it indicates that the actual psychological processes involved in everyday moral judgments of responsibility are likely to operate largely independently of theoretical views about determinism and mechanism." (Roskies, 2006, p. 422)

We have described worries that neuroscience poses to the humanistic image of the mind based on the considerations about determinism. But I suspect an equally strong threat comes from the understanding of neuroscience as supporting an eliminativist or epiphenomenalist solution to the mind body problem. The reason why these two views are intuitively threatening to freedom and responsibility is not because the past and laws are sufficient for our behavior, but because they suggest that our behavior is caused by forces that *bypass* our conscious mental life (Nahmias, 2006). Paul Churchland (1988) has argued that neuroscience shows that our folk psychological theory is radically wrong and thus deserves the fate of phlogiston and witches. It represents a radical scientistic understanding of scientific image according to which assumptions about the mental that we take for granted are just plain nonsense. Whether this is a viable possibility remains open. But I suspect that there will not be a massive mismatch between concepts at the level of the mind and lower levels, or as Horgan and Woordword argued, "Folk psychology is here to stay" (1991).

Eliminativism represents a theory which, if proved right, would in fact eliminate not only folk psychology but also philosophy. Sometimes people use the concept neurophilosophy as a substitution for philosophy. Patricia Churchland, who introduced the notion neurophilosophy as the name of her book (1986), rejects such a characterization. She attempts to combine both sides of the discourse in a kind of "co-evolution" of the disciplines. The motivation in neurophilosophy is to bring both disciplines closer together. As Henrik Walter says:

> "We can consider neurophilosophy as a discipline that moves in on the mind-brain problem from two opposite directions. Either we begin on the empirical side and happen

> upon philosophical questions, or we set out with philosophical puzzles and need empirical findings to solve them. ....It is best understood as a bridge discipline between subjective experience, philosophical theorizing, and empirical research." (Walter, 2001, p. 125)

I hope neurophilosophical framework will bring us more detailed knowledge about how control and volition are processed in the brain, together with an understanding of how these notions are connected to our subjective feeling of freedom. We all feel that there is a distinction between winking and blinking, so I think Farah's observation from the above quote that "everything we do is like a knee-jerk in the following important way: it results from a chain of purely physical events that are as impossible to resist as the laws of physics" misses an important point.

4. Conclusion

My suggestion is that we have to be careful not to jump too quickly from theories in neuroscience to the theories at the level of mind and common-sense understanding. I have tried to show that someone who comes to the conclusion that human beings are not able to deliberate, make decisions and act as free agents on the bases of neuroscientific evidence is committing inferential leaps bypassing two deep philosophical problems: the problem of free will and the mind-body problem. Namely, it is not straightforward that determinism leads to denying of free will and moral responsibility. For this, one needs to refute arguments for compatibilism and show that the compatibilist's free will is too weak and thus not worth wanting. Second, the exact relations between neuroscience, cognitive theories and folk psychology are still open. Eliminativism and epiphenomenalism are not the only options for naturalists, although I suspect they are tacitly presupposed when considering treats to humans as free agents. I think neither the dualistic humanistic image nor the scientistic scientific image of the mind is a viable option for the naturalist, but it is possible to retain the kernel of both. As research in neurophilosophy is informed by both sides, it may bring us some neurally informed reconceptualizations of the concepts associated with the dualistic image of the mind, for example control instead of freedom.

References

Bayne, T. (2006). "Phenomenology and the Feeling of Doing: Wegner on the Conscious Will" in S. Pockett, W. P. Banks & S. Gallagher (eds.) *Does Consciousness Cause Behavior? An Investigation of the Nature of Volition*. Cambridge, MA: MIT Press.

Churchland, Paul M. (1988). *Matter and Consciousness*. Cambridge, MA: MIT Press.

Churchland, P.S. (1986). *Neurophilosophy*. Cambridge, MA: MIT Press.

Churchland, P.S. (2002). *Brain-Wise: Studies in Neurophilosophy*. Cambridge, MA: MIT Press.

Damasio, A. (1994). *Descartes' Error: Emotion, Reason, and the Human Brain*. New York: G.P. Putnam's Sons.

Damasio, A. (2003). *Looking for Spinoza: Joy, Sorrow, and the Feeling Brain.* London: William Heinemann.

Denno, D. (2003). "A mind to blame: new views on involuntariness." Behav. Sci. Law 21, pp. 601–618.

Dennett, D. (2003). *Freedom Evolves*. Allen Lane, The Penguin Press.

Farah, M. (2005), "Neuroethics: the practical and the philosophical". *Trends in cognitive Sciences, Vol. 9 No. 1.*, pp. 34-40.

Flanagan, Owen (2002). *The Problem of the Soul: Two Visions of the Mind and How to Reconcile Them*. New York: Basic Books.

Frith, C. (2007). *Making up the Mind: How the Brain Creates Our Mental World.* Blackwell Publishing, Malden, Ma, Oxford, Carlton.

Glannon, W. (2005). "Neurobiology, Neuroimaging, and Free Will". *Midwest Studies in Philosophy*, XXIX, pp. 68-82.

Horgan, T., Woodward, J. (1991). "Folk psychology is here to stay" in J. Greenwood (ed.). *The future of folk psychology"*. Cambridge: Cambridge University press.

Kane, R. (1996). *The Significance of Free Will.* Oxford: Oxford University Press.

Kane, R. (2005). *A Contemporary Introduction to Free Will.* Oxford, New York: Oxford University Press.

Mele, A. (2008). "Recent work on free will and science". *American Philosophical Quarterly* Vol. 45, Number 2, pp. 107-130.

Nahmias, E. (2002). "When Consciousness Matters: A Critical Review of Daniel Wegner's The illusion of conscious will". *Philosophical Psychology*, XV (4), pp. 527–541.

Nahmias, E. (2006). "Folk Fears about Freedom and Responsibility: Determinism vs. Reductionism". *The Journal of Cognition and Culture* vol. 6 (1-2), pp. 215-237

Nichols, S., Knobe, J. (2007). "*Moral Responsibility and Determinism: The Cognitive Science of Folk Intuitions*". Nous, 41, pp. 663–685.

Pareboom, D. (ed.) (1997). *Free Will.* Indianapolis: Hackett Publishing Company, Inc.

Roskies, A. (2006). "Neuroscientific challenges to free will and responsibility". *Trends in Cognitive Sciences* Vol.10 No. 9, pp. 419-423.

Ryle, Van Inwagen, P. (1983). *An Essay on Free Will.* Oxford: Oxford University Press.

Strawson, P.F. (1982) Freedom and Resentment. In G. Watson (ed). *Free Will.* Oxford: Oxford University Press

Walter, H. (2001). *Neurophilosophy of Free Will: From Libertarian Illusions to a Concept of Natural Autonomy*. Cambridge, Ma., London: MIT Press.

Weber, M. (2005). "Indeterminism in Neurobiology". *Philosophy of Science 74*, pp. 663-674.

Wegner, D. (2002). *The Illusion of Conscious Will.* Cambridge, Ma., London: MIT Press.

Wegner, D. (2008). "Self Is Magic" in J. Bear, J. Kaufman and R. Baumeister (eds.). *Are we Free?: Psychology and Free Will.* Oxford, New York: Oxford University Press.

Wegner, D., Wheatley, T. (1999). "Apparent Mental Causation". *American Psychologist*, Vol. 54, pp. 480- 491.

PART B. BEYOND THE CURRENT KNOWLEDGE: A MORE HOLISTIC VIEW

In: Philosophical Insights about Modern Science
Eds: Eva Žerovnik, Olga Markič and Andrej Ule
ISBN: 978-1-60741-373-8

Chapter 9

WAYS TO SUSTAINABLE ENERGY SOURCES

Andrej Detela*
J. Stefan Institute, Ljubljana, Slovenia

SYNOPSIS

This chapter introduces a holistic approach to modern dilemma of energy balance in human society. The content is evident from respective subtitles. It starts with a description of the historical background of energy flows in human civilization, continues with explanation of the physical foundations of the law of energy conservation. Modern forms of energy conversion from sources to consumers are outlined; ecological and ethical limits to the growth in energy consumption are discussed. The problem of fossil fuels is explained in some more detail and known alternatives to classical fossil fuels (the nuclear energy and various renewable sources such as hydro energy, wind energy, and solar energy) are analyzed. Promises and problems connected with biomass and hydrogen economy are highlighted next. The "looking into the future" chapter considers three totally different scenarios. The need for energy saving is stressed. Nuclear fusion as a new hope with related questions, and the exotic energy sources like enigmatic "free energy", are described at the end. It seems at this moment that the solution to energy thirst is not straightforward but is rather a combination of many small solutions inside the complex network of energy sources and drains. This seems to be achievable only within the new social paradigm of ethical awareness. Therefore, ethical and ecological considerations are included throughout the text.

INTRODUCTION: HISTORICAL BACKGROUND

In older times when people were still living in breath-to-breath relationship with energy flows in nature, the human mind did not care about the problem of energy. *Lucy*, an ancient ancestor of *Homo sapiens* from the plateaus of East Africa, had no need of energy for heat, additional transportation, or industrial production. She dealt with energy in the same way as

* A. Detela is a theoretical physicist and inventor of scientifically-based ecological solutions. He also is a philosopher, poet, traveller and writer of fiction.

all other living creatures do. Energy flows provided by the intake of food were in complete harmony with other natural cycles.

Maybe the first awareness of *energy* originated on the occasion when, travelling to colder areas of Asia and Europe more than 50 thousand years ago, man learnt how to domesticate *fire*. This was already something different from the behaviours of the surrounding animals. Man was entitled to be proud of this early achievement; in which he recognized a clear sign of his mental and spiritual power beyond the animal kingdom. However, he was well aware that fire was not in his exclusive possession; on the contrary, fire was granted to him by some life-giving divine Being taking charge and caring for all individual beings in the wide Universe[1]. So, quite naturally, fire became an important symbolic instrument of ancient spiritual practices, in the form of the God of fire (for instance *Agni* in ancient India but still persisting to our modern times).

Ethnologists claim that in ancient "primitive" cultures (they can still be found in some remote places of our globalized planet) man did not imagine to be in possession of his delusive mental supremacy over Nature [Lizot, 1976]. Not at all, yet he was in constant struggle to find delicate balance with huge natural powers that surrounded him and at the same time filled him with fear and awe but also with hope to get "tamed" if dealing in the right way. Respect for Nature and for all forms of life was part of everyday life, not a bit less than it is to modern-day ecologists. For instance, speeches, stories, and poems of American Indians are full of such evidence[2].

As centuries passed on and human civilization slowly elapsed from the *matriarchal society*, this close (and in many ways beneficial) relationship with Nature has been slowly dissolved, dismissed and became forgotten. Great intellectual and social systems were constructed. Some 5000 years ago such far-reaching inventions as the *wheel*, or the first forms of *writing*, or rudimentary *arithmetic*, came into use. Powerful states and military systems did rise and fall. We could say that in this new historical paradigm the concept of *energy* (or rather, of what is today called energy) shifted to a different meaning: it became more abstract and detached from direct exchange with Nature. In military societies of the antiquity (e.g., in the *Roman Empire*) the idea of energy was in close relationship to the idea of supremacy of the strong over the weak[3]. Man was aware of energy predominantly when it was manifested as brutal military force working for particular interests. If, long ago in the matriarchal society, energy had been worshipped as a web of open flows in nature, now this same energy was caught in possession and imprisoned, for the sake of projecting it to some definite but very partial goal. It became a means of control.

This same trend continued through the following millennia, time and again supported by inventions of new energy forms. The muscular power (represented by bow and arrow) was replaced by the thundering roar of gunpowder. The rising economy of the modern world

---

[1] A clear sign of this attitude was relationship with trees that offered firewood to ancient people [Brosse, 1989].

[2] Collected in many books, in Europe for instance: K. Recheis, G. Bydlinski: Weißt du, daß die Bäume reden (1983, 1995 – Do you know, the trees are speaking), Auch das Gras hat ein Lied (1995 – Also the grass has a song) – both in German; I. Sernec: Utrip ravnovesja (1995 – The pulse of balance), F. Burger: Kako naj vam prodamo modrino neba (1988 – How can we sell you the blue sky) – both in Slovene.

[3] One gets this holistic impression most easily when reading the original Roman texts. One interesting example can be found in writings by Plinius sr.: In order not to devalue silver in his treasury, the imperator Tiberius killed an inventor who first extracted aluminum from alumina clay, and ordered to destroy his laboratory. This same process was reinvented not earlier than 18 centuries later (in 1825) by C. Oersted who used the chemical process of electrolysis. Presumably, electric energy had been used also by the unfortunate Roman inventor.

looked upon nature as something that has no value *per se*, but is here only to be dominated and put under our control. Described by the words of the early protagonist of this ideology, *Francis Bacon*, "we are to torture nature's secrets from her ... nature should be put in constraint and bound into service like a slave" [Capra, 1982]. That self-assertive economy was seemingly even making a good profit of such a conviction – at least for a certain period of time, maybe a few hundred years. Subtle balance of energy flows went out of fashion.

With the advent of *industrial revolution* at the beginning of the 19th century, renewable energy sources (like wood) were no more enough to provide for the rising needs. Now great quantities of energy, concentrated in the form of chemical energy, were necessary to propel that type of economy. *Coal* came into wide use to drive *steam engines* and to heat new block settlements for industrial workers. A century later, another fossil fuel, *oil*, was put into force in *internal combustion engines* and provoked a far-reaching revolution in transportation. This change in human mobility was in so many ways based on the myth that each and every one is legible to *American dream* of absolute freedom, and that we can neglect the "side-effects" of this imaginary freedom[4].

In the last two centuries, *fossil fuels* overran *renewable energy sources* and took on very important role in the world's economy. Without them, this economy would instantly collapse[5]. It is often heard that oil is like the blood of modern civilization.

LAW OF ENERGY CONSERVATION

It was in that industrial time when the modern scientific concept of *energy* has been introduced. (Only some special forms of energy had been considered earlier). So it was the climate of industrial revolution, of engines, of the rising economy, that was in need of this new abstract term. The first generalized use of this concept is found in works of *J. R. Mayer* who in 1842 introduced the idea of *energy conversion* (conversion between different forms of energy), and *J. P. Joule* who further elaborated this idea and in 1847 determined *quantitative equivalence between mechanical energy and quantity of heat* [Lightman, 2000]. Shortly after that *R. Clausius* successfully introduced this idea into his "*mechanical theory of heat*" [Clausius, 1850] and in 1865 was the first one to use the word *energy* (in German *die Energie*) coined from the Greek word *energeia* (activity). Maybe this term came into use a little bit late, but instantly it became widely accepted.

In physics, *energy* is one of the most fundamental quantities, together with *joule* (J) as the basic unit for energy. With *mechanical work* of 1 J we can lift the mass of 1 kg up to a height of approximately 10 cm. Energy can be expressed also in terms of *power* and *time*, so *joule* is equivalent to *watt second* (1 J = 1 Ws). A greater unit for energy is one *kilowatt hour* (kWh) which equals exactly $3.6 \cdot 10^6$ J.

Energy is manifested in many different forms, e.g., like *mechanical work, kinetic energy, gravitational energy, internal energy* (*elastic energy, thermal energy, etc.*), *energy of heat, electric energy, magnetic energy* (energy of electric and magnetic fields), *chemical energy, nuclear energy*, and so on. Most of these terms have also their own different symbols, but belong to the same superclass named *energy* and have the same basic unit: *joule*.

---

[4] Problem brilliantly outlined in the recent documentary film The End of Suburbia (2004).

[5] As in May 2008, an emergent Wall Street crisis is taking place and fuel prices are rising.

The concept of energy plays a fundamental role in *theoretical physics*, and because it is so fundamental, it crept to every pore of our human endeavour. Probably the most basic law in theoretical physics is the law of the *conservation of energy*, in rudimentary form formulated already by Joule. Briefly, it goes like this: different forms of energy (itemized above) can be converted from one to another, but the sum total of all these forms is remaining constant. If it were not so, energy could be produced from mere nothing, which means that a *perpetual motion machine* (providing us with unlimited amounts of energy) could be feasible. Many alchemists from the middle ages were trying to construct such a machine, but in vain. However, just through efforts of this kind they paved the way to early mechanics[6] and to evolution of the energy concept. Observing and analyzing different forms of energy, one can enter quite deep into the secrets of Nature and deduce one more thing from this example: Although the modern concept of energy was born as late as in the middle of the 19th century, the platform for it had been in preparation through many previous centuries, until finally the new climate of the industrial time opened the bud [Capra, 1982].

In every branch of natural sciences (especially in *physics* but also in *chemistry*, *biology*, and in applied sciences like *machinery*, *electrical engineering*, etc.) calculations of energy conversions fruitfully yield to straightforward results. Introduction of energy simplified calculations in applied science and paved the way for solutions to many theoretical problems.

A century ago in his *theory of relativity*, *A. Einstein* derived the famous equation $E = mc^2$, expressing *equivalence of mass and energy*. This means that every amount of energy has its own mass attached to it, and vice versa, mass can be converted into various forms of energy. For instance, if we could somehow "weight" the daily energy consumption of the whole humanity (this amounts to approximately $3.8 \cdot 10^{11}$ kWh in one day) we should get mass of 15 kg. We cannot weight it in practice because this mass is, like energy consumption, dispersed all around the globe. However, in nuclear reactors where energy is much more concentrated, we can actually measure a slight decrease in mass (weight) after nuclear fuel (uranium rods) is extracted from the reactor! This tiny mass leaked out from the power plant through transmission lines – in the form of electric energy.

In *quantum physics* which was in constant development from the beginning of the 20th century onwards, the role of energy is even more important, it has somehow even a "mystical aspiration". For instance, it can be shown that the law of energy conservation is closely related to the fact that things do not depend on *absolute time*. Energy is deeply related to *time*. Interactions between particles (between *quantum states*) are formally described by *Hamiltonians* which are *energy operators* – a mathematical term used in the formal language of quantum mechanics.

Developments in theory of relativity and in quantum physics brought into use another form of energy: *nuclear energy*. As we have just seen, it is the most concentrated form of energy that we use today. It is sad to know that, like with most other forms of energy, it was first used and misused for the means of massive destruction (in *Hiroshima* and *Nagasaki*). We are still learning our lessons: what will come next? Or, are we forced to turn the spiral down and find the exit from aggressive interactions of our unbalanced patriarchal society.

[6] for instance, the vector calculus with forces, S. Stevinus in the 16th century.

MODERN FORMS OF ENERGY CONVERSION

Total human consumption of energy (together in all various forms of *primary sources*) today reaches the value of $1.4 \cdot 10^{14}$ kWh per year. To have a slight imagination of this large number, this value means *average power* of 16 TW (*terawatts*) for the whole world, or 2400 W for each single person among the present population of 6.7 billion people[7]. This is already 20 times more than our "natural" (bodily) consumption of energy, since natural power that our body gets by intake of food is only about 120 W [8].

Global *energy consumption* must be in balance with global *energy acquisition*, and the latter is distributed among the following *primary sources* [World energy resources and consumption, 2008]:

- a Coal 25%
- b Oil 37%
- c Natural gas (methane) 23%
- d Nuclear energy 6%
- e Hydroelectric power 3%
- f Biomass (wood etc.) 4%
- g Wind power 0.3%
- h Solar power 0.5%
- i Other renewable sources (geothermal etc.) 0.5%

The items (e) to (i) are *renewable energy sources*, while others are not. The first three items (a) to (c) are known as *fossil fuels*. *Nuclear energy* (d) is somewhat special: as it is neither a renewable source nor it belongs to fossil fuels.

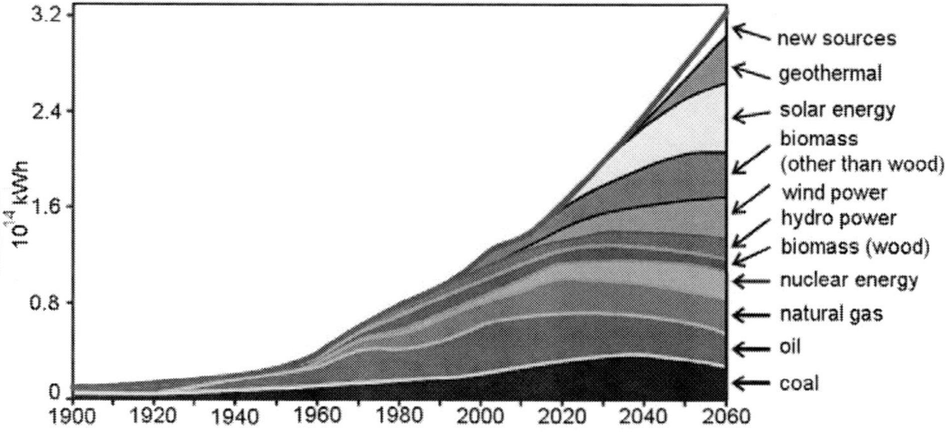

Figure 1. Global annual consumption of primary energy sources from the year 1900, together with projection into the future. The future scenario assumes 10 billion people in the year 2060 and annual economic growth of 3%.

[7] at the end of May 2008

[8] calculated from the recommended intake of "calories" (2500 kcal daily for a typical individual)

Figure 1 shows how our energy consumption has varied during the last century, together with the updated prognosis for the next 50 years. From the list, (f) has been used for millennia and (a) for a few hundred years, while all other sources are quite new, being in practical use for only 100 or 50 years (exceptions like hydro- and wind power used in some ancient mills do not change essentially the overall picture).

It is interesting to note for which purpose all this energy is used. *Industrial users* (agriculture, mining, manufacturing, and construction) consume about 37% of the total power. Personal and commercial *transportation* consumes approximately 20% (mainly as derivatives of oil in internal combustion engines). *Residential* heating, lighting and appliances use 11%. *Commercial uses* (mainly lighting, heating and cooling of commercial buildings) amount to 5% of the total. The remaining 27% of the world's energy is lost in energy transmission and generation.

From different forms of energy, *electric energy* can be used in most variable ways, so it is an important *secondary source* of energy. By power lines it can be transported over large distances and at the same time has no harmful emissions at the place of use[9]. Also, it can be regulated in many ways; there are numberless appliances running on electricity with nearly 100% *energy efficiency*, and this number of different applications is still growing, to a great extent due to new means of *electronic control*. Electric energy is thus the most precious form of energy, also because nearly 100% of electric energy can be converted to be used for any purpose. In contrast, only one part of energy hidden in heat can be converted into electric energy (the precise percentage depends on temperature difference in the engine[10]). That is why there is a slowly growing proportion of energy that is used in the form of electric energy. Today approximately one third of primary energy sources is consumed for production of electric power, however, due to energy losses in power plants only 13% of the primary energy sources are really converted into electricity. With regard to good applicability of electric energy, it would be ideal if all energy were used in the form of electricity – however, one must always ask how (using which primary source) do we produce electricity. As it stands now (2008) it is produced mainly from coal, from nuclear power, and from hydropower.

In the following chapters we shall see that all of these sources are limited in one sense or another. We do not dispose of an ideal solution providing us with clean electric energy because all primary energy sources are burdened with problems that are not yet successfully solved. This elementary problem goes quite deep, so at present one cannot see an ideal solution on the horizon.

LIMITS TO GROWTH OF ENERGY CONSUMPTION

Until recently, the limited amount of energy available for our disposal seemed to be a threat in principle, and not in everyday reality. Our Earth seemed to be so large and its wealth far from any danger of being exhausted. There were so many new lands yet to be discovered, so many visions of new hidden treasures, and so many possibilities! A dream age of

[9] However, there are some problems with electromagnetic emissions produced by high-voltage power lines and transformers, effecting on health.

[10] The best thermal power plants can reach 50% efficiency of energy conversion from chemical energy in fossil fuels to electric energy. Usually the efficiency is lower, up to 40% [Jäger et al., 2001; Lambertz, 2005; Sekavčnik, 2007].

technology ... Miracles like electricity, self-moving vehicles (cars), flying devices (airplanes), conveying of messages over great distances (telephone, radio, TV, internet), thinking machines (computers) and so forth were constantly popping out, anew and anew. There seemed to be no limitations to human creativity.

The twenty years following the Second World War were still full of prosperity. But then, in the sixties, a new awareness slowly began to creep in. It is difficult to say exactly which manifestation was the first, which announced the fundamental change. It was like a river that can be seen appearing from time to time from underneath the pebbles and disappearing again into the darkness until finally, in its full power and glory, it bursts into the light.

The age of prosperity had come to a kind of saturation. Tired of objective quantities, young people turned to new spiritual qualities. This was the golden age of trends like the *student revolts*[11] and the *hippie movement* [12]. Millions of young people listened to music and dreamed about a spiritually warm world without complicated economic and industrial systems. It was like a wave washing over a beach, overturning many pebbles, capable of bringing about many changes.

In the late sixties, the *Club of Rome* published something which sounded like a warning. It included curves, the results of many detailed numerical simulations, which showed the future development of several important variables, for example number of people living on our planet, rate of industrial production, quantity of food produced, reserves of natural resources (oil, coal, ores etc.), contamination of the Earth, or soil exhaustion[13]. The curves showed an exponential continuation of population growth and industrial production, but a growing crisis in all other areas. The world would gradually come to a state of saturation. Then, according to these calculations, something like a disaster should happen around the year 2015. Within a few years, the number of people would fall to a fraction of the previous number, industrial production would disappear completely and food production would barely be enough for the surviving part of humanity. As one might guess, the earth would be terribly polluted and exhausted, without many important natural resources.

The computers continued the calculations into a more distant future. Slowly, a new stability would be re-established, but this time by a smaller and invariable number of people, and still without any industrial production! Will this maybe present a restitution of the ancient forms of life-style?

There are possible objections to such a prediction. The computer did not take into account the new knowledge that this very analysis can provide, the new events occurring during our decades which span this critical period. People could say: Maybe we have enough time to change our habits and thus avoid the disaster [Gabor et al., 1981].

---

[11] Starting in Paris (May 1968) and instantly spreading to many other European and American universities.

[12] Music of that time opens a direct insight into these social movements. As explained in Abbie Hoffman's legendary book Woodstock Nation, or in the book And the voice to sing with by Joan Baez, the Woodstock music festival in August 1969 was a clear (but surely not the first) mark of the new awareness.

[13] The author first came across these curves in the French scientific magazine Science et Vie (approx. 1970) but they appeared also in many other serious publications.

Fossil Fuels

Now many people are asserting that this time of the great change has come. After a long period of worsening the environment of our planet, after many years of warnings in vain (numerous *ecologists* are quite active already for many decades), the recent signs of a *climate change* warn us that the situation is growing really serious. Finally the human impact on this dangerous process has been admitted and claimed to be scientifically confirmed, also by *Intergovernmental panel for climate change* (*IPCC*) under the umbrella of *UNO*. Detailed mathematical analysis shows that we are approaching the critical point beyond which this process would escape any control and become *irreversible* [IPCC, 2007].

The main mechanism behind the climate change is simple: 85% of energy that we use today is extracted from *fossil fuels* (coal, oil, and natural gas). Combustion of fossil fuels produces *carbon dioxide* (CO_2) and due to these emissions its concentration in the air has reached a dangerous value. Carbon dioxide is not transparent for the *long-wave infrared radiation* which escapes from the Earth and normally provides for cooling our planet to moderate temperature. This cooling process is now obstructed and global temperature is rising. Exactly this situation (called the *greenhouse effect*) is found on the *planet Venus*. Well, Venus is far from us, but *global warming* affects our planet. Glaciers and polar caps are disappearing; frequency of extreme meteorological events (droughts, floods, hurricanes, blizzards, etc.) is quickly growing[14]; ocean level is rising and huge areas located by sea are threatened to disappear under water waves. Higher temperatures may also induce great changes in *ecosystems*; therefore, new migrations of parasites would induce new immune diseases that we could not put under control so easily.

Certainly, there are enormous changes taking place on a global scale, among them a growing ecological awareness. All those who consider themselves important now speak about ecology: politicians use it to attract electors, producers to attract customers. It is almost the ideology of our time. But does this awareness really work? For instance, how are the conclusions of the *Kyoto Protocol* (1997)[15] reflected in our everyday life at this moment? Did we begin replacing gasoline cars with *hybrid electric cars* which consume much less energy? If not, we are in danger of suffocating ourselves in our cities, not to speak about global warming and other effects. Did we cease cutting down the rainforests, the lungs of our planet, swallowers of carbon dioxide? Without them, we cannot live. No, we did not cease to behave in such irrational ways, since each year the global forest area diminishes for a surface equal to the whole Greece. And now the basic question: Did we do enough to develop social institutions that would put under control the interests of capital over environment? Human sanity and health at first glance have nothing to do with the interests of capital, quite the contrary; however, for mere survival both should be in balance. Without green nature, we shall slowly lose our health and sanity.

With regard to ecology, a great deal of hypocrisy is evident. The *Third World* is striving hard to emulate the economies of the developed countries, often copying the problematic production processes which, forty years ago when nobody yet spoke about ecology, helped

[14] On 3rd May 2008 a cyclone swept away houses and trees in whole areas of Burma with 50.000 victims.

[15] The Kyoto Protocol is an international agreement linked to the United Nations Framework Convention on Climate Change. The major feature of the Kyoto Protocol is that it sets binding targets for 37 industrialized countries and the European community for reducing greenhouse gas emissions.

the rich world to strengthen its sovereignty. Now times have changed, but when the rich world is trying to make profit from the quickly rising economy of China or India, for the sake of profit nobody asks if these fast developing countries have the same ecological standards as the rich who had time enough to think about the privilege of ecology. In such a situation, ecology may even become a means of maintaining the superiority of the rich over the poor.

Things are far from being straightforward, so we must work hard to find appropriate solutions in time. Modern *economy*, understood as the science of optimal allocation of resources, should take into account new awareness about the real meaning of resources. Their complete value was not taken into consideration when, in the period of industrial revolution, economy has been established as a scientific discipline (by economists like *A. Smith* or sociologists like *H. Spencer*). If clean waters and air, large pristine forests, rich *biodiversity* of species, and other natural endowments leading to sustainable harmony of all living beings, were involved in economic calculations, then the activities of human society would be far different from what they currently are.

And, how is this present dilemma reflected in the area of energy consumption?

Serious problems connected with classical *fossil fuels* are already well known. They produce carbon dioxide, not to speak about other harmful emissions (*nitrogen oxides*, *sulphur oxides*, soot, *heavy metals*, dangerous *nanoparticles* ...). With regard to harmful emissions, *coal* is the worst, *oil* is better and *natural gas* is the best type among the fossil fuels. Fortunately, the trend of consumption actually goes in this very direction. The resources of oil will be fully exhausted in about 40 years (*peak oil production* may be reached quite soon, resulting in severe oil price increases) and the resources of gas in about 70 years. These numbers were calculated accordingly to actual trends in energy consumption and trends in discovering new oil-fields. The resources of coal will last longer, it is estimated that for about 165 years, but the peak of exploitation is probably approaching.

Natural gas is composed predominantly of *methane* (>98%) dashed with *heavier hydrocarbons* like *ethane*, so with high proportion of chemically bonded hydrogen it is the cleanest source among fossil fuels; hence with great hope also much money is invested to build the necessary infrastructure for its use – especially in Europe. However, its resources are also limited. Another difficulty is discussed below.

A problem with the fossil fuels is the fact that their resources are not evenly distributed; in the world they are concentrated predominantly in certain "hot" regions. The problem of oil is well known. Some of these regions are politically quite unstable, also because of relentless interests of the world's greatest consumers of oil whose economy is largely dependent on this precious liquid. Greed for oil (present in the rich world but also many new emerging economic tigers) on one side, and corrupted regimes (especially in the undeveloped world) on the other side, work hand in hand as the prevailing two hidden reasons for many local wars. It is difficult to discern one reason from another, especially because quite often greed promotes corruption and *vice versa*. International prices of oil are rising quickly and one dares to foretell not even the situation of the following year. Every prognosis is put on an extremely slippery ground. In general, we cannot rely on fossil fuels; from several reasons it is obvious that we have to find some alternative source of energy, and most definitely, very soon.

KNOWN ALTERNATIVES TO CLASSICAL FUELS: NUCLEAR ENERGY

Is *nuclear energy* such a promising alternative? After the disaster in *Chernobyl* (in April 1986) with more than 100.000 victims (estimated number counted over 15 successive years), there was a long halt in proliferation of *nuclear reactors*. There is still a lot of discussion whether reactors are safe enough or not. Here one should not put all the reactors into the same rang: some types of nuclear reactors are really far from being safe (like those in Chernobyl) while modern technologies offer much safer solutions. Naturally, in the present fear of CO_2 emissions, many lobbies who advocate nuclear energy have caught new wind into sails, since production of nuclear energy has no side-effect of CO_2. Thus, to a considerable part of experts, nuclear energy seems to be a better alternative [Lovelock, 2006]. Here we speak about most common nuclear reactors, running on *enriched uranium*. There are also reactors running on *natural uranium* (the so-called *breeding reactors*) and those are more dangerous because they produce huge amounts of *plutonium*, an extremely dangerous chemical element that remains highly radioactive for thousands of years.

No matter which type of reactor we consider, several unsolved questions remain. The first question is how to consider the annoying "side-effects" such as the problem of *cooling* and the problem of *nuclear waste*. Every nuclear reactor has to be cooled and one needs huge amounts of cold water for this purpose – but shortage of water is a growing problem in most parts of the world.

Nuclear waste is another question without a clear answer. After extraction from the reactor, nuclear waste is highly radioactive still for many years and must be deposited for a certain period beside the reactor. After this first period, one tries to store it permanently in a safe place – the best solution is to put it into protected shafts very deep under the Earth crust. But it is not so easy to obtain legal permissions for a location just everywhere, since nobody wants to have nuclear waste "in his own yard". Transports over long distances are often necessary, and who can guarantee that these transports will suffer no unexpected accidents? The whole nuclear system and all other logistics connected to it are very complicated. The problem is not only technical; it exists also on the social level. At present nuclear waste is often transported to undeveloped countries far-away, which is, from ethical and ecological viewpoint, totally inadmissible. If situation is so confused, who can protect nuclear systems from terrorist attacks, without fail? When a reactor is finally extinguished, it remains radioactive for more than 1000 years – who can guarantee that for the next millennium there will be no upheavals in that region? In former Yugoslavia we have seen that crazy militia often attack the most vulnerable points. Whether we like it or not, it is a total illusion to expect that people will always behave in a sane manner. Notwithstanding the human factor there exists possibility of an earthquake which also could endanger nuclear safety[16].

Another question related to safety of nuclear power plants is, in a certain way, the social extrapolation of the problem outlined above. A nuclear power plant is a concentrated technology and at the same time an extremely *complex system* with millions of different parts that must function together in harmony. One single man cannot have a complete view over all this, so a great deal of confidence among different actors in play is required. But today when

[16] The earthquake on 12th May 2008 in Sichuan (China) was a serious threat to nuclear reactors in the area. Luckily, none has been seriously damaged.

the interests of capital are prepared to manipulate with reality this confidence is heavily threatened: whom can we really trust? Can we know if this or that man speaks sincerely and takes into regard all the necessary facts? One man cannot know all the facts himself, therefore, are all other actors embroiled into the play also sincere? One can never get a clear answer to such complex raw of question – simply because of the *system complexity*, and nuclear systems *are* complex.

KNOWN ALTERNATIVES TO CLASSICAL FUELS: RENEWABLE SOURCES

(A) Hydro Power and Wind Power

Let us turn now to *renewable energy sources*. What can we say about items from (e) to (i) on the list above?

Hydroelectric power plants are still in favour but it is worth to note that today people have become much more sensitive to possibility of devastating large natural areas by turning them into water reservoirs. A better solution that can get broader public support is a decentralised network of many small hydro power plants ("micro hydros"), each one producing several kilowatts or even less than 1 kW of electric power. We can use brooks in the same way than the constructors of the old mills did, namely without considerable earthworks. A lot of electric power is still hidden in these unexploited water streams. But decentralisation is essential. Nature is sensitive and vulnerable; if our intervention into nature follows the same lines of sensitivity and vulnerability, if we restrict ourselves to extremely careful and limited size of our interventions, then harmony can be sustained and protected.

Large hydro projects are much more problematic. The hydroelectric facilities of the great European or North American rivers are mainly exhausted. There are some facilities left, but public opinion is against turning them into stale lakes. In the economically developed world the times of sacrificing everything for the sake of "progress" seem to be over forever, although interests of the capital are time and again trying to undermine the awakening eco-awareness of public opinion.

A more unfortunate situation is found around the rest of the globe. Eco-awareness has not grown to such an extent as in the developed world. In Africa there was an ecological and cultural *fiasco* with *Asuan dam*, or more precisely, with the politically imposed way how the dam had been prematurely constructed without a serious analysis of all possible alternatives. Great dams on South American rivers (e.g., *Itaipú dam* on *Paraná river*) have destroyed not only large rainforest habitats (extremely rich in *biodiversity*, therefore a great part of our global gene-bank), but also many Indian tribes living there for thousands of years and whom nobody asked for permission. A comparable event is now threatening India (*Narmada river*) and China (*Yangzi river*). In such huge political projects, often millions of people are pushed away from the lands of their ancestors. Even more crazy case is the Chinese controversial project of "taming" several great Himalayan rivers like *Brahmaputra* (even changing their flows eastwards to populated parts of China). This project is extremely problematic and risky; it is already strengthening international tensions between India, Bangladesh (both dependent on Himalayan water), and China[17]. From such examples, one ventures to say that the world's supply of hydropower is already approaching its reasonable limits.

From more or less conventional energy sources, we are now going to describe those that are still under intensive development. There is another possibility of hydropower: it could also be extracted from *ocean currents*. This is something new. Great Britain, endowed by several strong and steady sea currents (e.g., in *Pentland Firth* south of the Orkneys), has plans in this direction. Presumably all actual needs for electric energy in Britain could be covered this way. Great turbines are meant to be implanted at the sea bottom. This is an interesting alternative but further studies must show whether it is feasible, ecologically acceptable, and economically profitable. Power plants using *tidal currents* (e.g., in the French *estuaries*) could also be ranked into this class.

Wind power is, next to solar energy, actually the fastest-growing energy source in the world (with the rate of increase of about 25% per year). It is becoming increasingly interesting in those countries that are endowed with strong *and* steady winds – for instance in plains lying by open sea. At present, it seems that wind power is worth of investments only in such favourable places, still, it is highly useful and promising. *Wind turbines* have been improved a lot during the last decade; now they are more efficient, more endurable, and cheaper. The amount of electric energy produced by wind power plants is rising exponentially and will soon reach 1% of the total electric power produced in the world. At present ¾ of the world production belongs to Europe, and one half of the European production belongs to Germany, the rest to Spain, Denmark, France, etc.

When one evaluates if locations are convenient for *wind fields*, one criterion to consider is economic profitability, another criterion is ecological acceptability: the wind fields should do serious damage neither to natural habitats nor to landscape. We mention this because there are many cases of corruption observed. Producers of wind technology and men of construction business are often lobbying mayors, politicians and other responsible people to get licenses for construction, even in such cases when the wind field would squeeze out much more invested money than electricity. We must be aware that ecology is often only a seductive façade, an excuse for behaviour without any ethical attitude. In the author's opinion, in such cases "eco-business" is leading us into a blind alley.

Dealing with Complex Systems

A great knowledge is needed to make someone capable for decisions that contain also the ecological aspect, because ecology is always dealing with *complex systems* where only a great deal of interconnected data can give the overall picture of reality in such systems. In any evaluation of energy projects, we need at least the so-called *life cycle assessment* (*LCA*) that takes into account every possible ecological impact on the environment [ISO 14040, 1997; Frankl et al., 2000]. What is even more, sincerity and ethics are also needed – the same precious qualities that we have met in connection with nuclear safety or responsible decision-making in dealing with any other complex system.

In our creative work we are dealing more and more with the issue of complex systems, for example: a natural ecosystem or technological system or human society or any

[17] Can this be interpreted as a warning that in the near future wars for water may supersede present-day wars for oil? Lack of water is a serious global issue; today more than a billion people do not have access to water, suitable for drinking – and this number is quickly inflating (accordingly to actual trends: 2.4 billion in 2025).

combination of all these. In our approach to the complex systems, human values like *ethics* and *sincerity* are of a paramount importance. If we loose these values, we also loose every *criterion for the truth*. It is a total illusion to believe that observation of merely technical facts will give a complete answer to our questions. Our human *instrumental mind* that is so much adored in modern times is simply not enough: We also must involve the genuine *human sensitivity* that is functioning only through *intuitive, holistic approach* to what we dare to call *reality* [Bohm, 1998; Krishnamurti et al., 1977; Thakar, 1971; Burden, 1975].

(B) Solar Power

Let us continue our analysis of renewable energy sources with *solar power*. Solar rays are unceasingly bringing 1370 watts of radiation power per every square meter located slightly above the Earth atmosphere. This number is called the *solar constant*. On a good day without clouds, an odd half of this energy reaches the land below. Well, there is day and night and there are days of bad weather, but one can still collect several *kilowatt hours* of solar energy per day and per square meter of Earth (about 3 kWh, this number being largely dependent on geographical latitude, season, and weather).

This energy is useful especially when it is converted directly into *heat*, with purpose to warm our living interiors in *solar architecture* or to warm sanitary water in *solar collectors*. Systems of this kind are rather cheap and have very good efficiency (nearly 100%) of energy conversion. Solar energy is free and produces no side-effects, so it is clever to replace classical heating systems (running on oil or other fossil fuels or even on electricity) by new systems that use solar power. Every sane government supports investments in this direction. It is necessary to note that, like with any new technology, a lot of knowledge is needed to make these systems function properly. For instance, precise calculations of time-dependent energy flows in the whole complex thermal system (for instance, a private house) are necessary, taking into account also new materials (like special *greenhouse glasses* or modern insulating materials or *heat-storing materials*). On this condition, solar systems are efficient and therefore fully worth of investments.

If solar radiation is concentrated by an array of mirrors, we can reach very high temperature in the focal area of such an optical system. Water can be easily vaporized and used either for cooking or for powering steam turbines and thus producing electric power. Many small experimental setups of this kind already exist in the world. One of them is the *solar kitchen* for 2.000 people in *Auroville*, South India.

Solar energy can also be converted directly into electric energy. This is done in the *photovoltaic cells*. They are still rather expensive (due to present lack of producing facilities in *silicon refinement* technology) and have energy efficiency below 20%. Therefore it is wise to convert solar energy directly into heat wherever heat is needed in our buildings, and to use photovoltaic cells only for energy supply of those appliances that cannot run without electricity (lights, electric motors, computers, audio and video systems). The sector of *solar electricity* has annual average growth of 35% over the past ten years and is now the fastest growing sector of energy conversion [Worldwatch Institute, 2007].

A house covering most of its energy needs by solar power is called an *eco-house*, while in a *passive house* practically all energy needs are covered by sun. Further support to this kind of architecture could reduce our total energy consumption for at least 20%, and this without very great investments. This is an example of an appropriate *systemic approach* – energy flows inside our human society are woven into a complex network and it is worth to select those knots in the web (and then act upon them) which maximally improve the long-term stability of the network, with the minimal input from outside.

Although they do not belong to this section of solar power technologies, let us now briefly mention also some other renewable energy sources. *Geothermal energy* is welcomed wherever it comes close enough to the Earth's surface (in Iceland, for instance), otherwise investments into shafts and pipes are usually still too expensive[18]. One can use also the *energy of the ocean waves* but these systems are still in development. The same holds for electric power plants running on temperature difference between cold water at the ocean bottom and warm water at the surface. Here the upper temperature never reaches 100°C, so in such a power station steam is not produced from water but from some other chemical agent. Since a power plant of this latter type has very low efficiency (below 5%) great amounts of warm and cold water are to be mixed together. And this may again be a great disturbance in global ecosystems (oceans are already damaged quite a lot), not to mention huge installations needed.

(C) Biomass

Let us conclude our survey of practically used energy sources at the point where we have started – with wood and other *biomass* that has been used by people for many thousands of years up to this day. Can this conventional source furnish us with some new possibilities? In principle *yes*, but again only in a limited sense. Surely we can use dry wood in our sacred fireplaces like our grandmothers have done for so many years. Surely we can incinerate most of organic trash from home and industry. We can also organize some new logistics in order to collect one part of wood (and other naturally grown organic material) that is today left over to rot in natural environment. Sincere ecological sense can always tell us how far we can go in order not to deteriorate healthy *natural ecosystems*. Finally, we can also explore if biomass can be grown in a new way, for instance by species like some special kinds of *algae*, etc.

But surely there are certain very clear limits that we should not transgress [Lovelock, 2006]. If one is devastating natural forest areas in order to turn them into plantations for biomass of any kind (like it is done so much in these days in many countries that have been so quickly enslaved by "modern" economy, from Chile to Indonesia), then this does not have any fragrance of eco-awareness but only of human greed with catastrophic consequences.

Today there is a lot of discussion about bringing *biofuels* into use. In many countries it is planned (or it is already in use, like in Brazil) to replace a certain part (10% or more) of classical fossil fuels (petrol or diesel oil) by naturally grown alcohols or oils from various plants (cane, sugar-beet, corn, oil-rape, hemp, sunflower, etc.). All of these plants are not equally convenient for the purpose. It seems that we spend more fuel in production of some biofuels than we get back, while other plants seem to turn back more than we put in. So again a lot of knowledge (and sincerity!) is needed here in order to make a reasonable decision. And especially, one must be extremely careful not to wrest plantation area from healthy habitats

[18] We speak about geothermal energy in its strict sense, therefore excluding heat pumps that are not energy sources but only devices using a special thermodynamic process of energy conversion.

and against the ways of people living there. Such miserable cases are many; they are done only for the sake of profit, disregarding all the rest (ecology, ethics). In long term such actions lead to barren land, to uncontrollable erosion, to misery of all those living there. So there is no place for hypocrisy in the name of "progress".

In brief, we cannot say that use of biomass is bad in principle, but one must be careful and should sensibly observe reasonable limits to our allowed interference into natural habitats. These limits are probably quite near (much nearer than it seemed to insensitive mind even a few years ago), so all the classical fossil fuels that we use today surely cannot be replaced by biofuels. If something like that happened (from any reason, for instance from manipulative interest of a certain GMO company), the prices of human food would quickly rise up so high that great famine and social instability would certainly follow. In fact, this is already happening: In the last three years the prices of food on global market have risen for 83% (and even much more for three basic nutrients: rice, wheat, and corn)[19].

(D) Hydrogen Economy

In the last few years, several visionaries are promoting *hydrogen economy* [Rifkin, 2002]. The main idea is to build a decentralized network of small power plants running on pure *hydrogen gas*. Chemical energy of hydrogen can be converted directly into electric energy inside a *fuel cell*, an electrochemical device similar to a battery stack, therefore with very few moving parts. The energy efficiency of this conversion is reasonably good (above 50%) and the only emission is water vapour which is already a natural part of our atmosphere.

The basic mechanism of reaction is simple. Hydrogen diffuses through a *semipermeable membrane* and then reacts with oxygen from the surrounding air; the product of this reaction is pure water plus electric energy. Electrodes that collect electric charge generated in this process are placed on both sides of the membrane. It seems that after 165 years of development, fuel cells are now approaching the stage of maturity.

Hydrogen is *not* a *primary energy source* available in nature. Therefore a good question is where to get hydrogen. There are several possibilities. The first one is to produce it from oil or natural gas (by a chemical process called *reforming*), and although here we are returning to oil or natural gas again, this solution (combined with use of electric vehicles) has at least two advantages over classical use of oil: much better efficiency[20] and clean urban areas [Lampič, 2006].

But there are also other ways to hydrogen. It can be extracted from *biomass* (wood etc.) by a similar reforming process. Let us repeat once again that biomass is all right if it is gathered from natural environment in moderate quantities that are provided by natural cycles, but its exploitation may become a human disaster if large natural areas (like forests or fields with crops for human food) were turned into organized plantations for biomass. For example, we can criticize one proposed scenario: gradual replacement of all gasoline cars by electric vehicles, powered by hydrogen from wood and other biomass. This scenario is already above

---

[19] Moreover, just now (in April 2008) these prices are spiralling out of control.

[20] Energy efficiency (from chemical energy of fuel to mechanical energy of movement) of conventional cars in urban driving cycle is usually below 10%, while energy efficiency (from hydrogen to movement) of electric cars in urban driving cycle is in the range between 40% and 50%.

the limits of what Nature can yield in a balanced way, not to speak about the fact that vehicles consume only one fifth of the overall energy consumption.

The third possibility is production of hydrogen by *electrolysis of water*. For this purpose a great amount of electric energy is needed, so this way is all right under condition that we develop a network of "clean" electric sources (nuclear power or renewable sources like solar power, wind power etc.). Without doubt, hydrogen economy would amply stimulate development of clean primary sources; both go together hand in hand [Sekavčnik, 2007].

So we see that hydrogen economy is inseparable from other solutions to our energy problems. Within a sane network of clean energy sources, hydrogen can surely play a significant role, since it can be easily stored (much easier than electric energy). Emergency of hydrogen-based economy will also promote global public participation in the energy network; this new social context is highly welcomed. Consumers of energy may become also energy producers – just like in case of information technology, where each computer is both a source and user of information [Rifkin, 2002]. Vivid human activity may automatically promote responsible attitude towards issues of energy and ecology. So let us consecrate our endeavour towards edifying such a sane network.

Looking into the Future

Among energy sources that we are using now, not even a single one is quite free of negative side-effects. Fossil fuels are responsible for global warming, for political instabilities, and so on. Nuclear power, with its complex technology and retarded radioactivity, seems to be suspicious (dangerous, not well controlled) in our manipulative society. Hydro- and wind power can deteriorate large natural areas. A similar problem arises with excessive exploitation of biomass. Solar power, when used for production of electricity, is functioning only in daylight and requires an expensive investment. Other renewable sources can be used only in certain rare spots of our globe.

The whole world is crying for development towards a noble standard of living and human dignity. Countries in development have the same right to progress as those who have already consumed this right and are now prosperous. But it seems that this world process can never bear fruits of fulfilment as long as it is subtly controlled by economic system based on unreflected human greed – an economic system that has itself escaped every human control. Nowadays it has enslaved even the richest individuals on our globe: no one can affirm to be joyfully released on the rack of dehumanized economy.

Our Earth has certain capability of restoring its eco-stability (by chemical and biochemical processes in natural cycles, etc.), but today the ecological burden that we have produced is definitely above the allowable limit. In the end, this may soon lead to a total collapse. So, what can we do? If we are quite sincere, there are three different possible scenarios for our future [Detela, 2002]:

1. We shall not respect natural limits to energy consumption, but shall continue to expand artificial needs (in the name of production and profits), waiting for a miracle. But this miracle will not come. Accordingly, a collapse on the global scale will follow.

2. We shall return to the lifestyle of the pre-industrial age, with very moderate energy consumption. However, this is the least likely possibility because human pride is so strong that man would accept this drastic discipline only after the onset of a total collapse.
3. We shall continue to profit from science and technology, but in a very different sense. A new awareness of the fact that unbalanced economic "progress" does not make people even a bit happier will finally emerge. So everyone will turn deep into questioning what human body and soul really need. In time, we shall develop such new activities and production processes that use no more energy than necessary, and we shall combine this new attitude with very careful choice of those products which are really necessary. If we follow these new rules that regulate sustainable societies, a global collapse is avoidable. Our pride will admit this change, because even more knowledge, greater technological perfection and spiritual maturity will be needed to make such a change possible.

Every reader will agree; that of the three possibilities stated above, the first one is not really a solution to our problems. And the second possibility is very improbable, because it goes against one of the basic rules of human history: the same path can never be trodden twice. Even if there is an attempt to do so, there are obviously different *initial conditions* and they will necessarily lead to different *end results*. We cannot pretend that nothing has happened in the last few hundred years; we cannot hide all that we have learned during that time. We are also conditioned by our recent past and so we must learn to live with it and digest it in the best way we can.

By the author's and many other people opinion only the third possibility takes these new initial conditions into account. This solution is equally optimistic and realistic, and it is the only possible one. We see that it is very heterogeneous, a combination of many small *partial solutions* – but what is uniting them, is new ethical awareness emerging in this turbulent era. We are going to have some more look at this last scenario.

Let us assume the obvious transformation in global consciousness, let us assume that humanity is striving for happiness along the way of wisdom and of new global ethics. Let us assume that everyone is aware of a simple fact that true happiness and serenity lie at the very source of our existence, not at the end of human endeavour. On condition of accepting the obvious transformation, our life gets simplified and new insights follow quite naturally. We shall resume them in the conclusion.

SAVING ENERGY

Since our energy resources are limited, an important factor in any kind of energy use is *energy efficiency*. It is extremely important to use energy in such a way that energy loss is minimized at each step; that always and for every purpose *the minimum possible amount of energy* is consumed. It can be shown by general analysis based on theory of *dissipative systems* that *energy saving* is indispensable for long-term stability of every self-sustaining complex system [Prigogine et al., 1984; Davies, 1995; Capra, 1996].

Surely this requirement dictates many novel technological solutions, like design of *power-saving devices* or "intelligent control" of energy flows or better thermal insulation. There can be numerous small changes made, each one reducing our energy consumption just a little bit, but all together they contribute quite a lot. It is estimated that, by various careful adaptations of this kind, we could reduce total energy consumption to one half or even less of the actual value. This is so easy as soon as we brush aside the modern myth of all-perfect, almighty man – yet totally unreflecting and irresponsible, capricious with limitless material possessions (by the author's and many others opinion). A great part of our consumption is just for the sake of showing-off, such as using landrover cars to drive in the cities.

Such an artificial, inflated reality doesn't bestow on one even a smidgen of wisdom. *Wisdom* (as understood and practised in many primordial cultures, even the ones which survived till today) demands a genuine contact with Earth. We must not deceive ourselves with the attractive illusion that modern technology, which today's world so blindly trusts, has liberated us from the trials that we have to go through (as human souls). Mystification of science and technology can be dangerous, because it blinds us, and we're no longer alert to the true problems of our time and push the "progress" for too long in the wrong direction. Of course, we don't need to reject all of the achievements of the previous centuries, but they alone do not suffice. If we illuminate them with the insight of wisdom, it is soon revealed that we, human beings, have actual need for a very few of the innumerable possibilities at our disposal. The rest are only a seduction and a burden. This is clear to everyone awakened in the dawn of the new (forgotten) wisdom.

On this condition, and *only on this condition*, it is permissible and also quite possible that totally new energy sources will be discovered, developed, and put into everyday use. Without this condition, there is a probability they would be misused again and again, infinitely.

So let us be optimistic. Let us assume that people are basically sane and wise. Let us enumerate several new possible energy sources that are now only in the stage of development, with hope to give energy in some more or less distant future. Some of them are in the focus of strong international groups working in the mainstream of modern scientific research, while some are on the level of solitary speculations by researchers pushed out of the "orthodox science" to an obscure margin. As learnt from the past, both possibilities should be taken seriously.

PERSPECTIVE FOR NEW OUTBREAKS: NUCLEAR FUSION

First, let us jump into the mainstream science. Among scientists, there is a great hope of taming such a kind of nuclear reaction that is continually taking place in the inner core of the Sun or any other active star. In this reaction which is called *nuclear fusion* two hydrogen nuclei fuse together into one single nucleus of helium. There are several possible paths of this reaction, in laboratories the path usually starts with a pair of *deuterium atoms* (deuterium is a heavy *isotope* of hydrogen). Nuclear energy liberated in this process is about one million times greater than the energy of typical chemical reactions. Natural resources of deuterium in our biosphere are practically unlimited. Another advantage (in comparison to nuclear reaction in uranium that we use today in our reactors) is much lower *residual radioactivity*, therefore also absence of those hard problems with nuclear waste that we face in up-to-date nuclear power plants.

The process of nuclear fusion is taking place (as much as we know today) only at very elevated temperature (at least 40 million degrees centigrade) and only at high density of hydrogen. One method to cope with these extreme conditions consists from a continuous series of minute nuclear explosions (similar to those in a hydrogen bomb although much smaller), produced by intense laser beams focused onto small pellets of solid fuel containing hydrogen.

Another method, with even much more research done on it, is trying to tame nuclear fusion inside a larger volume of space. Scientists, for instance those united under the umbrella of the international projects *ITER* and *DEMO*, are trying to reproduce the above-mentioned extreme conditions inside a special *fusion reactor* [ITER, 2008]. Reactor of the type that is in the focus of research has an interesting shape of a *torus* (shape like a doughnut), with walls made of special ceramic materials resistant to high temperature and to the obvious radiation. High-temperature plasma is kept asunder from the walls by very strong magnetic fields. *Magnetic confinement* of plasma inside the torus is a great problem but seems to be solvable. An arrangement of *superconductive coils* (again of an interesting shape) is designed for this purpose. The whole system needs also a *cryostat* that cools these large coils to very low temperature of liquid helium, in order to keep the coils in the quantum state of *superconductivity*. From both temperature extremes and from other details, one can imagine the amazing complexity of the whole assembly.

An ever increasing series of such experimental reactors has already been constructed. With the latest (and the biggest) experimental setup it seems that the amount of energy produced by nuclear fusion is already close to the energy spent to run the whole system. So there is hope to fulfil the obvious prerequisite for practical use of this new energy source: to produce more energy than invest it in the first place. This does not mean that nuclear fusion will provide us with energy in just a few years. The whole system is so complex that at least 20 or 30 more years (according to actual expectations) are needed to develop all the details up to the point of practical use. However, in any case this direction of research is promising since the goal is an elegant new technology. We are not yet at home with it, so only future experience will be able to tell us whether our hopes are realistic or not. A complex system assembled of numberless parts that are still in development may always be a hiding place for inconvenient side-effects – effects that are not yet visible but later can spoil the feasibility of the whole system. So, like every daring project, also this one represents a great challenge and simultaneously a certain risk of final success.

Now let us see what trials are taking place outside the "mainstream" science. In 1989, news spread from several laboratories[21] that nuclear fusion had been successfully realized even at room temperature, so this reaction was nicknamed as *cold fusion*. The hypothetical reaction took place in *palladium* (a precious metal similar to platinum) saturated with *deuterium nuclei* so palladium crystal lattice acted as the *catalyst* for nuclear reaction and high temperature was no more necessary. The physics of this catalyzing reaction was not totally understood but knowledge from other branches of physics (like *semiconductor physics*) hinted that several important parameters in reaction (like the *effective mass* of

[21] Starting by M. Fleischmann and S. Pons from the University of Utah

hydrogen nuclei) may change for several orders of magnitude, and thus trigger the nuclear reaction.

Deuterium nuclei are brought to palladium electrode simply by *electrolysis of heavy water*, so the whole system is in fact a simple *electrolytic cell.* Amazing simplicity of the new "nuclear reactor" that can be assembled in every kitchen or garage would be an enormous advantage of cold fusion over the conventional "hot fusion". So quite obviously, cold fusion has instantly become the "hot" subject of research in many laboratories around the globe. Although the success of the reported experiments was mixed, in the following two years 92 groups from 10 different countries reported that they liberated energy and some other products of the nuclear reaction. However, it seems that this cold reaction (if it exists at all) is anomalous in many ways, since it does yield some of the expected products, but not all of them. So the certainty of cold fusion remains ambiguous [Collins et al., 1993].

If we search only through the most "serious" and respected scientific magazines, we find more than 100 papers on this issue up to this day, regardless of the fact that several times (for instance in 1991, at a regular conference on cold fusion in Italy) cold fusion was declared to be "a typical example of premature enthusiasm without adequate scientific basis". Later it was again whitewashed, especially after approval of the *American Dept of Energy* in 2004 [ADOE, 2004].

A new branch of science named *condensed matter nuclear science* was established and is still in progress – although without a convincing theoretical background, without a complete understanding what is going on, and especially, without any firm experimental proof. Such a proof would be if one knew all the necessary experimental conditions for cold fusion and be able to repeat the experiment. However, we do not have this knowledge today, thus we do not know whether the "successful experiments" are only an experimental error or, on the contrary, a genuine *exothermal nuclear reaction*, but with uncontrolled (somewhat accidental) conditions of success.

EXOTIC ENERGY SOURCES: FREE ENERGY?

In description of our human quest of new energy sources, with hypothetical *cold fusion* we are already transcending the well-defined solutions based on conventional knowledge (knowledge based on existing scientific theories). Now let us peep for one more step further, into even "wilder" alternatives.

We often read about many unconventional proposals, usually quoted as sources of "*free energy*". Everyone who types "free energy" into some search engine will get hundreds of hits on the screen. We shall mention some of them in the following paragraphs. One is confused since most of these hits are, at least at the first glance, against the basic *physical law of energy conservation*. An obvious question arises here: Where does *free energy*, available everywhere and to everyone (and presumably without any impact on environment?), come from? However, many inventors are convinced that they have designed a free energy machine, even more; they claim to have one functioning. Sometimes they call it "over unity machine" since energy efficiency is presumably more than 100% – such a machine should give out more than it gets in. So, is there a flaw in each and every one of these wonderful machines (or, more precisely, in the inventors' understanding of them), or is our present-day knowledge of nature inadequate?

Closer observation of free energy devices shows that all of them cannot be thrown into the same pot. Let us first look briefly into the first pot – the one with suspicious proposals. Quite sure, many proposals in this area are not very serious; they reveal lack of scientific knowledge and appropriate method. This means, many "inventors" without long scientific training are quickly convinced of a vaguely apparent success and plunge into self-deceit of experimental error, before working out all necessary details and before testing (theoretically and experimentally) the device accordingly to numerous influences. Usually such unripe devices are presented to public as "*perpetual motion machines*" and the "inventors" are not able to give the answer where the energy comes from; or if it really comes from "nowhere", in which respect does this sensational discovery change the whole structure of theoretical physics – quite definitely, perpetual motion machine would shatter it completely!

From large collection of proposals in so many different directions and based on so many different theories (in case of theoretical background), it is not easy to evaluate all of them and to separate the husk from the grain. The case of cold fusion discussed above is more familiar with our conventional knowledge but, as we have seen, even this one escapes clear evaluation. Thoroughly honest arbitration needs both detached and daring approach, and this takes time, so the choice of "promising proposals" (among hundreds of others) to be sorted into the second pot, is always somehow subjective.

It is not always clear what the term *free energy* does mean in every specific case; but often it means *zero-point energy of quantum vacuum*. Such proposals have therefore some background in modern physics – often as sophisticated scientific theories of the highest rank. So, although in this case we have at least some analytical tools, analysis is often far from trivial.

Zero-point energy may be explained in various terms like *gravitational vacuum energy* or *electromagnetic vacuum energy*. One experimental proof of oscillations in quantum vacuum is the *Casimir effect* in quantum physics[22]. However, we still do not know how to suck this enigmatic energy to our everyday world – if this is fundamentally feasible at all. Quantum physics as we understand it today does not allow sucking *zero-point energy* – hence the name!

Let us mention a few interesting remarks on *"electromagnetic free energy devices"*, the most "popular" subclass of free energy devices. The theory behind them is not simple, especially when we know that even the basic theory of electromagnetism that we use today (in the form of *Maxwell's equations*) is only a simplified (and pruned) version of what ingenious *J. C. Maxwell* used in his original papers. Several decades later *N. Tesla* did a lot of effort to bring the electromagnetic "energy of ether" down to Earth; most of his legacy (200.000 sheets of paper) has remained above comprehension up to this day [Tesla, 1978; Cheney, 1981][23]. In modern times, there are many inventors who claim that they have designed and constructed an electromagnetic free energy device [Childress, 1995][24].

[22] A minute force acting upon a tiny plate has actually been measured when the plate was exposed to "vacuum oscillations" asymmetrically from both sides [Ballentine, 1990].

[23] An early pioneer of ideas on the subtle structure of the Universe, and who amply influenced Tesla, was Ruđer Bošković, a great 18th century scientist from Dubrovnik [Bošković, 1745, 1758].

[24] For each case mentioned in the following paragraphs, the reader can get more information from the web, simply by typing "free energy" + respective name.

A frequent example that one can often read about is a hypothetical *magnetostatic motor* running on free energy (*Takahashi engine*, *Kawai engine*, *Johnson engine*, etc.). Its vital part is a complicatedly arranged set of permanent magnets in both rotor and stator; some versions have also auxiliary electromagnets. The inventors claim that positive mechanical energy can be extracted from the magnetic energy of the system as rotor magnets make one complete cycle of movement. However, inventors usually remain silent about where the magnetic energy is extracted from – or hint that the classical formulation of Maxwell's equations is inadequate.

Similarly, the machine constructed by *EBM technology* in Budapest is a rotary electromagnetic machine, and presumably it is running after being disconnected from any external power input. There are more hypothetical inventions of this kind: *T. Bearden* has patented a "motionless electromagnetic generator" and published a scientific paper on it; and so on.

Another device (named *methernitha*) was developed and constructed in a remote Swiss village; the inventors are laying great stress upon ethical awareness of their endeavour and close relationship with Nature, our great teacher. This sounds very sane, promising, and serious. Methernitha is not a magnetostatic machine but an electrostatic one. It seems to be a special (and highly advanced) kind of a *Wimhurst machine* running on *atmospheric electricity* – in this case it is not a real "free energy" device but is still fully worth of respect and consideration, inseparably together with its spiritual message.

Still remaining in the class of hypothetical electromagnetic free energy devices, we can mention also those based on exotic effects in plasma. *P.N. & A.N. Correa* and *H. Aspden* claim to have made use of *pulsed abnormal discharge* in plasma, *N. Graneau* used water-plasma explosions in cold fog, *T. Mizuno* tamed an anomalous kind of plasma electrolysis in water, etc. In the device developed by *Blacklight Power* hydrogen atoms yield energy presumably by transition to a quantum state with energy *under* the conventionally recognized base state. All of these proposals are supported by respective patents and scientific papers[25].

The author of the present paper could not find a clear answer whether any of these machines can really function without any external power source – where "functioning" means "continuous supply of electric or mechanical power to an external load". Usually we need some additional power supply in order to provide for appropriate temperature and other working conditions (at least initially). Besides, a lot of electronic equipment is always needed in indispensable testing procedures; and all these measuring boxes also use some electric power. Was it unambiguously assured that some of this power did not leak back to the tested device? Did the inventors try to cut the whole system off from any external source, just to see if it can run by itself?

Despite the confusion with all these seemingly queer proposals that cannot be sorted into any ordinary theory, it is not very clever to push all of them into dark shadow of "*marginal pseudoscience*". If they are not standing on even terms with "*official science*" then this is not yet a legitimate reason for *disqualification a priori* [Kuhn, 1962]. A famous quotation from *T. H. Huxley* reads: *"It is customary fate of new truths to begin with heresies and end as superstitions"*. Here we are standing in untrodden land without signposts and no one knows

[25] They were not cited here first because of space limitations, second as this paper is not primarily meant as a scientific overview but more of a personal conclusion of the author – who nevertheless is sincere in his studies of the field.

definitely the right way. Therefore, each separate case should be observed and analyzed seriously and with all due respect, not only by solitary inventors but also by the global scientific community – as far as it is possible in this delicate situation. If we proceed in this open manner, unserious proposals will be quickly spotted and discarded, whereas the serious ones will be given deserved attention. This light of open consideration can dispel the darkness of the present situation.

CONCLUSION

Within modern human society, *energy sources* and *energy drains* (consumers) are of many different types; they are interconnected into a coloured *energy network* that is today much more complex then it has been ever before. Every source or drain in this network is functioning accordingly to its own nature represented by its own characteristic laws (with regard to numerous economic, ecological, and social requirements). It is good to know these laws well enough, otherwise we cannot provide for the *long-term stability* of the whole network.

At present we do not know an ideal solution for our energy thirst, we cannot rely only on one or two types of energy sources (for instance on *fossil fuels* or *nuclear power*); so we must take into account many different solutions to energy problems. Energy crisis that we face today was produced mainly by excessive use of *fossil fuels* (coal, oil, natural gas). Blind faith in conventional *nuclear energy* is also not the best solution. The way out of the energy crisis should combine many small solutions inside the complex network of energy sources and drains. A great knowledge about characteristic nature of every particular network element is needed in order to provide for the *optimum balance* of the whole network. *Renewable sources*, especially the *solar energy*, by side of the *wind energy* and *decentralized hydro energy*, should be given deserved attention. *Biomass* is good but only in moderate quantities and under strict *LCA standards*. These standards should be considered for any candidate of possible energy sources.

In the same time it is wise to investigate also the new possibilities of energy sources (like *nuclear fusion* or even hypothetical sources of "*free energy*"); although today they look more or less exotic.

Living beings in nature do not waste energy. *Saving energy* is indispensable for long-term stability of life. Regretfully the idea of saving has been treated nearly like a *taboo* in our modern consumption oriented society, but now we are enticed and forced to find a new balance with the environment. Sparing use of energy should be incorporated into all forms of everyday life. There are numberless small practical adaptations in this direction that can be made. Considerate dealing with energy problem means numerous practical arrangements towards energy-saving solutions, together with development of sustainable energy sources.

In the present-day crisis, it is good and worth to remember of the long-tested values cherished by our ancestors and by followers of great spiritual cultures. Pure, unalloyed happiness is the fundamental nature of our deepest *self*, claimed the wise men of all times, in all human cultures. Adopting such an attitude makes things very simple. Material world cannot offer us much more than food, clothes and a warm shelter. Our activities in this world are needed only for the preservation of that fundamental harmony with the world, which enables us and all the beings that we love to remove the barriers preventing the experience of our true nature. Only if we assume a simple and innerly free life stance, when we are no longer slaves to fulfilling ever new material needs, is it possible to look at reality with curious, child-like purity, to experience a heartfelt connection with the world, to pulsate in the joy of existence. Only then can we relax and experience the heart and self of the beloved beings – of ourselves and of all those close to us. Our activity is like a tree, which grows, blossoms and bears fruit. The fruit is our gift to the unknown birds, but happiness is more like the sap, which flows into the roots and veins from the mysterious depths.

One thing is already certain: It is pointless to believe that solution will be achieved merely by efforts in the realm of science and technology. We all already know today that ecology is posing strict limits to our behaviour, since exponential growth of production and consumption cannot continue infinitely. Time and again, we can continue to develop new sources of energy, but if simultaneously we do not learn to subdue the beasts of spiritual ignorance roaming in jungles of human mind, then every intellectual effort is in vain.

What remains to be learnt is that ecological behaviour is not so much a matter of technology but much more a new *ethical awareness*: to realize that human happiness is at the (still veiled) source of our existence and can never be achieved through uncivilized human greed. Modern manipulative economy is continually supporting this harmful illusion, but in our hearts there is a place for liberation – not only for our personal security but also for other fellow beings. Sages of all times (also of ours) know that simple and balanced life is quite enough. In our age of communication this means also a new social paradigm of *spiritual and ethical awareness*. On this basis one day we shall discover that the energy problem is solvable much easier than it seems in these turbulent times.

Acknowledgment

I feel grateful to Thomas C Daffern, Andrej Ule and Gorazd Lampič, who read the manuscript and helped to make many points clearer and easier to understand.

References

ADOE: American Department of Energy Report on Cold Fusion, released on Dec. 1, 2004

Ballentine L.E., Quantum Mechanics (Prentice Hall, 1990)

Bohm D., On Creativity (Routledge, 1998)

Bošković R., Theoria philosophiae naturalis, reducta ad unicam legem virium in natura existentium (Theory of natural philosophy, reduced to a single law of natural forces, Vienna, 1758); De viribus vivis (On living forces, 1745)

Brosse J., Mythologie des arbres (Librairie Plon, 1989)

Burden V., The Process of Intuition – a psychology of creativity (Theosophical Publ. House, 1975)

Capra F., The Turning Point (Simon and Schuster Publ., 1982)

Capra F., The Web of Life – a new scientific understanding of living systems (Anchor Books, 1996)

Cheney M., Tesla, Man out of Time (Dell Publ. Co., 1981)

Childress D.H. (compiled by), The Free Energy Device Handbook – a compilation of patents and reports (Adventures Unlimited Press, 1994)

Clausius R., Űber die bewegende Kraft der Wärme (Annalen der Physik, 1850)

Collins H., Pinch T., The Golem (Cambridge Univ. Press, 1993)

Davies P., The Cosmic Blueprint – Order and Complexity at the Edge of Chaos (Penguin, 1995)

Detela A., Magnetni vozli (Magnetic Knots, Ljubljana, 2002)

Frankl P., Rubik F., Life Cycle Assessment in Industry and Business – adoption patterns, applications and implications (Springer, 2000)

Gabor D., Colombo U., Beyond the Age of Waste – a report to the Club of Rome (Franklin Book Co., 1981)

IPCC Fourth Assessment Report (The AR4 Synthesis Report, Cambridge Un. Press, 2007)

ISO 14040 "Environmental Management – Life Cycle Assessment – principles and framework", Int. organization for standardization, Geneva, Switzerland (1997)

ITER: http://www.iter.org/

Jäger G., Theis K.A.: Increase of power plant efficiency, VGB Power Tech, *Int. Journal for Electricity and Heat Generation,* Vol. 81 (2001)

Krishnamurti J., Bohm D., Truth and Actuality (Krishnamurti Trust, 1977)

Kuhn T., The Structure of Scientific Revolutions (University of Chicago, 1962)

Lambertz J., Efficient power generation in coal- and gas-fired power plants, VGB Power Tech, *Int. Journal for Electricity and Heat Generation,* Vol. 85 (2005)

Lampič G., Analiza uvajanja električnih pogonov v različne vrste vozil in zasnova pogona za sodobni mestni električni hibridni avto (Analysis of implementation of electric drives into different kinds of vehicles, especially in a modern urban hybrid electric car, University of Ljubljana, 2006)

Lightman A., Great Ideas in Physics (McGraw-Hill, 2000)

Lizot J., Le cercle des feux: Faits et dits des Indiens Yanomami (Editions du Seuil, Paris, 1976). I feel grateful to late Andrej O. Župančič, a great soul of Slovene science, who among many other researches in his rich life lived a certain time among this tribe (inhabiting upper Orinoco river basin in South America) and told me about their culture.

Lovelock J., The Revenge of Gaia (Penguin Books, 2006)

Prigogine I, Stengers I., Order out of Chaos (Bantam Books, 1984)

Rifkin J., The Hydrogen Economy (Tarcher/Putnam Books, 2002)

Sekavčnik M.: Vodikove tehnologije – utopija ali resničnost (Hydrogen technologies – utopia or actuality, University of Ljubljana, 2007)

Tesla N., Colorado Spring Notes 1899 - 1900 (Tesla Museum, Belgrade, 1978)

Thakar V., Totality in Essence (Motilal Banarsidass, 1971)

Worldwatch Institute: www.worldwatch.org/

World energy resources and consumption, 2008: http://en.wikipedia.org/

In: Philosophical Insights about Modern Science
Eds: Eva Žerovnik, Olga Markič and Andrej Ule
ISBN: 978-1-60741-373-8

Chapter 10

EVIDENCE FOR BIOFIELD

Igor Jerman*, Robert T. Leskovar and Rok Krašovec
BION Institute, Ljubljana, Slovenia

SYNOPSIS

Advances in science are not achieved merely through the accumulation of knowledge but also through the development of new concepts. In physics, the inclusive concept of the field was used centuries ago with great success, while in life sciences we are still waiting for a similar universally acceptable term. For the time being, in biology we only use the field concept in its strictly physical meaning, e.g., cellular transmembrane electrical field. Nevertheless, an increasing number of research groups are making surprising discoveries that will require a new conceptual and empirical breakthrough also in the field of life sciences; one of the promising concepts is the concept of an emergent and potentially all-encompassing biofield. Many open questions still remain: first, what is the physical basis of the biofield; second, what is the scientific evidence for such a field; third, what is the biological meaning of the biofield and fourth, what would be the significance of the biofield for our understanding of consciousness.

INTRODUCTION

Centuries ago physicists noticed that energy is interconnected, structured and ordered (in other words, integrated) throughout space, and this deep insight they shaped into the field concept. Treating of the forces of nature through the application of this inclusive concept represented a great leap forward in physics. However, in contemporary biology and medicine the concept of long-range forces has not been integrated that well, at least as the mainstream science goes. As it stands now, organisms are regarded as physical objects, as complex aggregates of their independent or synergistic parts (macromolecules, organelles etc.), which are believed to interact mainly by short-range forces. Even the physico-chemical fields of

* Stegne 21, 1000 Ljubljana, Slovenia, e-mail: igor.jerman@bion.si

these entities are regarded as short-range in their scope; such as the electric fields of molecules and cells, e.g., the cellular transmembrane electric field. This by itself could do perfectly well, if it reflected the whole picture. However, an increasing number of research groups throughout the world are making surprising discoveries which strongly indicate (if not prove) that organisms' energy – at least some part of their total energies – are integrated into a sort of an all-inclusive, *long-range* and to a certain degree *coherent field.* This field is assumed to provide a long-range order within and around the organism. It is a new concept – considered by some authors as the biofield (e.g., Rubik, 2002) – that can account for many different biological (and even physical) phenomena.

The concept of biofield may remind us of vitalism – a stream of thought according to which life is led by a principle different from the standard, measurable physical forces. Classical vitalism, well established in certain circles around the time of Newton, saw the living principle (or the living force) as something apart from nature and its forces. It was assumed as immeasurable and even undetectable, thus invoking a supernatural principle, inaccessible to empirical scientific research. If this inability of research and scientific explanation is the mark of vitalism, then the concept of the biofield is definitely not vitalistic. Rather, the biofield should be considered as a normal, measurable and scientifically explainable natural field, even if currently beyond standard physical concepts. As such it should not be limited only to living beings; it should rather be a fundamental constituent of nature, at least in principle.

Let us be clear, in physical terms it is not yet known what exactly the biofield would correspond to. The first relevant question about the biofield is what its constituent particles are and how it is connected to the already known fields (in analogy to particle physics). Second, in a broader meaning of the term it is not called the biofield because it would be so unique to organisms, but because it is essential for the living process – for its long range coordination, regulation etc. Therefore, as we shall explain in the following, on one hand the biofield should be regarded as an ordinary element of nature and in this larger connotation it should perhaps be called the subtle field[1]. On the other hand however, it does not demonstrate properties of an ordinary field accessible to the accustomed physical research – it demands more subtle techniques of research and new explanatory modes.

VARIOUS CONCEPTS OF BIOFIELD

Electric and Electromagnetic Fields

According to the prevalent notion of the biofield the latter is conceived as connected with the endogenous electromagnetic (EM) fields of organisms. The electric and EM fields of organisms are known and are partially dealt with by the conventional biological science. It is well known that every living cell has an electric field of a very high intensity (around 10^7 V/m) though of a rather low voltage (~ -70 mV) that is the basis for excitability – one of the basic features of life. According to a very strong theoretical consideration started by the biophysicist Herbert Fröhlich this resting electric field represents the basis for coherent endogenous EM oscillations within living systems on a cellular level (Fröhlich, 1975, 1988).

[1] The term already used by some authors.

This oscillatory field has not yet been proven experimentally at the level of proposed frequencies (high microwaves: $10^{11} - 10^{12}$ Hz) and for its coherency. Yet, the Czech group of Prof. Pokorny succeeded to prove the endogenous field at radiowave frequencies (8–9 MHz) in a direct way (Pokorny et al., 2001; Pokorny, 2004). There are also indirect indications for the endogenous electromagnetic field from the research of rouleaux formation of living erythrocytes (Rowlands and Sewchand, 1982) and the dielectrophoretic research (Pollock and Pohl, 1988). Other indirect indications are coming from the research in the field of biophotonics, more specifically, its area of exploring the (statistically) coherent ultra-weak emissions of photons from living cells (Popp and Nagel, 1988; Popp et al., 1992). Formal theories and empirical evidence exist, which suggest that biological systems use long-range communication with biophotons and optimise it according to the well known relation of coherent and squeezed light (Popp et al., 2002). The idea that not only coherent states but also squeezed states may play a role in biological regulation is a consequent and progressive conclusion since biological optimization may make use of quantum effects just in the ultra-weak range of intensities where squeezed states exist (Bajpai, 1999).

Therefore, as experimentally proven, we have a strong quasi-static transmembrane electric field (the so-called resting membrane potential) within the organism and, a high frequency EM field presumably formed on its basis (which also has been detected as will be detailed below) and is more or less coherent. This coherent EM field is supposed to organize many processes on the cellular level. Its putative role is more thoroughly elaborated in the theoretical extension of the Fröhlich's quantum field, worked out by the Italian group of quantum field theoretical physicists (del Giudice et al., 1984, 1985, 1988). Their theory proposes a special, network like, organisation of the coherent EM field that could coordinate countless chemical reactions within cells. It could be at least partially responsible for the organisation and orientation of cellular microtubular network (Hameroff and Penrose, 1995) – its skeleton that plays very important role in cell's division and thus in morphogenesis. Similar to this endogenous EM field is also the before-mentioned photon field of Popp. According to Popp and many other research groups the emitted coherent photons of this field are not a by-product of cellular chemistry, but are a marker of the photon field involvement in the cellular physiology (van Wijk and van Wijk, 2004). The ultraweak photon emission was found to be strongly correlated with the cell's cycle (Popp et al., 1992). Experiments involving transparency of tissue to such radiation, as well thorough physical analysis of the emission, strongly indicate the coherent state of the photon field inside organisms and if properly stimulated, the biophotons may coherently spread into the environment (Popp et al., 2002). Such a subtle field could emerge from networks of macromolecules and their structural organization. Perhaps signalling between such complex patterns of macromolecules (with the corresponding dipoles) might be taking place via photon fields.

The related phenomenon, also covered by biophotonics, is called delayed luminescence, in which after excitation of a living system, in particular in the optical range, the system emits a stronger photon current than without illumination. Though biophotonics and delayed luminescence have been researched in different ways, in essence both contain all the features of the interaction of biological matter with photons. They are different from common photobiology in so far as they are confined to the quantum nature of these physical and biological processes. This confinement allows distinguishing biophotonics from ordinary photobiology, fluorescence techniques and similar classical, well-known fields of science and engineering.

The phenomenon of biophoton emission does not result from the thermal radiation in the infrared range (Popp, 1999). At present it is well known that the biophotons are emitted also in the range from the visible up to the UV range. The intensity of biophotons can be registered from a few photons per second and square centimetre surface area on up to some hundred photons (per second and per cm^2) from every living system under investigation. The spectral distribution never displays small peaks around definite frequencies. Rather, the quite flat distribution within the range of at least 300 to 800 nm has to be assigned to a thermodynamic system far away from equilibrium, since the probability of occupying the phase space is on average almost constant and exceeds the Boltzmann distribution in this spectral range by at least a factor of 10^{10} (in the red) up to 10^{40} (in the UV-range).

After excitation by white or monochromatic light, every biological system increases the photon emission up to about 5 times higher in orders of magnitude. Then it relaxes in darkness to its original biophoton emission quite slowly and never in an exponential manner, yet with high accuracy as a *1/t*-function, where *t* is the time after excitation. This phenomenon of a delayed luminescence is the basis for all kinds of exciting the biological matter, irrespective of whether one takes ordinary lamps, LEDs, laser or even other forms of excitation like ultrasound. The spectral distribution of delayed luminescence corresponds to that of biophotons (Chang et al., 1998).

The biophoton field importance can be seen in the cell division process. A surprising fact is that almost never an error occurs in the distribution of the molecules, which are partitioned exactly into two equal fractions by the daughter cells. According to the "random walk" theory we should expect about 10^5 out of 10^{10} molecules to be located at incorrect positions. Most likely explanation is the presence of a force which repels molecules that are in erroneous positions back to their correct place. The possible solution of describing such a force has been made by calculating the cavity resonator waves of a single cell (Popp et al., 1979.).

The questions that still remain unanswered are where the biophotons come from and how are they sustained. In order to find the answers, Popp with co-workers (Popp et al., 1981) have shown that biophoton emission can be traced back to DNA as the most likely candidate for the (main) source, and that delayed luminescence corresponds to excited states of the biophoton field. At the same time, one should not neglect the possibility that proteins also may have potential for bioluminescence and electron transfer.

In addition, all the correlations between biophoton phenomena and biological functions such as cell growth, cell differentiation, biological rhythms, and cancer development, turned out to be consistent with the coherence hypothesis but could be only very poorly explained in terms of radical reactions (Popp, 1999).

To summarise, on the physically established (and measurable) level, we have two fields: one (static electric – transmembrane potential) that has been known for long, and the other, a high frequency oscillating and more or less coherent EM field. The latter, at least as the physical research goes[2], has two components: a) one in the microwave and lower frequency range (Fröhlich's field) and another (b) in the visible (including near IR and near UV) light range (Popp's photon field). The Fröhlich's field was shown empirically, although, as previously mentioned, at lower frequencies than predicted. The other, Popp's component is

---

[2] In reality of living beings there could be only one very extended EM field. But causally viewed, we should treat it as if of two components, since the Fröhlich's field stems form the transmembrane electric field and vibrations of long macromolecules, while the photon field should stem from DNA photon-electron dynamics.

(for the time being) well researched only theoretically and is thought to be mainly connected to DNA electronic oscillations[3]. A strong empirical support to the Popp's photon field theory is provided by the finding of the statistical coherence of biophotons that, consequently, are separated from the photons originating as a by-product of biochemical reactions (Popp, 1979; Popp et al., 2002). If we consider this finding and a recent one by Engel that proves the existence of coherence in the photosynthetic system (Engel et al., 2007), the biofield may be considered as consisting of three components: the electric field, the Fröhlich's regime and the photon field. Each of these fields is believed to have either an energizing or informational role. The electric field (membrane potential) has a well established energy component whereas the two EM fields are regarded as providing organization (having mainly an informational role). Theoretical predictions indicate that the high-frequency EM field could resonate with the EM frequencies of biological molecular interactions on one hand, and cover larger areas through coherency and quantum mechanical entanglement on the other (Ho, 1993). A recent finding suggests that the Fröhlich's field and the Popp's field are mutually interconnected because of rather strong mode couplings in living systems (Popp et al., 2006).

The two fields (actually the three components) are scientifically more or less well established, at least in theory, therefore, certain scientists are convinced these three alone could represent the complex electromagnetic biofield; see for instance Rubik (2002). But some others think that it would be very difficult, if not impossible, to cover many diverse phenomena connected to the biofield concept only with EM fields, however coherently or extraordinarily they behaved.

Beyond the EM Field and the Mainstream Physics

There are numerous studies that prove there is more to the biofield than just classical and coherent fields known to contemporary physics (Rein, 2004). Many of the researchers attempted to prove that consciousness could influence the physical phenomena. One of the most elaborated and prominent efforts in this direction is Tiller's empirical research as well as his theoretical model. In contrast to some physicists who are trying to prove the beyond-physical levels of existence and are in strong opposition to contemporary physics – Einstein's relativity theory and quantum mechanics (Correa and Correa, 2002; Correa, 2001; Aspden, 1969) – Tiller's theoretical model is firmly rooted in the established physics, yet at the same time shows the way beyond its framework. In his theory Tiller proposes that the so-called physical vacuum is not an absolutely empty space at all (this is concordant also with quantum mechanics and vacuum field fluctuations). Namely, according to Tiller the physical reality should consist of two unique levels with a variable degree of coupling between the two: (1) our ordinary electric and atom/molecule level functioning in the conventional spacetime, and (2) a coarse physical vacuum that represents the magneto-electric wave level functioning in a reciprocal-space time (Tiller et al., 2005). The basic particle of this second level of existence would be the magnetic monopole and it would function in a different symmetry space (the so-called SU2 symmetry), in the so-called R (wave) space. Because of this it would not be easily detectable, but would still interfere with the ordinary physical reality (functioning in the U1 symmetry) under certain conditions, among them the ones that would include the human

[3] For more detail (since it is about exciplexes) see Popp 1984a, b.

conscious intention. The conscious intention would work on a different level of reality than the known forms of energy. Tiller's numerous experiments have shown that even his special device (called IIED), when influenced by a directed human intention, can have empirically demonstrable effects, for instance in an increase or decrease in the acidity (pH) of a solution (Tiller et al., 2001). The prerequisite for repeating the experiments with the IIED is an adequate "conditioning" of the lab, where the experiments are performed. According to Tiller, conditioning means an adequate establishing of subtle field conditions by means of cultivating mental quietening for a few days in that lab. One of the major findings of Tiller's experiments is that the processes in the R space are interconnected (entangled) in a similar manner, though differently from the processes in the realm of quantum mechanics. Therefore, even two events far apart in space may be closely interconnected, interdependent; there is no separation in space as we are accustomed to in our ordinary physical reality.

Tiller's ideas stem from the accepted physics and therefore we treat them as superior to certain other theoretical attempts already mentioned in the beginning of this subsection. If we connect them to our main theme of the biofield we may extend its purely electromagnetic nature to Tiller's R space or to something similar that transcends our known and scientifically well described physical reality and demonstrates characteristics of waves and long-range interconnectedness, a sort of holism. Whatever the source (magnetic monopoles, ambipolar massfree electricity (Correa, 2001), unknown dark matter particles - like postulated massive WIMPs or light axions, zero field energy etc.), it should have a deep relation to the ordinary matter-energy coupling, while at the same time its influences would not be easily perceivable. Since the electric and electromagnetic (EM) fields of organisms are covered by contemporary physics and can be detected (even thought not easily in certain cases), we might reserve the term biofield only for some equivalent of the Tiller's R level, something that goes beyond the established physical knowledge. In such a case the notion of biofield could correspond to various traditional layman terms such as Chinese Qi, Hindu prana, Reich's orgone, Reichenbach's Odic force, health practitioner's bioenergy, etheric double (energy) or subtle field etc. From this standpoint, the electric and EM fields of organisms may be regarded just as the fields mediating between the biofield and the material structures of life.

The Dual Nature of the Biofield

Based on experimental and theoretical work of others as well as on our own experience we assume that the biofield would manifest itself a) under strong electric fields, for instance such as demonstrated across the membrane of living cells, b) under the weak, partially coherent EM field as demonstrated within living cells, c) under conditions involving conscious intention and d) under other special conditions involving strong magnetic fields, plasma state of matter etc. As already said, we may assume that in organisms the biofield comes into interaction with organic matter through the strong transmembrane electric field. If the latter is destroyed its coupling with the holistic and wave like biofield vanishes: life comes to its end. When the electric field is normal the coupling between the latter and the biofield results in the resonance between the intricate processes within the supposed R level of the biofield and the Fröhlich-Giudice-Popp endogenous EM field that directly organises countless molecular processes within the cell.

We assume that the more the EM field is coherent, the more closely it is connected (or better say coupled) to the biofield. The biofield can therefore be assumed as either closely coupled or almost decoupled to the material level of the organism. In the first instance we would have a healthy, biochemically and physiologically well coordinated organism (be it a bacterium or a whale) while in the second case we would have illness or sleep or coma etc. In this light we may understand that what most traditional medicine (healing touch, homeopathy, acupuncture) tries to do would be to bring the two misaligned levels – EM and the biofield's – into as perfect alignment as possible; health would then be the natural result of this process.

In the energy aspect of life, the EM field directly wrestles with the molecular processes, thus assuring an indispensable long-range order on the cellular level (Popp et al., 2006), while the integral biofield would establish an even more extended long-range order, perhaps spanning over the entire multicellular organism and beyond. Namely, a long-range connection (presumably without energy or matter transfer) was demonstrated between human individuals (Wackermann et al., 2003, Standish et al., 2004, Grinberg-Zylberbaum et al., 1994) as well as between isolated neurons (Pizzi et al., 2004). The experimental system that includes humans was a simple one; two subjects in two separate Faraday cages at a distance of several meters. When subject one was exposed to light, the brain of subject two instantly demonstrated waves that correlated with the applied photo stimulus.

The concept of biofield is also very important for a deeper understanding of illness. It is broadly believed that the biofield might be closely connected to the consciousness of an organism (not necessarily a human being). Namely, consciousness is much too synthetic, holistic to be directly expressionable through countless fragmented biochemical reactions or even individual electric fields of brain cells (see other Chapters dealing with Consciousness). On the other hand the biofield is deeply integrative and can be responsive to conscious intention as shown by Tiller's and some other experiments (like Wackermann et al., 2003). In this light we could understand the strong influence of our psyche on our health and also many so-called parapsychological phenomena.

Measurements of the Biofield

If the biofield is at least partially a physical phenomenon then it should be accessible to scientific research, perhaps not directly but by measuring its indirect effects. The EM level of an organism is measurable, as we have already seen. Even the endogenous EM field was successfully measured in a direct or indirect way. The coherent photon particles of the predicted photon field can also be readily detected (Popp 1984, 1999 etc.). For the latter (Popp et al.., 1992) there are still debates pro and contra. However, as for the biofield's R level, many researchers think that it is and will stay - immeasurable.

Our experiences and other researchers' studies show, however, that the biofield is accessible to various indirect measurements. One – more crude – method is computer-controlled electrophotography. In one of its applications it was improved by Korotkov, who invented the so-called GDV camera (Korotkov, 2002). Here, through well controlled electrophotography of human fingers and their computer analysis, one is able to get a more or less accurate supplementary diagnostic overview of a human organism. A very similar technique is used at the Bion Institute (Škarja et al.., 1998): the electrophotography of water drops, previously exposed to various fields or radiations. It has been subject to severe computer analysis. According to our general model of the biofield, during electrophotography subtle interference occurs between the R level of either water drops or fingers and the applied strong electric field. The form of the corona discharge brings information about the quality and intensity of the applied biofield and is fashioned according to this interference.

Early-stage research called Digital Visualisation of Biofield (DVB), currently going on at the Bion Institute, detects and analyzes subtle variations of the background light in proximity of an organism. It seems that the organism's biofield interferes with the surrounding light at very small distances, occasionally in a coherent manner. Namely, occasionally such interferences follow the shape of an organism, even after the shape has been physically modified (e.g., after cutting an apple, interferences still formed for a few seconds in the shape that resembled the whole apple) (Leskovar et al., 2005).

Analysis of the conductivity distribution functions of the skin is also a possible technique for an indirect biofield measurement. Recent observations showed that biophoton emission of the human body does not only correlate with biological rhythms but they both correlate also with the electrical parameters of the body's skin (Popp et al., 2006).

Another type of measurements can be achieved by measuring the absorption and transmission of the externally applied electric field penetrating the human being, larger space, plant etc. (Škarja, 2007). Here it seems as if the near electric field and the biofield interfered.

We may assume that the so-called bioresonance diagnostics also detects the biofield of the examined person, even if the theory of the bioresonance speaks only about ordinary EM fields (but is consequently inconsistent with existing physical theories).

Possible Biological Implications of the Biofield

The mainstream contemporary biology is based on molecules and their interaction, more specifically, on the famous DNA molecule. The latter should possess information about everything that concerns an organism – its vehicle for expression, as for instance, seen by Dawkins (Dawkins, 1976). One of the so-called organicist streams in biology on the other hand stresses the importance of the whole living state and proposes the morphogenetic field[4] to implement the ontogenetic (embryogenetic) plan of the organism (Goodwin, 1985; Sole, Goodwin, 2001; Goodwin, Webster, 1996). While not denying the importance of genes, this biological line of thought suggests that they can only determine the boundary conditions for the actualisation of the organism's form, while they cannot actually build it. This could mean that an organism is not built on the basis of some fixed information and a computer-like program, but on the basis of a law-governing field.

What would be the relation between the morphogenetic field proposed by organicists , and the biofield? First, the two fields, at least according to suggestions, cannot be regarded as equal. In its most elementary meaning the morphogenetic field can be identified as the adhesive intercellular force field with its tensions and pressures. These forces come to fore in formative embryogenetic processes. Here the morphogenetic field works according to the known laws of physics and chemistry. But it could work also on a deeper level – on the level of cell's division, therefore on the subcellular level. Namely, the morphogenesis is not only a

[4] In organicistic circles (not vitalistic!) the term morphogenetic field is mainly understood as a field of all the forces that assist in morphogenesis.

matter of intercellular mechanical forces but also a matter of the orientation of the cellular division plane (orientation of the mitotic spindle). And here the morphogenetic field comes together with the coherent cellular EM field and the biofield. According to Popp the whole cell division situation corresponds to a cavity resonator in the visible range of the electromagnetic field and is therefore closely connected to Popp's biophotonic field (Popp et al., 2006). In this respect the morphogenetic field is connected to the endogenous EM field and consequently, also to the biofield. If the EM field level organized countless molecular reactions and energetic transformations, mainly on the (sub)cellular level, than the biofield with its larger area of interconnectedness would be responsible for morphogenesis, not in the sense of a mechanical force (where the already mentioned connotation would be sufficient), but in the sense of a formative cause (see the last section), not actually shaping the form of the organism or its organ but providing their structure and/or position. However, it would also have other roles, for instance an integrative role, and therefore cannot be identified as the morphogenetic field. In conclusion, we could say that the morphogenetic field can be regarded as triple in its nature: on the most elementary level it would involve mechanical forces needed to actually shape the developing organism. On another, EM level, it would mainly influence the orientation of the cellular division, and on the R level (biofield) it would function as the macroscopic morphogenetic field, pervading the whole organism.[5]

Embryogenesis could be regarded as a very dynamic and mutual interaction between the biofield in its morphogenetic aspect on one hand and the DNA of the organism on the other. According to Popp, the DNA is not only a carrier of genetic information but also an important carrier and a possible originator of biophotons – at regular distances (true also for proteins, especially bundles of microtubules, receptors at synapses, patterns of neurons firing). In this regard the macromolecules such as DNA and proteins could function also as one of the connecting factors between the molecular plane of the organism and its (supramolecular) integrative field. In any case the signals from DNA or protein assemblies could modify the morphogenetic aspect of the biofield so that it would begin a series of symmetry breakings. A gradual shaping of the organisms follows the initial form – more or less spherical (totally symmetric) – and continues into its adult form. During this process three elements might be dynamically interconnected: DNA with a myriad of its genes, the morphogenetic field on all its three levels and the actual form of the developing embryo.

As already said, the biofield may have an important integrative function, providing the wholeness of the organism. We may assume that the biofield could be related to the well known network of acupuncture meridians and points as well as other subtle interconnections within the organism (reflexive points); electroacupuncture practice (for instance Voll's diagnostic and healing) strongly supports the connection between the electrical field and holistic network throughout the organism that does not have a satisfactory scientific explanation yet. In an extension of this integrative function the biofield could also be responsible for the connection between the organism's psyche, consciousness and intelligence on one side and the operative physical body (brain) on the other. From this complex integrative role a completely new – at least as far as the official medical knowledge goes – perception of illness becomes possible. For instance various psychological stresses could be imagined to disturb the biofield so that it could no longer integrate the physiological

[5] Contemporary science latest developments in the field of cell differentiation and division should not be overlooked; some of this is dealt with in Part A of these volumes; Chapters 1,2,4.

processes in a satisfactory way. The field disturbances would then be pronounced on special points that would reflect a particular psychological state – as a consequence, an illness in that specific area would result. We may speculate that a part of this knowledge is delivered by homeopathy, acupuncture, reflexology and various psychosomatic treatments, though without true scientific background. We may expect still more knowledge and more effective healing practices when the biofield is better understood and thoroughly researched.

Possible Connection of the Biofield to Intelligence and Consciousness

One of the pronounced marks of living beings is not only the high organisation of energy and matter, but also their flexibility. This means that organisms are sensing external signals, and before the actual response takes place, 'messages' from the environment are compared with the organism's inner state defined by the overall status at that moment (organism's physiology, genetics, energy or social status, etc.) and by previous experience. Then a decision is made and as a rule, organisms seem to execute the most economical response. Such flexibility is a characteristic seen by many researchers (Trewavas, 2005) as biological intelligence and it is regarded as something opposed to automatism where there is no true choice. Biological intelligence requires at least some freedom of choice, in other words, organisms should have ways to manipulate and control their own information flow, at least partially.

As far as the molecular level goes, some very elementary intelligence[6] can be found on the quantum level of activity (Conrad, 1993) and is expressed in the sensitivity to the adjacent possible (Kauffman, 2000,) before the collapse of the wave function (decoherence). The 'adjacent possible' means the set of possible events, implicit in the wave function of the system under consideration, whose probability is unconditional, i.e. the events are possible directly, without any intervening events. Self-organized systems are supposed to travel towards this unknown territory with a maximum speed that still allows them to preserve organisation gained during evolution. Through this elementary form of flexibility (intelligence), enzymes may find their functioning conformation at the level of many superposed possibilities in the wave function (Hameroff, 1987). In their famous paper Engel and co-workers (2007) describe for the first time the preservation of coherence in biological systems. The research object was a photosynthetic complex whose function is to capture solar light efficiently, and then transmit the excitation energy to reaction centres. The energy transfer mechanism includes proteins that may promote coherence transfer which means that proteins might be capable to sense many states simultaneously and 'always' select the correct answer; in an isolated photosynthetic complex the correct answer is the lowest energy state.

An extension of this elementary photosynthetic-like process may be found on the cellular level, where countless molecular interactions must be integrated, coordinated and adapted to the need of a larger whole, be it a cell, an organ or an organism (see for instance Albrecht-Buehler, 1985). For example, DNA helix does not possess any special chemical robustness, therefore every living cell must constantly pay attention that the DNA is replicated as accurately as possible and that the nucleotide sequence is adequately repaired when

[6] Of course, it is not intelligence in the human or animal sense, but just a capability to choose the most economical alternative.

spontaneous or induced mutations occur (Friedberg et al., 1995). Only one base substitution within millions of DNA pairs is enough for the induction of the DNA repair mechanism. How does an individual cell place all these million base pairs under such a strict and constant surveillance? A similar dilemma occurs in protein/DNA interactions. Namely, structural and thermodynamical studies show that proteins known to bind particular DNA sequences are very unlikely to have enough time to test (by using their reading heads) every single base pair and find out if the sequence is indeed a proper one (Bruinsma, 2002). Also, plants have a continuous control of the distribution of roots, shoots or leaves (Trewavas, 2005). By having this control over the root distribution they sense the volume of the soil which in turn affects the level of plant growth. In biological sciences there are many phenomena demonstrating an extraordinary and highly dynamic order that is much more organised than it would be expected if only Brownian motion was in force (Vitiello, 1992). Here we find Popp's and Fröhlich's electromagnetic coherent regime as an important if not indispensable pillar of this order. And again, every cell, be it prokaryotic or eukaryotic, has to adapt each moment to countless external signals, and, when conditions are met, prepares itself for eventual division. Actually, the intelligence (i.e. sensing, adapting & flexibility) of the cell should be on a much higher level than the intelligence of a molecule. It could comprise the ordinary quantum level of all its molecules and also the quantum level of its coherent multi-frequency EM field.

When we come to a whole multicellular organism, not only the molecular and the cellular organizations apply, but also the possible intelligence exerted through the biofield that would permeate the whole living being. Through its integrative power the biofield would assure the entireness of information that is needed for best decisions in the concrete living challenges and situations of a higher animal. Of course, organisms' bodily intelligence works mainly through the brain where the biofield's essential information (integrated reflection of brain's excitation states) would stand as an integrative whole, while on the ordinary material level the information content of the brain seems fragmented into countless excitations of its individual cells (neurons). Through its integrative power the biofield would stand not only as the basis of the organism's intelligence but also as the basis of organism's consciousness, which however, is difficult to objectively access. To the best approximation, a conscious state is experienced in an abstract and channelized manner by means of cellular and electrical events in the brain (neural substrate).

And what could be found beyond the R level? According to Tiller, whose theory is based on contemporary electromagnetics under SU2 symmetry conditions, there are also other, even more subtle levels of reality (Tiller et al., 2005). They may be connected with emotions and thoughts as well as with their corresponding intelligence and consciousness. Of course, at the moment this connection of the biofield and consciousness is only a hypothesis. However, this is a testable hypothesis that invites to a new and fresh scientific endeavour that can lead us into new realms of understanding consciousness and intelligence, perhaps far reaching from the contemporary views. Partially, it has already been supported experimentally, by experiments such as the already mentioned long range connection between humans or influence of consciousness on the Tiller's IIED system.

Biofield and Inanimate Nature

According to our considerations the biofield is not limited only to organisms as already said; it also is part of the so-called inanimate nature, such as water and crystals. And like every physical field, it should also have its elementary particles, its material carriers. One could speculate that these particles represent (at least partially) the so-called dark matter that interpenetrates the visible matter and is very unwilling to interact with the latter, at least in the ways that involve friction, collision or direct magnetic or electric effects, as implied in standard measurement techniques (Griest, 1996). Even in Tiller's interpretation, the R level of reality has some special quanta called magnetic monopoles that work in SU2 symmetry, as already discussed. After another suggestion, which is partially experimentally confirmed, the particles may be electric dipoles (dions, see Marshall, 2006). To some degree the biofield carriers may be identified with Reich's orgone; the main difference is that he treated orgone as a massless substance, while we assume that the R particles should have some mass – even if small – and are consequently in accordance with Einstein's famous finding expressed in the formula $E = mc^2$.

Another interesting concept of the possible biofield's function in the inanimate nature is the so-called "memory of water", as for instance demonstrated in the homeopathic practice[7]. As far as the established understanding of water goes, the physicists are right that water does not carry long-term ordering for long. Water molecules namely erase (forget) their previous molecular or supramolecular imprints in a matter of picoseconds, however, serious researchers of the phenomenon claim the molecular organization (imprinted by other molecules once dissolved in water) lasts much much longer (Rey, 2003, Elia and Nicolli, 1999, Rao et al., 2007, Chaplin, 2007). There is a thorough quantum field theoretical consideration that speaks in favour of the "water memory" (del Giudice and Preparata, 1998). The field that could impose the imprints of molecules dissolved in water in the past – especially in cases where the homeopathically prepared solution (with strong mechanical stiring) comes as low as below one sole molecule of the originating substance – is similar to Fröhlich's coherent field in organisms. According to Giudice and co-workers, the field in water should also be organised in coherent domains and should remain very stable since it is estimated to be on the lowest energy level; from this stability the long lasting organization (memory) follows spontaneously. We have therefore a situation analogous to that in organisms: the coherent EM modes and the biofield component connected to them. While the sub-molecular and molecular organisation of water itself could be covered by the EM level alone, some ways for imprinting molecular structures into water without a molecular contact could involve the biofield. At the Bion Institute two such methods of imprinting are being practiced: one involving a high voltage electric field (for more detail see Jerman et al., 2005) and the other without such a field. In our experience, not only the phenomenon of water memory exists, but also the memory of other liquids and even solids, indicating the possibility of the existence of a large spanning memory of matter.

[7] The research of "water memory" is a branch of science and has nothing to do with homeopathy understood as a healing practice based on Hahnemann's doctrine.

Conclusion

Biofield and Quantum World

Many modern thinkers and healing practitioners that propose (by experience or intuitively) some deeper level of reality, through which we are inherently connected and through which we can manifest our thoughts and intentions, speak about the quantum world (Stenger, 1999). The interconnectedness that quantum mechanics has uncovered is undoubtedly valid for small quanta that were in relation prior to the collapse of their joint wave function, but according to quantum mechanics, every collapse brings total oblivion of all past connections. And since the manifested, experienced, sensed world means constant flux of collapses through countless quantum interactions, its quantum wholeness is being constantly interrupted; it has no continuity. There are theories that the quantum reality at least in its field aspect (see Vitiello 2001) can represent the deeper level of reality that joins this (physical) world with thoughts, intentions, emotions and consciousness on one hand and assures wholeness of everything on the other. The biofield with its R level and possible quantum characteristics is also a good candidate for this dual role. But what could be the relation between the biofield and the quantum level of reality? First, the biofield should obviously work also at the quantum level since it should have its own quanta (e.g., magnetic monopoles or axions or whatever still to be discovered). Second, if we take Tiller's model, its quantum level is different from the ordinary one for the simple reason that it works in a higher symmetry level of reality. And third, even when "classically" treated (i.e. not on the quantum level) the biofield exemplifies wholeness as well as the connectivity with thoughts and other mental phenomena; as supported by numerous experiments throughout the world. Therefore, in conclusion, we may assume that the R level of the biofield functions in a similar way as other quantum processes. In addition, it seems to have the needed continuity of long-range correlations that the quantum world is lacking. If these theoretical derivations and partial experimental evidence for the biofield are taken as serious hypotheses, science may be able to discover new elementary particles, fields and laws, which govern the macroscopic reality and the wholeness of all beings and organized matter.

Biofield and Life

What the concept of biofield teaches us about life can only be touched very briefly, since firstly, life is a very abstruse and complex problem, normally viewed only in terms of molecules and their physicochemical interactions, and secondly, we still know very little about the biofield. From the standpoint of our main theme life can be regarded as a multilevel process. On the most superficial level we see organisms with their life cycles full of various activities that can be described in terms of physiology, ontogenetics and/or ethology. On a deeper level that has fascinated most biologists since 1953, we found DNA and the related biochemistry – Watson and Crick were for instance convinced that with DNA they discovered the outmost secret of life (Shapiro, 2007)[8].

On a still deeper and largely neglected level we see life as an electromagnetic phenomenon in line with Fröhlich's and Popp's theories. It is a much deeper level than the preceding one, but most probably not the last. This deeper level could be the R level of the biofield or its equivalent. We are able to describe life on the first two levels to a considerable extent; on the third one we are beginning to grasp some of its fascinating secrets (see for instance Popp et al, 2006). On the level corresponding to Tiller's theoretically predicted R level a vast void has opened; we are still tapping in the dark, not even knowing the exact nature of the field and its particles. But we can expect that this new horizon will unravel fresh and deep secrets of life, perhaps unveiling the mystery of the beginning of life – that still upsets scientists (Shapiro, 2007) – and some further mysteries of its evolution. It might disclose the role and nature of intelligence in the living world and many other things that we are perhaps still unaware of. As a result of all this we may expect a new and well grounded theoretical biology with some clear and generally accepted definition of life that we lack today, and a completely new and much broader outlook on life.

To speculate further: maybe we shall see that on the level of the biofield everything is alive and imbued with some basic intelligence that cannot be reduced to any Turing machine-based algorithm however complex it may be; and that on the ordinary level of our perception only those entities that are capable of representing the intricacies of the biofield are perceived as being alive. It is possible though, as said, that other physical theories and experiments, especially the physics of elementary particles and astrophysics, will also add in the search for understanding the reality behind the biofield.

Philosophy Behind Present Day Natural Sciences: Question of Formative Causation

Since Aristotle our civilisation has been enriched with some light regarding the causation. Aristotle found four basic types of causes: material (conditional), efficient (mechanical), formal and final (teleological). Of these four, natural sciences admit only the first two, while the last (for Aristotle the most important) is forbidden and believed to be non-scientific. But natural science excludes also the formal cause that can be interpreted as the formative or informational[9] and reduces it to the efficient cause. It seems as if the contemporary science is still fixed into the Newtonian world of short-range forces, in spite of many serious thinkers being dissatisfied with the constricted scientific world of only material and efficient causes. It is true that the quantum mechanics opened some new perspectives, and especially the Bohm's interpretation explicitly and emphatically included the formative causation into the scene through the so-called quantum potential (Bohm, 1982); yet his illuminative interpretation is regarded as superfluous, if not heretic to the mainstream Copenhagen school.

As already said the biofield should primarily work in a subtle informational way. It traces the paths for more concrete forces to express themselves. This can be seen in

[8] New developments in gene regulation and the role of the so called "junk" DNA are showing that this "dark matter of molecular biology" may well have numerous regulatory roles needed for a complex organism to develop. Especially strong is this regulation by non-coding RNA's in the brain (see, Ch1 of Ule J et al, Part A)

[9] In the sense of Bohm's concept of informing (see Bohm, 1982) and not Shannon's information theory.

electrophotography where the electric streamers (driven by strong electrical forces) are at least partially directed and shaped according to the stored information in a water drop or a finger (or the brain). In this way the biofield (on its R or equivalent level) would work similarly to the magnetic vector potential as envisaged in the famous Bohm-Aharonov effect and also empirically demonstrated (see Imry and Webb, 1989). Here the potential (it has no force!) changes only the phase of the wave function of some elementary particle in the vicinity, which, of course, may result in a vastly changed behaviour of a collection of such particles. In other words, though originally weak and subtle, the formative cause can have profound influences on events in the ordinary (macroscopic) world.

REFERENCES

Albrecht-Buehler, G. (1985). Is Cytoplasm Intelligent too? In: *Muscle and Cell Motility VI* (ed. J. Shay) p. 1-21

Aspden, H. (1969). Physics without Einstein, Southampton, Sabberton Publications.

Bajpai, R.P. (1999). Coherent nature of the radiation emitted in delayed luminescence of leaves. *Journal of Theoretical Biology, 19,* 287-299.

Bohm, D. (1982). Wholeness and the implicate order, Routledge & Kegan Paul, London, Boston.

Bruinsma, R.F. (2002). Physics of protein-DNA interaction. *Physica A: Statistical Mechanics and its Applications, 313*, 211-237.

Chang, J.J., Fisch, J., Popp, F.A. (Ed.). (1998). Biophotons. Kluwer Academic Publishers, Dordrecht.

Chaplin, M. F. (2007). The Memory of Water: an overview. Homeopathy, 96, 143-150.

Conrad, M. (1993). Emergent computation through self-assembly, Nanobiology, 2(1): 5-30.

Correa, P.N. (2001). Electroscopic demonstration of reverse potentials of energy flow able to draw kinetic and electric energies. Biophysical research Series S2-04, Publ. by Correa and Correa.

Correa, P.N., Correa, A.N. (2002). Fundamental measurement of biological energies I, Akronos Publishing @ Aetherometry.com, Canada.

Dawkins, R. (1976). The selfish gene, Oxford University Press, Oxford.

Del Giudice, E., Doglia, A., Milani, M. (1984). Order and structures in living systems, in: Nonlinear Electrodynamics in Biological Systems, Plenum Publ. Corp., 477-487.

Del Giudice, E., Doglia, A., Milani, M., Vitiello, G. (1985). A quantum field theoretical approach to the collective behaviour of biological systems. *Nuclear Physics B251(FS13)*, 375-400.

Del Giudice, E., Doglia, A., Milani, M., Vitiello, G. (1988). Coherence of electromagnetic radiation in biological systems, *Cell Biophysics, 13,* 221-224.

Del Giudice, E., Preparata, P. (1998). Coherent electrodynamics in water. In Schulte, J., Endler, P.C. (Ed.) Fundamental Research in Ultra High Dilution and Homoeopathy, Netherlands, Kluwer Academic Publishers, 89-103.

Elia, V., Nicolli, M. (1999). Thermodynamics of extremelly diluted aqueous solutions. *Ann. N.Y. Acad. Sci.*, 879: 241-248

Engel, G.S., Calhoun, T.R., Read E.L. et al. (2007). Evidence for wavelike energy transfer through quantum coherence in photosynthetic systems. *Nature, 446,* 782-786.

Friedberg, E.C., Walker, G.C., Siede, W. (1995). DNA repair and mutagenesis. Washington, D.C., ASM Press.

Fröhlich, H. (1975). The extraordinary dielectric properties of biological materials and the action of enzymes. *Proceedings of the National Academy of Sciences USA,* 72, 4211-4215.

Fröhlich, H. (1988). Theoretical physics and biology, in: Biological Coherence and Response to External Stimuli, Springer Verlag, Berlin, 1-24.

Goodwin, B.C. (1985). The causes of morphogenesis. *Bioessays, 3,* 32-36.

Goodwin, B.C., Webster, G. (1996). *Form and Transformation: Generative and Relational Principles in Biology*, Cambridge Univ Press.

Griest. K. (1996). The Nature of the Dark Matter, http://web.mit.edu/afs/ athena.mit.edu/user/ r/e/ redingtn/www/netadv/specr/012/012.html

Grinberg-Zylberbaum, G., Delaflor, M., Attie, L., Goswami, A. (1994). The Einstein–Podolsky–Rosen paradox in the brain: the transferred potential. *Physical Essays, 7,* 422-428

Hameroff, S., Penrose, R. (1995). Orchestrated reduction of quantum coherence in brain microtubules. In: Proceedings of the international neural Network Society, Washington DC.

Hameroff, S.R. (1987). Ultimate computing: biomolecular consciousness and nanotechnology. Elsevier, Amsterdam.

Ho, M-W. (1993). The Rainbow and the Worm. World Scientific, Singapore.

Imry, Y., Webb R. A. (1989). Quantum Interference and the Aharonov-Bohm Effect, *Scientific American*, 260(4).

Jerman, I., Ružič, R., Krašovec, R., Škarja M., Mogilnicki, L. (2005). Electrical transfer of molecule information into water, its storage and bioeffects on plants and bacteria. *Electromagnetic Biology and Medicine, 24*, 341-353.

Kauffman, S.A. (2000). Investigations. Oxford Univ. Press.

Korotkov, K. (2002). Human Energy Field: Study with GDV Bioelectrography. Fair Lawn (NJ). Backbone Publishing Co.

Leskovar, R.T., Jerman, I., Škarja, M. (2005). Near-field influence of organism´s endogenous electromagnetic field on environmental light particles. In: Pokorný J (Ed.). Coherence and electromagnetic fields in biological systems. Prague, 74-75.

Marshall, I.N. (2006). Elementary electric dipoles. Frontiers Perspectives, 15: 22-25.

Pizzi, R., Fantasia, A., Gelain, F., Rossetti, D., Vescovi, A. (2004). Non-local correlations between separated neural networks. In: Donkor, E., Pirick, A.R., Brandt, H.E. (Ed.), Quantum Information and Computation II. Proceedings of SPIE 5436, 107-117.

Pokorny, J. (2004). Excitations of vibrations in microtubules in living cells. *Bioelectrochemistry, 63,* 321-326.

Pokorny, J., Hašek, J., Jelinek, F., Šaroch, J., Balaban, B. (2001). Electromagnetic activity of yeast in the M phase. *Electromagnetic Biology and Medicine, 20,* 371-396.

Pollock, K.J., Pohl, D.G. (1988). Emission of radiation by active cells, In: Fröhlich H. (Ed.), Biological Coherence and Response to External Stimuli, Springer Verlag, Berlin, Heidelberg, New York, London, Paris, Tokyo, 141-147.

Popp, F.A. (1979). Coherent Photon Storage of Biological Systems. In: Popp, F.A., Becker, G., Konig, H.L., Peschka, W., (Ed.), Electromagnetic Bio-Information. Urban & Schwarzenberg, Munich, Vienna, Baltimore.

Popp, F.A. (1984). Biologie des Lichts, Paul Parey Verlag, Berlin.

Popp, F.A. (1999). About the Coherence of Biophotons. In: Sassaroli, E., Srivastava, Y., Swain, J. Widom, A. (Ed.), Macroscopic Quantum Coherence, World Scientific, Singapore-New Jersey-London-Hong Kong, 130-150.

Popp, F.A. and Nagl, W. (1988). Concerning the question of coherence in biological systems. *Cell Biophysics, 13,* 218-20.

Popp, F.A., Beloussov, L., Klimek, W., Swain, J. (2006). Coupling of Fröhlich-Modes as a Basis of Biological Regulation. In: Hyland, G.J. and Rowlands, P. (Ed.), Herbert Fröhlich FRS: A physicist ahead of his time. The University of Liverpool Press, Liverpool, 139-175.

Popp, F.A., Chang, J.J., Herzog, A., Yan, Z., Yan, Y. (2002). Evidence of squeezed light in biological systems. *Physical Letters A,* 293, 98-102.

Popp, F.A., Li, K.H., Gu, Q. (1992). Recent Advances in Biophoton Research, World Scientific, Singapore.

Popp, F.A., Nagl, W., Li, K.H., Scholz, W., Weingärtner, O., Wolf, R. (1984). Biophoton Emission: New Evidence for Coherence and DNA as a Source, *Cell Biophysics, 6,* 33-52.

Popp, F.A., Ruth, B., Bahr, W., Böhm, J., Grass, P., Grolig, G., Rattemeyer, M., Schmidt, H.G., Wulle, P. (1981). Emssion of visible and ultraviolet radiation by active biological systems. *Collective Phenomena 3,* 187-214.

Rao, M. L. Roy, R., Bell, I. R., Hoover, R. (2007). The defining role of structure (including epitaxy) in the plausibility of homeopathy. Homeopathy, 96, 175 – 182.

Rein, G. (2004). Bioinformation Within the Biofield: Beyond Bioelectromagnetics. *Journal of Alternative and Complementary Medicine, 10,* 59-68.

Rey, L. (2003). Thermoluminescence of ultra-high dilutions of lithium chloride and sodium chloride. Phisica A., 323, 67-74.

Rowlands, S. and Sewchand, L.S. (1982). Quantum Mechanical Interaction of Human Erythrocytes. *Canadian Journal of Physiology and Pharmacology, 60,* 52-59.

Rubik, B. (2002). The Biofield Hypothesis: Its Biophysical Basis and Role in Medicine. *Journal of Alternative and Complementary Medicine, 8,* 703-718.

Shapiro, R. (2007). A simpler origin for life. *Scientific American, 296,* 25-31.

Sole, R, Goodwin, B.C. (2001). *Signs of Life: How Complexity Pervades Biology,* Basic Books

Standish, L.J., Kozak, L., Johnson, L.C., Richards, T. (2004). Electroencephalographic evidence of correlated event-related signals between the brains of spatially and sensory isolated human subjects. *Journal of Alternative and Complementary Medicine, 10,* 307-314.

Stenger, V.J. (1999). Bioenergetic fields. *The Scientific Review of Alternative Medicine, 3,* 14-21.

Škarja, M. (2007). Near field measurements of biofield of organisms and in nature. in: Kononenko, Igor. (Ed.). Proceedings of measuring energy fields: International scientific conference, Kamnik, Tunjice, 13-14.

Škarja, M., Berden, M., Jerman, I. (1998). The influence of ionic composition of water on the corona discharge around water drops. *Journal of Applied Physics, 84,* 2436-2442.

Tiller, W.A., Dibble, W.E., Fandel, J.G. (2005). Some science adventures with real magic, Pavior Publishing.

Tiller, W.A., Dibble, W.E., Kohane, M.J. (2001). Conscious acts of creation: the emergence of a new physics. Pavior Publishing, Walnut Creek, California.

Trewavas, A. (2005). Green plants as intelligent organisms. *Trends in Plant Science, 10*, 413-419.

Van Wijk, R., Van Wijk, E.P.A. (2004). Human biophoton emission. *Recent Research Developments in Photochemistry & Photobiology, 7*, 139-173.

Vitiello G. (2001). My Double Unveiled. Amsterdam: John Benjamins.

Vitiello, G. (1992). Coherence and electromagnetic fields in living matter. *Nanobiology, 1*, 221-228.

Wackermann, J., Seiter, C., Keibel, H., Walach, H. (2003). Correlations between brain electrical activities of two spatially separated human subjects, *Neuroscience Letters, 336*, 60-64.

In: Philosophical Insights about Modern Science
Eds: Eva Žerovnik, Olga Markič and Andrej Ule
ISBN: 978-1-60741-373-8

Chapter 11

HOW TO RESEARCH EXPERIENCE?

Urban Kordeš*
Faculty of Education, University of Ljubljana, Slovenia

SYNOPSIS

The exploration of cognition, consciousness and everything else connected to these two areas is one of the biggest adventures modern science has undertaken in the last decades. The exponential progress of neuroscience has allowed us to take a scientific look at areas which we were until recently unable to observe *in vivo*. It is an interesting coincidence that it was the growth of objective (third-person) cognition research which encouraged research of direct lived human experience (first-person). It was neuroscientists who first began to notice that it is not that simple to collect the so-called "phenomenal data". It does not suffice merely to ask for data or prepare appropriate questionnaires. The questions arise: "If I want to situate experience in a theoretical framework – how can I think about it if I do not even know how to observe it? And how can I examine experience, if even the person reporting to me about it does not know how to observe it?"

In the areas where we are faced with a direct contact with living, concrete, unique, human beings, the ignorance of their (our) experiential landscapes merely suggests the loss of an important, perhaps even a crucial element. Thus a new area of research has recently been conceived: an attempt to systematically observe experience. We are talking about the so-called *phenomenologically inspired research approaches.*

The article briefly introduces this rather new approach in cognitive science. It is a mostly qualitative research project dedicated to the examination of experience, based on the assumption that phenomenal data cannot be reduced to or derived from the third-person perspective. Rather than aiming at objective (third-person) descriptions or theories it focuses on the lived experience as it is given.

The paper outlines the state-of-the-art in this field. Since phenomenological (or first-person) research is a common name for a wide variety of approaches, it also tries to introduce and classify the most important research directions.

The outline concludes with a *pro et contra* discussion of phenomenological research as a (potential) scientific endeavor: It is surely an open-ended, rigorous and systematic

* Kardeljeva ploščad 16, 1000 Ljubljana, Slovenia; e-mail: urban.kordes@guest.arnes.si

study. But on the other hand it cannot promise intersubjectively verifiable results, the data is not objective and its researcher is anything but detached from the "object" of research.

The text, of course, does not offer any concluding answers about the future of phenomenological inquiry, it merely reflects motives and hopes of a growing research community.

WHAT IS THE AIM OF THE SCIENCE OF CONSCIOUSNESS?

It is probably no exaggeration to say that the exploration of cognition, consciousness and everything else connected to these two areas is one of the biggest adventures modern science has undertaken in the last decades. Not so long ago this field of research was still in the hands of mystics, theologians, and philosophers at best. And some 100 years ago these were joined by psychologists and psychotherapists, who had felt the lack of appropriate scientific approaches since the very beginning (it is well known that Freud took up his research only because he had no empirical measuring techniques available which would allow him to approach his research in a similar way as physicists research the "behaviour" of inanimate nature).

It was not until the last decades that science has made its way into the human mind. The word *consciousness* has only recently been accepted into serious scientific texts. One can even talk about a "Science of Consciousness", thanks to the progress of computer technology which, during the sixties, brought about the development of artificial intelligence and resulted in the emergence of cognitive science. After the so-called computer metaphor of the functioning of the brain has proved to be partially disappointing, the ball was taken up by neurologists with new, non-invasive approaches to exploring the brain. Tremendous progress of cognitive neuroscience has recently virtually overshadowed the endeavours of other disciplines engaging in the research of cognition. Neuroscientists today are similarly enthusiastic as computer experts were some twenty years ago. They feel that a path has been opened to understanding the functioning of the brain in a way similar to understanding the "functioning" of the physical world – the descriptions of the dynamics of neural networks are becoming more and more accurate and the connections between neurophysiological processes and experience even more obvious.

We appear to be close to the moment when *reason* will be able to explain the functioning of *reason* (just like it already explained the "functioning" of inanimate nature). Of course we have not yet reached that stage. Cognitive neuroscientists still have much to learn using the trial-and-error principle. Nevertheless, an explanation of mind in the sense of formulas and variables appears to be possible.

A SCIENTIFIC EXPLANATION OF EXPERIENCE?

Let us consider what theories have to offer or what the term "to explain" actually means, e.g.: when a stone hits the ground, we say: "Oh, look, gravity," and – if we are skilled in physics – in this way we explain the observed phenomenon. We have found a sublime background of a subjective observation (experience): a process common to a falling stone, an orbiting planet, or a mighty rotation of entire galaxy. The explanation makes us see what lies *underneath*, the common pattern independent of this stone (planet, galaxy) and, most importantly, *independent of my perception*. If my perception happens to be mistaken, that is too bad for my perception (or me). If I think that I see white colour – wrong! In reality what I am looking at is just a dense and intertwined patchwork of all colours of the spectrum. A scientific explanation tells me what things are really like, objectively – independent of my experience. Such explanations can occasionally even be expressed in formulas, enabling me to predict and thus control the observed phenomenon.

What about consciousness? How does an explanation work in this area? Is the ultimate goal of the sciences of consciousness a grand unified formula describing (predicting?) behaviour and/or experience in the same way as physical theories describe the inanimate world?

If we put aside the major questions concerning such a goal (complexity of living beings, autonomy, and, last but not least, ethical dilemmas), we still have to face the fact that knowledge (the view "from outside") has its opposite – experience (the view "from within"). What about experience? How to examine it and what is the goal of such research? And even more importantly: *to whom* does such research make sense?

In the case of observing one's own experience, situating it into a theoretical framework (i.e. into a general, statistical context) does not have the same effect as in the case of phenomena in which I do not existentially take part. I do not understand the phenomenon any better by attaching a few scientific (or formulas) terms to a certain experience, at least not in the existential sense. Quite the opposite – this way I only distance myself from understanding it or knowing it, as I get carried away from observing experience to indulging in theories, statistics (experiencing a theoretical soliloquy so to speak). I pass from observing the experiential here-and-now to thinking about concepts. Above all, so-called "explanation" in turn makes part of experience. Thus it has not (merely) explained the experience but changed (or even replaced) it!

While the final goal of explanation in science is to achieve a description that will clearly determine connections between variables and thus allow for prediction, this is clearly questionable in the case of examining one's own experience. As mentioned before, by trying to predict one's own experience one arrives at a new experience (and quite a schizophrenic one at that!). It appears that knowing (one's own) experience means something completely different: something that cannot be properly defined in a propositional way, but rather makes whatever happens in consciousness more transparent (sometimes we say that we become more "aware"), so that we can *act* with greater ease.

Sometimes it is possible to say – once I have properly situated my experience into a theoretical framework – that I "understand" it. But this is completely different from letting our experience present itself precisely as it presents itself, instead of contextualising it, abstracting it, or "explaining" it. This only happens when I succeed to observe it in a way that expands the space of experience instead of dividing it. It would appear that in the area of experience algorithmic-propositional explanations do not have the same power or meaning as in other areas.

Is It Sufficient Merely to Explain the Connection between Physiology and Experience?

If I drink a cup of strong coffee, the world becomes nicer. I perceive everything in a more optimistic way, my relationship to the world is suddenly more intensive, erotic... Neurologists would say: "Of course, your adrenalin level has increased, your brain produces more neurotransmitters etc." Such explanations are in concord with the demands of natural science: they clarify and make predictions.

Neuroscience is making an epistemological mistake assuming a direct connection between the state of neurophysiological variables and mental states: here we are talking about the results of correlation research which is trying to determine the neurophysiological variables connected to (for example) a specific kind of experience (e.g., adrenalin – excitement). The mistake based on these correlations lies in assuming that a given variable is "responsible for" (or that it "causes") a given mental state. But even putting the mind-body problem aside, it is pretty clear that experience is connected to the *gestalt* of the entire state of the cognitive system. It is more a kind of convergence of numerous factors than a reflection of one single factor.

If we take the established neurological correlations as causes of mental states, we can imagine that one day we will be able to control consciousness at our free will (conscious will?). We will just take a pill and reach the desired state. But is this possible? The answer is: we do not know. Once I have asked Paul Haller, a Zen master from San Francisco: "If there was a pill assuring you enlightenment – would you take it?" "No way!" he cried even before I was able to finish the sentence. (He had apparently thought about this before.) He explained that this way he would be deprived of all the experience he had to undergo (of all his life, so to speak) and it is exactly this experience that matters. Experience in its entirety, awareness (its sharpness and broadness) etc. appear to be crucial attributes in "understanding" experience. The way is the goal.

In the field of consciousness, understanding appears to be not a place, but a way of walking. It is not an achievement, but doing. It is becoming clearer and clearer that in exploring consciousness (experience), it is not enough to simply describe it ("from the outside"), but that we must get to *know* it intimately ("from within"). We have to learn how to open up its space.

At first sight, such statements appear to belong to the new-age repertoire rather than science. This article intends to show that this is not the case. It is a fact that neurological third-person explanations ("from outside") have a clear and strong position in science, while it completely ignores research of experience or the view "from within". For Copernicus and Newton, giving up the admiration for the complexity and entirety of private lived experience was a small price to pay compared to the advantages brought about by scientific procedures. Even today the situation remains the same: giving up observation of the uniqueness of individual experiential landscapes appears to be a small price to pay for achieving general, objective results by following the standard way of observing. It is nevertheless interesting that it was the progress of neuroscience which brought about the growth of the research area described in the article. It was neuroscientists who first started to notice that it is not that simple to collect the so-called "phenomenal data". It does not suffice merely to ask for data or prepare appropriate questionnaires. Even if I want to situate experience in a theoretical framework – how can I think about it if I do not even know how to observe it? And how can I examine it, if even the person reporting to me about it does not know how to observe it?

In the areas where we are faced with a direct contact with living, concrete, unique human beings, the ignorance of their (our) experiential landscapes suggests the loss of an important, perhaps even a crucial element. Thus a new area of research has recently been established: an attempt to systematically observe experience. We are talking about the so-called *phenomenologically inspired research approaches.*

Before I embark on presenting this methodological novelty (or recently rediscovered classic, to put it more accurately), let me describe what appears to be a kind of an identity crisis experienced by many researchers of experience. Any enthusiasm about the meaning and/or research of one's own experience can fade away in sight of the success neurological "explanations" have achieved in soothing pain and illness. I believe it is no exaggeration to say that researchers of experience often tacitly envy neuroscientists. No epistemological doubts! Ever new, fascinating discoveries, many times even widely applicable! Research of experience is often accompanied by a kind of a split between dr. Jekyll (the "true" scientist observing "from outside", deciphering causal connections, searching for diagnoses and cures) and Mr Hyde (attempting to open up space and listen, hidden from the eyes of his colleagues, naively wondering at things and letting experience be what it is). Two personalities which are as of yet incompatible: the existential version of the so called "hard" (mind-body) problem.

It is by no means my intention to say that psychological theories or neurological explanations are meaningless.[1] But it is probably necessary to build up a certain opposition for the young and still fragile area of phenomenological research to gain some ground. Let us therefore start by describing the foundations upon which experience research is based.

PHENOMENOLOGY – THE BASIC ASSUMPTIONS

One of the foundations is the phenomenological insight that experience, more accurately my first-person lived experience, is primary. Not only it comes first, it is all I have – I cannot get to know anything outside the field of my experience. Thoughts, meanings, descriptions, visions, images, feelings, emotions… nothing exists outside one's conscious experience. The experiential world can be organised (e.g., to "inner" and "outer" or by classifying it into feelings, thoughts etc. or by scientifically explaining the observed phenomena), but I cannot get to consciously know anything outside my experience.

Phenomenological research therefore focuses on the observation of direct lived experience.

The term "phenomenological" comes from a philosophical movement established around 1900, which posed the above mentioned statement as its basic epistemological credo. But long before the official foundation of phenomenology, the poet Johann Wolfgang Goethe suggested his research programme which, if it had survived, could today be classified as a phenomenological one. He considered the experiential properties of light and how we experience colours, he proposed techniques to direct attention to the observation of a given light phenomenon so as to experience it as fully as possible. Goethe was not interested in

[1] The major approaches to phenomenological research suggest a balanced conjunction of both views (e.g. Varela's neurophenomenology.)

explanations but in the experience. He was concerned with the connections between different kinds of experience. In a truly poetic manner he suggested procedures of "throwing" oneself into experiencing colour. He also described how to reach the experience of an *Urphänomen* through gradual research of an entire range of experiential possibilities.

Goethe's "poetic" perspective met with disapproval and quickly faded into oblivion. It was no match for Newton's approach, which was also based on observation, but on the observation of repeatable experiments in controlled conditions. Here the depth and quality of the observer's experience no longer mattered. The experimenter had to determine and measure specific quantities. The essence of Newton's perspective was to "discover" hidden elements common to *all* phenomena involving light. He "discovered", for example, that white light is composed of a whole spectrum of colours. Newton's message read: white colour is an illusion. In truth what we are looking at is a multitude of different wavelengths.

Some 200 years later the mathematician and philosopher Edmund Husserl realised a simple truth that – despite everything – it is experience that has a necessary priority and, in the final instance, it is all we have at our disposal. There is no way we can escape its scope. He proclaimed that we should "return to the things themselves!" This proclamation symbolically marks the beginning of phenomenology, one of the strongest philosophical movements in the 20th century. Husserl himself was well-trained in procedures of natural science, but he found it unacceptable that we neglect direct lived experience on account of getting carried away into theoretical explanations. He founded phenomenology as a strict science which systematically explored experience: things (or phenomena) as they appear to us (instead of hidden truths and laws considered by other sciences).

Some of the basic methodological guidelines of Husserl's research programme were[2]:

- To focus on phenomena (things as they appear in our experience) and *epoche* – bracketing. Husserl suggests putting the usual early judgement calls about things aside (or putting them into "brackets") and applying phenomenological reduction – the reduction of the observed phenomena "as the only thing given and certain in experience". As Kotnik (2003, p. 102) says, the emphasis is on "the research of what is given in experience only, but it is imperative to include experience in its entirety". The first step towards achieving this is to recognise the complexity and infinity of this field.
- Rule: "Never explain, just describe!" This is the single most important methodological guideline of phenomenological research. This instruction may seem simple at first sight, but it is extremely difficult and complex to carry out in practice. It takes a great deal of reflection and skill. Only once we try to merely describe experience without classifying or explaining or situating it into theoretical frameworks in any way, we become aware of how deep our need to explain is and how hard it is to give it up. Kotnik (ibid.) quotes Ihde in talking about "how difficult it is to distinguish between the actually describable experience as it shows itself directly, and non-experiential elements such as assumptions or presuppositions. Any kind of theory, idea, notion or construction tending to go beyond the phenomena is already an explanation."

[2] The guidelines are recapitulated after a simplified, but very concise review of phenomenology in Kotnik (2003, str. 102) – which also includes quotations of Ihde (1977, str. 32 do 45).

- Refrain from convictions or evaluations of the "reality" of the observed phenomena. Without intersubjective verification (typical of the scientific method) we cannot distinguish between "illusion" and "reality". From the stance of the priority of direct experience such distinctions are merely one way to classify the experiential world (and as such no more valid than other possible distinctions). That is why Husserl recommends that even this – at first sight primary judgement – be bracketed and we observe the field of experience as it shows itself, without judging it.

As Husserl believed that only by abiding to the above stated guidelines we would get "lost in phenomena" (ibid.), he added a fourth guideline recommending the search for structure and invariable properties of the observed phenomena. Husserl believed that this way it would be possible to create a "transcendental" science – a system surpassing the ephemeral uniqueness of concrete experience by extracting the essential (transcendental) elements from it.

Phenomenological Research

To make a long story short, phenomenology advocates the research of experience, without indulging into metaphysical or theoretical speculation.

I must ask the educated reader to pardon my simplification – of course there is much more to phenomenology than that, especially if we take into account its variations and derivations (Merleau-Ponty, Sartre, Heidegger, Pfänder, Schütz, etc.). What phenomenological research (as a qualitative-methodological approach) takes after phenomenology as a philosophical discipline is the epistemological stance described above. Phenomenological research can generally be defined as a methodological category encompassing all empirical research methods, approaches, ways of data collection or data analysis based on above stated (phenomenological) epistemological foundations and guidelines.

This chapter intends to sketch this relatively new research approach.

The Chart

Phenomenological research is a very wide notion, and – since it is still in its infancy – rather undefined. Mescht (2004, p. 2) says: "Here I face a dilemma, since phenomenology has come to mean different things to different people. The particularly subjective and qualitative character of phenomenological research has led researchers to refer to any example of highly interpretive, qualitative research as *phenomenological*".

There is as of yet no general consensus about what the subject of phenomenological research actually is. Not so much due to disagreements among researchers in this field, but because – as I have mentioned before – different types of reporting about experience make up various areas of research. Usually researchers do not spend time in discussing how to gain phenomenal data or about what this data actually is. The fact that the "just ask" attitude does not suffice at all (just like in physics it is not enough to merely look at natural phenomena) is usually overseen. Phenomenological data is usually treated as a kind of unavoidable evil in the process of collecting objective data.

Phenomenological research can be defined as an attempt at serious, systematic collection (and sometimes analysis) of experiential data. Before we consider different approaches, let us first see what kind of data we are interested in. Varela describes the so-called phenomenal data as:

> "In spite of the variety of terminology being used, a consensus seems to have emerged that Thomas Nagel's expression 'what it is like to be' succeeds in capturing well what is at stake. Clearly 'what it is like to be' a bat or a human being refers to how things (everything) looks when being a bat or a human being. In other words this is just another way of talking about what philosophers have called phenomenality since the Presocratics. A phenomenon, in the most original sense of the word, is an appearance and therefore something relational. It is what something is for something else; it is a being for by opposition to a being in itself independently of its apprehension by another entity." (Varela et Shear, 1999, pp. 3)

Phenomenological research does not take its area lightly. Besides, phenomenological tradition is based on the results of early psychological experiments in introspection (Wundt, Titchener etc.) – and has learnt from their mistakes. Many phenomenological researchers are also inspired by the so-called *mindfulness* practices (such as Zazen or Vipassana meditation and their derivations in modern-day psychotherapy).

As mentioned in the introduction, the subject of phenomenological research is not exactly right up the alley of the scientific (let alone reductionist) view of the world. Nevertheless, there have recently been several attempts at systematic observation of experience. Schweitzer (2002, p. 1) classifies them into two groups: the so-called Husserlian type and the "What It's like for them...?" type.

The first group encompasses – roughly speaking – studies following Husserl's model of phenomenological reduction. Here a researcher is sitting in a chair, introspectively delving into his experience or aspect of it (e.g., the experience of time). Then he tries to describe the characteristics of the observed phenomenon as clearly as possible.

The second group covers all methods which try to grasp the experience of others: individuals or – more often – specific groups (e.g., schizophrenic teen-agers etc.). The article mainly focuses on this category as it is by far most widespread.

The described classification is of course just an approximation. It does not encompass all the nuances of phenomenological research. This goes especially for the recent radical experiments (Černigoj, 2007). In order to present a larger picture and to be epistemologically consistent, we will try to sketch a chart of different approaches in phenomenological research. I will present it as a three dimensional model, each dimension representing one important feature along which various approaches differ:

1. Research perspective

This axis classifies research approaches according to whose experience we observe. First-person perspective means that we observe our own experience. Third-person perspective means that we are interested in the experience of others. But there are also second-person approaches based on dialogue.

2. Intersubjectivity or the "scope" of theory

This axis classifies approaches according to the general value of their theoretical results. On the one hand we find approaches offering general results valid for the experience of anyone. On the other hand we have observations offering results valid only for me alone, here-and-now (there are also approaches in which propositional insights are not part of the research process at all).

The article mainly considers approaches which offer at least some degree of general insight. The scientific status of other approaches can be deemed doubtful, due to that they are mentioned merely as a curiosity. However, it needs to be pointed out that the fact that in some versions of phenomenological research there is no expectation of intersubjective verification of the obtained results does not necessarily mean that general insights are impossible. We just do not expect them because we allow for the uniqueness of individual experience or the dialectical process.

3. Emphasis on...

What is it that we observe? The structure of experience, the contents we use to interpret or make sense of a given situation? There is of course a full range of possibilities between the two extremes. One can focus on certain parts of the experiential landscape, on its entirety, on a certain activity, a lived situation etc.

The third axis is a patchwork of different emphases determining different approaches rather than a linear spectrum. Different goals call for different methods: if a researcher wants to explore an entire experiential landscape, for example, he must find a way to make a representative sample of experience; but if one is interested in a specific experience (e.g., the experience of intuition), one must repetitively delve into certain selected moments.

It would appear that all known forms of phenomenological research today can be characterised by their position in the 3D space determined by the three axes. But of course, this classification is just as incomplete as any other and it is unable to grasp the vivaciousness of the emerging area in its entirety.

With the help of these basic classificatory tools we can now take a look at some of the most interesting landscapes of the area.

Husserlian Research

1. Research perspective	2. Intersubjectivity or "scope" of the theory	3. Emphasis on...
First-person	Intersubjective, general (wide scope)	Structure of individual (common) patterns of experience

This approach has already been mentioned before. It involves research approaches closest to the original idea of phenomenological reduction as conceived by Husserl. We talk about first-person research as the researcher observes his experience through systematic introspection.

The researcher focuses on a specific experience in a phenomenological manner. One can focus on the perception of shadows and light (Goethe's favourite subject), or the observation of some object, one can listen to music etc. Observing is hermeneutically rounded up into experiencing. According to the idea of phenomenological bracketing, the observer intends to isolate a "pure" experience – that which comes before interpretation and is common to the whole category of experience. Thus Husserl, for example, observed the experience of time and temporality. He believed that he succeeded in defining the structure and invariable properties of the observed phenomena (the third axis of our research space), while in the area of observing light Goethe felt that he found the way to the so-called *ürphenomenon* – a kind of experiential essence and smallest common denominator. He believed that the way to this experiential essence can be described in a manner which anyone would be able to follow.

Thus we can see that it is an attempt to give an intersubjective, general theory, valid not only for the observer, but for everybody (second axis). Many researchers feel that the latter is a matter of discussion (in spite of the fact that Husserl dedicated time and energy to proving that phenomenological reduction is an intersubjective science) – the question of the possibility of deducing the characteristics of other people's experience based only on one's own experience remains open.

Here it must be pointed out that it is exactly in this field that phenomenal research has not yet established itself as an empirical science. Philosophers often imagine that the access to experience is simple – and that research (i.e., collecting empirical data) is therefore unnecessary or rather trivial (we just need to classify what we already know). At this point phenomenological researchers attempt to distance themselves from philosophy and establish their own direction as a science based on empirical fact – regardless of how subjective and elusive it is.

Descriptive Experience Sampling

1. Research perspective	2. Intersubjectivity or "scope" of the theory	3. Emphasis on...
Third- and second-person	General or valid for a specific group (partial to wide scope)	Structure of (common) basic patterns of experience

Descriptive experience sampling is methodologically speaking the most sophisticated kind of phenomenological research. It focuses on examining the "pure" structure of experience – the basic constituents or patterns composing an individual's experience. The method was founded by the psychologist Russel Hulburt as early as 1980, but it only gained worldwide acclaim at the turn of the century. Its derivations were mostly accepted by psychotherapists and the so-called "positive psychology" researchers (Csíkszentmihályi). Most researchers involved in this kind of research do not use the term "phenomenological" to describe it (this research is not based on philosophical phenomenology, but similar conclusions were reached independently).

The basic idea of descriptive experience sampling (DES) is to describe the structure of experience as precisely as possible, regardless of the content and meaning ascribed to it by the experiencing person. The goal of DES is to compose a kind of encyclopaedia of basic experiential elements and connections between them. DES can be compared to geological surveying of the ground: we take samples in random places, purify them and analyse them in a laboratory. Similarly in DES we survey (sample) experience in randomly selected moments.

Practical application of such surveying involves a research subject carrying a device which emits a discrete signal 7 to 10 times per day at randomly selected moments. The subject attempts to "freeze" the experience just prior to the signal by writing it down into a notebook as precisely as possible. (There are of course updated versions using palmers and/or cellphones.) When a subject gathers 6 to 8 good "samples" (usually after one day of sampling), he meets the researcher and together they "purify" and analyse the notes. Researchers emphasise that they are not interested in explaining the thoughts or other experiences – the subjects are encouraged to focus only on the structure (the question is "how?" not "why?"). They repeat the process several times, until the researcher (usually in agreement with the subject) decides that the gathered "samples" give an appropriate depiction of the structure of the entire experiential landscape. Afterwards, researchers extract the most prominent characteristics of the basic experiential structures according to the above mentioned principles of phenomenal reduction[3].

The result of analysis is a list of the subject's basic experiential elements ("phenomena") and a description of connections between them (the structure of experience). Each element is named and defined according to five categories. Let us look at *worded thinking* as an example[4]:

Pure phenomenon: Worded thinking is the experience of thinking in particular distinct words, but those words are not being (internally or externally) spoken, heard, seen, or voiced in any other way.

Example: "I was thinking, 'I should give him the letter.' Those exact words were somehow present in my awareness, but I can't tell you how. They were not spoken, and I did not see them. But somehow they were there, one after the other."

Variants: Sometimes but not always the phenomenon will include a hint of visual experience.

Discriminations:

Inner speech: Words experienced in inner speech have definite vocal characteristics (pitch, timbre, inflection, etc.) and timing (rhythm, sequence, etc.). Words in worded thinking are not vocalized.

Unworded speech: is the experience of speaking in one's own inner voice, except that there is no experience of the words themselves. Thus the characteristics of the voicing are present in unworded speech but absent in worded thinking.

---

[3] The DES method has been meticulously described in several books by Russel Hulburt, most concisely in Sampling Normal and Schizophrenic Inner Experience (Hulburt, 1990).

[4] Cf.: http://www.nevada.edu/~russ/codebook.html#target4

Unsymbolized thinking: Unsymbolized thinking does not include the experience of words.

Images: A clearly seen visual image of a word or sentence is classified as an image (of words).

Considerations: . . ."

The DES researchers are aware of the fact that their method is basically idiographic – and thus do not expect the results to be generally valid. It is quite interesting to see how differently we experience phenomena which are assumed to affect everybody in the same way. On the other hand, researchers tend to look for common categories shared by several groups of researched subjects.

The experiential structures they describe as very common are e.g.: Inner Speech, Partially Worded Speech, Unworded Speech (UWS), Worded Thinking, Image, Imageless Seeing, Unsymbolized Thinking, Inner Hearing, Feeling, Sensory Awareness, Just Doing, Just Talking, Just Listening, Just Reading, Just Watching TV, Multiple Awareness, etc. Each individual subject of course experiences idiosyncratic phenomena, but Russel Hulburt principally focuses on examining certain groups of people attempting to find common characteristics of their experiential world. His most famous cases are the studies of eating disorders and schizophrenia (Hulburt, 1992).

Let us mention also that researches involved with DES emphasise the importance of research interviews (through which they "grasp" and "purify" the samples). It is recognised as an essential element of the method – the element distinguishing "naïve" informal observation of one's own experience from scientific, systematic (and, according to them, much more successful) research.

Researches of the *"What it is like to be...?"* Type

1. Research perspective	2. Intersubjectivity or "scope" of the theory	3. Emphasis on...
Third-person	Wide or limited scope	Content of experiencing, making sense of lived situations etc.

Researches of the "What it is like to be...?" type are the most widespread and most diverse type of phenomenological research. The expression "What it is like to be...?" comes from the article of the cognitive philosopher Nagel "What it is like to be a bat?" in which the author points out that we can never really know how other people experience the world (let alone other creatures). This we can only assume.

This does not mean of course that we are unable to examine it. The conclusion that we can never fully know other peoples' experience merely induces a certain amount of modesty to third-person phenomenological research and reminds us to insist on the "I don't know" point of view. The biggest problem of this type of research is also the source of its beauty and vitality.

Researches of the "What it is like to be...?" type intend to shed some light on phenomena through the way in which agents experience them. They are primarily interested in how agents make sense of the lived (experiential) situation, what are the experiences connected to it etc. They try to compose as detailed, profound and vivid picture of the experiential space linked to phenomena as possible. It might be said that it is a type of radical qualitative research which abstains at any cost from generalising its results.

This category consists of very diverse types of studies, from examining children's experience of traumatic events to describing the feeling of professional competence in helping professions. The areas in which this phenomenological approach is most successful belong to psychotherapeutic research, all types of pedagogical sciences and other helping professions.

Phenomenological research of the "What it is like to be..." type (others call it "empirical phenomenology") originated in the seventies, when the experimental psychologist Amedeo Giorgi (1970) published the initial interpretative guidelines of this approach in the *Journal of Phenomenological Psychology*. In several articles and monographs that followed he also described very precise and practical methodological guidelines. The four volumes of *Duquesne studies in phenomenological psychology,* published by the Duquesne University Press are also very important (even today we can read about the "Duquesne school of phenomenology"), as they present the ontological and epistemological background of this approach. For several decades one of the most quoted references in the field of phenomenological research (especially in pedagogy) was the *Phenomenology + Pedagogy* review. Unfortunately it no longer exists. It has been partially substituted by the *Indo-Pacific Journal of Phenomenology* – an online publication trying to present as wide a range of phenomenological research as possible.

In the follow-up I will attempt to list some of the principal characteristics of the "What it is like to be...?" approach (according to Mescht, 2004, pp. 2 – 3). But first I would like to stress that none of the described characteristics is exclusive "property" of this approach. I mention them merely to give the reader an impression of what this research looks like.

- An acknowledgement that research participants' 'reality' is not directly accessible to the researcher, and that the researcher's focus is thus neither on the phenomenon nor on the participants, but rather on the 'dialogue' of individuals with their contexts, the sense-making as constructed by the agents and the link between experience and behaviour;
- A focus on 'lived experience', an obsession with the concrete; verbal data are interrogated for how they 'language' participants' physical, emotional and intellectual being-in-the-world. Other data – such as data collected through observation or filming – are rarely used;
- An insistence on description, rather than interpretation; while the line between these – description and interpretation – is thin and perhaps contestable, the drive to stay with description until a holistic picture of the issue emerges is fundamental to phenomenological research. It embraces the notion of Husserlian bracketing and works against the tendency to make early judgment calls based on pre-conceived notions.
- Participants are purposively selected on the basis of experience of the phenomenon under investigation, as well as their linguistic proficiency in the research language. Since participants' reports are usually the only data researchers work with, it is essential that participants are verbally fluent and expressive. One of the weaknesses

of phenomenological research is exactly its focus on eloquent and verbally expressive subjects;
- The researcher adopts a position of "conceptual silence" (Stones, 1988, p. 124 in Mecht, p. 5), or trust (Kordeš 2004), bracketing *a priori* theories, hunches and suppositions. This position can be described by Freud's term of "floating attention". In an attempt to honour all data equally (and not be tempted to analyse and thus set aside what appears to be irrelevant) the interview protocols are reduced to natural meaning units, in which each unit represents a statement that makes complete sense, expressed in the words of the participant. The researcher explicates the natural meaning units, and then *describes* what is presented, thus attempting to capture the *lived-world* of the participant. Only when a holistic sense of the participant's lived world is obtained through description does it become appropriate to extract themes and compare findings with other sources, such as literature;
- The participant's lived experience of the phenomenon is then set within its context, both locally (usually the organisation) and more broadly (perhaps in terms of national or international circumstances, such as policy).

Examples of the "What it is like to be...?" type of research are (some of the themes of researches published in the *Phenomenology + Pedagogy* review[5]):

- Waiting for a diagnosis (Ferguson)
- How does a child experience things? (Langeveld)
- Women endeavouring to look attractive (Szekely)

See also the very important work done by Daniel Stern (2004), who gained wide reputation for his microphenomenological (or micro-analytical, as he calls them) researches of the experience of a moment.

Neurophenomenologists' studies (Varela, 1996) could also be included into this category, especially those by Claire Petitmengin, who describes her work as second-person approach, as it – in addition to research of subject's experience – emphasizes the need for the research subject to attain self-awareness.

Enacting States or Roles

1. Research perspective	2. Intersubjectivity or "scope" of the theory	3. Emphasis on...
First-person	General or valid for a specific group (partial to wide scope)	Experiences (content and structure) of specific existential states or roles

Example: The researcher intentionally (temporarily) disables her sight in order to get to know (some) features of the experiential world of the blind. Thus she attempts to find out what it is like to be blind. She is of course aware of the fact that she will never be able to experience blindness in its entirety. But she can hope to gain insights which the blind are

[5] The studies are available at: http://www.phenomenologyonline.com/articles/articles.html

unable to transmit verbally (propositionally) as the fully sighted simply do not possess the experiential categories that could match their experience.

Enacting existential positions (states) or roles is – as can be seen – the most existentially involved or participatory type of phenomenological research. It distinguishes itself from the previous category in that the researcher describes her own experience. It is therefore a true first-person research.

The goal of research is not necessarily to get acquainted with extreme experiential situations (such as in the above-mentioned case). Enacting encompasses a wide range of areas from examining special circumstances or states in the researcher's life to intentional changing of one's own experiential world in order to better understand the world of others. The spectrum ranges from studies similar to the above-mentioned third-person group ("What it is like to be...?") to extreme attempts at changing one's own experiential world. What they have in common is the fact that researchers attempt to personally experience a selected (often unusual) experiential "situation".

There is a small but very interesting group of studies in which researchers attempt to enact the experiential world of others. The above mentioned "self-induced blindness" could be listed among them. Others include examining the experience of the handicapped, the homeless, the imprisoned etc.

This group (partially) also includes studies of altered or special states of consciousness. A radical example is the first-person studies of the effects of drugs (common in the seventies and eighties: Leary, Lilly etc.)

Participatory phenomenological research is not very common. Understandably so, as such research demands a high degree of participation, while results are not necessarily generally valid. Nevertheless, it is necessary to pose the question: can one speak about a given existential position if one does not experience it by oneself at least in the above-mentioned degree?

This type of phenomenological research can only be carried out by means of systematic and planned introspection following the rules of phenomenological reduction (up to a point). Often it helps to have an assistant helping the researcher with questions and taking notes of the researcher's observations (e.g., some of the researchers of the effects of psychoactive substances who carried out their experiments accompanied by assistants).

This one and the "What it is like to be...?" approach can be interpreted in two ways:

- As a holistic research method which includes collecting "raw" empirical data and its scientific processing. In phenomenological research it is allowed to finish a study without the final analysis – it suffices to clearly present the selected phenomenon from the point of view of the agents' experience. Some authors do attempt to classify the achieved data into phenomenological categories, while others give merely an orderly description fully reflecting the agent's words.
- As a method of collecting empirical data. An account of the agents' experiential worlds can be used as empirical input for any other type of qualitative research (qualitative text analysis, case study, material for an ethnographic study etc.)

This is also true of other phenomenological methods.

Phenomenological Case Study and Phenomenological Interview

1. Research perspective	2. Intersubjectivity or "scope" of the theory	3. Emphasis on...
Second-person	Limited scope	Researcher allows the respondent to select the focus

Our review of phenomenological research approaches is now almost complete. Let me just touch upon a phenomenological case study – a method very rarely used but nevertheless personifying the very idea of phenomenology: dialogue.

A phenomenological case study could also be called "freestyle phenomenology". It is based in the dialectical method in which the researcher through his questions attempts to open up the respondent's space. The researcher has no fixed conception of themes he would like to address, but rather attempts to direct the dialogue towards the themes most relevant to the respondent. This way he allows the respondent's awareness of his experience to widen. Through questions he helps him to "discover" areas of experiential world he might earlier not have been aware of. The questions merely offer support – we use them to show our interest, our participation. It is imperative for the interviewer to maintain an open mind and to persist in the "I don't know" position. The more a priory judgements, ideas and beliefs we manage to get rid of, the more space we create for new insights. The respondent must feel that his words "hit the spot", that he is being heard and seen. Questions beginning with "why" have no place in phenomenological dialogue, as it is not intended to encourage explanations and interpretations. Its primary interest is "how".

Of course it is useful to have an idea about which cognitive categories are to be examined before starting the interview, but this is merely a "plan B" in case that the respondent loses track and it is necessary to break the silence. As much as possible, the respondent should determine the direction of the interview.

Phenomenological dialogue can be used as a research method helping us to describe the experiential world of the respondent. Such a description, the result of several interviews, is called a phenomenological case study. The result of such a study is as clear, graphic and systematic description of the research subject's experiential landscape as possible – without attempting theoretical comparisons, classifications etc.

But it has an even bigger role as an instrument in another, even more extreme phenomenological technique – *self-research*. From the point of view of self-research the roles in the interview are reversed: the researcher here is actually the one who is answering. The interviewer is there merely to support him in examining his own experiential space. As it is extremely difficult to strictly and systematically examine oneself or one's experience, the (self-) researcher asks a fellow researcher to help him in observing the field of his experience.

It is not the goal of such phenomenological dialogue to gain propositional insights about one's experiential world (as in the case of the Husserlian method), but to change (expand) one's experience. Jack Petranker (2003) says that any study of (one's own) experience changes this experience (we become "conscious differently"). This results in a newly formed experiential world and, even more importantly, in improving the skill of self-observation.

At this point – when addressing research which makes sense exclusively to the researcher himself – we are stepping beyond the limits of science, at least in the present-day sense of the word. But as we can see, it does not necessarily mean transgressing the limits of possible research. Can a research that brings no intersubjectively verifiable insights be considered scientific? A research, in which the principal area of research is one's own experiential field and where the bond linking researchers into a community is their research orientation rather than their results?

FIRST-PERSON RESEARCH

"Certainty", as we can see, is not exactly a word associated with the area of phenomenological approaches. Can the sideways of exploratory curiosity still belong to science? Have we lost track in this quasi-spiritual experimentation or are we on the threshold of reaching something epochal? We do not know.

Before finishing, let us just enumerate some relatively solid facts about what is happening in this young (and fragile) branch of cognitive science:

1 The basis of phenomenological approaches is the epistemological insight that our first-person experience is primary. Not only it is primary; it is all we have at our disposal – nothing can be perceived outside the field of one's own experience. The experiential world can of course be organised (e.g., through scientific explanation), but phenomenological research prefers raw description, without categorising, filtering, judging.

2 Simple observation: despite my experience being existentially very close to me, I am often incredibly unaware of it. This appears to be quite common. (Thus we can put it into plural.) We very rarely consider observing our experiential world. Most of our time is spent on inventing stories or explanations – but very rarely do we take a wider view of the entire process of "invention". We usually live our stories without reflecting upon them. Only exceptionally do we take the time to observe them.

3 Phenomenological research DOES NOT belong to science because (despite some opposing views) intersubjective validation cannot become a base for validating truth-value of findings. The fact that my observations are incompatible with the observations of others does not reduce their value as phenomenological data.

4 Phenomenological research DOES belong to science, because it is open-ended. It is thorough and systematic research with no a priori set goals (unlike in spiritual training, here we do not know where we are going). If I take a "scientific" view of my experience, I intend to merely see and not to change it. The absence of intentionality, the absence of judgement, the absence of escaping the *here-and-now*, the observation of that which is there – these are the basic features of scientific work. If we apply them in examining experience, they create a meta-cognitive dimension, from which experience can be viewed almost uninterrupted.

5 One of the principal goals of the phenomenological enterprise is to create a research community or rather a culture of experience researchers. We are not bound by our

results (as mentioned before, intersubjective verification is not a condition for the validity of experience), but our intentions (to explore experience).

One of the turning points for first-person science was an article written by Francisco Varela in 1996 in which he suggested the foundation of a new field of research – neurophenomenology – which would, beside neurological research, direct its attention to the research of experience. In the last decade this field has made some headway, but it is still far too early to try to extract any methodological guidelines. Personally I can only say that I have to marvel again and again at how fulfilling it is to open up the space of experience when I do nothing but observe or listen. I am familiar with the feeling of freshness, presence, insight that wash over me – and others sharing my path – in such moments.

As for the rest, it is mainly hopes and open questions. Could something more like a science emerge from this? Perhaps a new 'science of being' (Černigoj, 2007)? Even if it came to that, researchers in this field will always be aware that we can never reach the end. We can do nothing but keep examining, and hope that one day we will be able to gaze into the unknown without fear, without expectations, without filters, without blinking.

References

Bradfield, B. (2007): Examining the Lived World: The Place of Phenomenology in Psychiatry and Clinical Psychology. *Indo-Pacific Journal of Phenomenology*, 7, 1

Černigoj, M. (2007): *Jaz in mi – raziskovanje temeljev socialne psihologije.* Ipsa, Ljubljana

Giorgi, A. (1970): Toward phenomenologically based research in psychology. *Journal of Phenomenological Psychology,*1.

Hulburt, R. (1992): *Sampling Normal and Schizophrenic Inner Experience.* Plenum Press, New York

Ihde, D. (1977): *Experimental phenomenology.* G. P. Putnam's Sons, New York

Kotnik, R. (2003): *Pouk filozofije kot osebno doživljanje: izkustveno poučevanje filozofije kot aplikacija načel Gestalt terapije*, (Knjižna zbirka Učbeniki, 5), Pedagoška fakulteta v Mariboru

Mesch, H., Van der (2004): Phenomenology in Education: A Case Study in Educational Leadership. *Indo-Pacific Journal of Phenomenology*, 4, 1

Petranker, J, (2003): Inhabiting Conscious Experience: Engaged Objectivity in the First-Person Study of Consciousness. *Journal of consciousness studies*, 10, 12

Stern, D. (2004): *The Present Moment in Psychotherapy and Everyday Life.* W. W. Norton & Company, New York

Varela, F (1996): Neurophenomenology – remedy for a hard problem. *Journal of consciousness studies*

Varela, F. et Shear, J. (1999): *The View from Within.* Imprint Academic, Exeter

In: Philosophical Insights about Modern Science
Eds: Eva Žerovnik, Olga Markič and Andrej Ule
ISBN: 978-1-60741-373-8

Chapter 12

Consciousness and Vision

Mitja Peruš[*]
Laboratory for Cognitive Modeling, FRI,
University of Ljubljana, Ljubljana, Slovenia

1. Synopsis

Visual experience is to a large extent conscious. Unconscious visual perception, which is mainly automatic or involves less important contents, will not be considered here. Conscious vision requires bigger collaboration of science and philosophy than unconscious vision, although they both share the bulk of the neural substrate up to extrastriate cerebral cortices. There is a gap between our objective or rather intersubjective quantitative knowledge on the brain and visual processing on one hand and our introspective qualitative knowledge or rather feeling of visual phenomena (i.e., on how things look to us, what *experience* we have) on the other hand.

As the intrasubjective (first person) aspect will be found to remain unexplainable for now, we will mainly present research on the "objective" (third person) aspects, i.e. neuropsychological, neurophysiological and subneurophysical ones. The processes of pictorial and color experience, imagery, extraction of figure from background, synchronization and binding of perceptual elements into wholes (objects, scenes) and clinical phenomena like blindsight will be discussed.

Although we present the hypothesis that quantum processes are essentially involved in conscious experience, we emphasize that consciousness is a complex multi-level and multi-aspect phenomenon where physical, biochemical, dendritic, neuronal, network-dynamical, informational, psychical processes are *all* essential and irreducible.

Keywords: consciousness, visual, experience, attention, qualia, color, quantum, coherence.

[*] Žibertova 1, SLO-1000 Ljubljana, Slovenia; mitja_perus@t-2.net

2. Introduction to Consciousness

Holistic Aspects of Consciousness

There are two aspects of the holistic brain–mind process (Arbib, 2002; Roth, 2003; Oakley, 1985; Luria, 1973) which manifests consciousness (Velmans & Schneider, 2007):

1 *information-processing background* emerging from the so-called neural / physical correlates of conscious process (the third person aspect);
2 subjective, qualitative, *phenomenal experience* of the irreducible *I* (the first person perspective).

The first aspect incorporates roots of consciousness in a "sea" of unconscious and subconscious background processes. Together with (phenomenal) consciousness they constitute the mind, i.e. mental processes. Let us list, in a broad context, some mental functions that can be mediated by consciousness: *emotion, motivation, intentionality, planning, goal-achieving, I / self-schema, self-reflective representations and thoughts, judgements, control in novel circumstances, volition, creativity, spiritual "dimensions"*, but also ordinary *feelings* and *(felt) needs* (Damasio, 1999), and *qualia* (details in: Velmans & Schneider, 2007; Marcel & Bisiach, 1988; Pribram, 2004, 1998; Peruš, 1998a). *Qualia* are subjective, qualitative, phenomenal experiences ("how things seem to us"). Examples are experiencing yellowness of a lemon, feeling pain in one's own elbow, and in general also *what it is like to be* a person, etc. Qualia are "felt in the first person" only, not (directly) in the third person. A blind person cannot imagine precisely how it is to see; person A does not know precisely how person B feels (Flanagan, 1992). Qualia stay unexplained by now.

A special ("second-order cybernetics") case of the first-person perspective is *awareness*. It is self-reflective (recursive) and, since it is *Self*-based and entails self-consciousness, it is also *Self*-reflective, i.e. self-aware. *Introspective* awareness is a deeper or active (self-monitoring) case of self-awareness ("third-order cybernetics").

Pribram (1998) considers the *Self* having two aspects: the *I* which is a first-person self-perspective (e.g., self-visualizational, -episodic, -narrative – as in telling about one's own actions: e.g., "*I* have taken…"), and the *Me* which is a third-person self-perspective (intentional – as in referring to oneself, i.e. to "*my*self": e.g., "This belongs to *me*." "This is *me*, photographed thirty years ago.").

I will use mainly *the holonomic theory* by Karl Pribram (1991) because it is holistic, it is especially suitable for modeling vision, and it does not exclude relevance of quantum processes. He uses "*conscious experience*" instead of the usual too-reified (inappropriately substantive) term "consciousness", thus unifying the subjective and the objective aspects of the phenomemon.

Pribram distinguishes the following aspects of *conscious experience*:

1. *conscious process itself* (e.g., conscious thought, attention, intention),
2. *contents of consciousness* (percepts, or mental representations of, usually, external objects), and

3. *neural correlates of consciousness* (neuro-chemically determined states or patterns of neural activity that, presumably necessarily, accompany conscious processes) (Pribram, 2004, 1998; cf., e.g., Davies & Humphreys, 1993; ASSC, 1998; AuxilRef 1).

Neural Correlates of Consciousness

"One can no more hope to find consciousness by digging into the brain than one can find gravity by digging into the earth, " wrote Pribram (2004, p. 11). We can merely trace neural processes that are correlated to the conscious process itself. Conscious experience is based on physiological *arousal* which is, at least partially, controlled by the *reticular activating system* of the middle part of the brainstem. In this context, locus coeruleus is important for vigilance, and damage to intralaminal nuclei of the thalamus can cause irreversible coma and vegetative states (Frith et al., 1999).

Both wakefulness and dreaming (Jasper et al., 1998) are manifested by similar activity-patterns in the thalamo-cortical circuit, and both include subjective experience, but in REM-periods of dreaming we are usually not aware of it. Dreams are characterized by vivid, but unreflected, imagery, similar to a sort of delirium (rather than psychosis), in spite of lower arousal. (Experiments show that higher animals also have such dreams.) (Frith et al., 1999)

States of conscious experience are in general accompanied by increased neural and metabolic activity comparing to unconscious states. It is the same with superliminal versus subliminal stimulation and with paying attention (to an object encoded in a neural state) versus being non-attentive. Attention regulates access to conscious experience, but is, in turn, also often controlled by it. (Baars, 1997)

3. Theories of Conscious Experience

3.1. Focuses

Hypothetical theories of conscious experience can be *roughly* divided into several rival groups which focus on the following positions (reviews in: Velmans & Schneider, 2007; Marcel & Bisiach, 1988; Flanagan, 1992; Hameroff et al., 1996; Newmann, 1997; Rakić et al., 1997; Železnikar & Peruš, 1998; AuxilRef 7-10):

1. conscious process emerges from collective dynamics in specific networks (e.g., neural nets);
2. conscious process is a result of attentional scanning circuits (e.g., thalamo-cortical, reticular);
3. conscious process arises from coherent 40(-80) Hz firing of far-apart neurons belonging to the same brain area (e.g., V1) or to different areas;
4. conscious process is essentially a quantum phenomenon modulated by (sub)neuronal parallel-distributed processing (e.g., in dendritic, or/and microtubular, or/and biomolecular networks);

5. conscious process is entirely non-material or even non-natural, or it is entirely mysterious.

Ad 1. Emergence theories are usually bottom-up physicalist theories emphasizing hierarchical structures of assemblies of assemblies (attractors of attractors) of neurons or other units. I agree that such processes take place, but they operate top-down as well as bottom-up, and are not exclusively responsible for consciousness.

Ad 2. Conscious process is identified, usually by proponents of the main-stream neuroscience, with (an emergent process arising from):

a ERTAS (Extended Reticulo-Thalamic Activation System): *thalamo-cortical-loop*-driven (Jasper et al., 1998) voting-like competition, modulated by *nucleus reticularis thalami*, of modules for attention (Baars, 1997; Newmann, 1997);

b processes of the *intralaminar complex* of the thalamus in cooperation with the cortical areas interconnected with it (but also the reticular formation of the brain-stem);

c activation of the *theta-system* in the hippocampus (O'Keefe in Oakley, 1985); other (auxiliary) structures.

These views base on findings that, in a rough sketch, reticular activating system regulates wakefulness, the nucleus reticularis of the thalamus mediates processes of directing attention to various brain-areas (Bickle et al., 1999), and the superior colliculus of the tectum[1] mediates integration of sensory information.

A critique of these hypotheses can be given: They say more about the physiological origins of arousal, alertness and attention than about conscious experience. These arousal centers are, of course, necessary for conscious process, but are not sufficient to describe the conscious processing of mental *contents*. For the latter, neocortex is needed — although coma happens with damage to the intralaminar complex and (usually) not with damage to neocortex. The intralaminar nuclei trigger the cortical EEG characteristics. (Baars, 1997)

Ad 3. The scientific majority thinks that global processing and unification (*binding*) of information-"parts" *that belong together* is realized by coherent oscillations in neural units that encode the "parts". Cooperative phase-synchronization[2] was found, by many independent multi-electrode recordings[3], 1. among assemblies of neurons in the same cortical column where similar features are processed, 2. among distant assemblies belonging to different columns of the same cortical area, and 3. also among assemblies in different areas of the visual cortex when they simultaneously process some *common feature* of the stimulus (Gray et al., 1989, 1990).

However, Gray et al. (1990, p. 94) acknowledge that there is a question how can distant cortical columns become coherent often even with *zero phase-delay* in spite of considerable transmission-delays in the synchronizing connections. Quantum (or, in Bohm's terms, sub-

---

[1] Multiple inputs to the superior colliculus show that it should even be needed to generate a meaningful and coherent sense of selfness and its relations to experience – as argued by: B. Strehler (1991): Synapse 7, 144-191.

[2] I.e., oscillations "go up and down" simultaneously or always with the same delay.

[3] Multiple micro-electrode recording is necessary for measurement of phase-dynamics. Coherence is in-phase oscillation, i.e. waves go up and down together.

quantum) non-local processing with *instantaneous*, noise-less intercommunication (Ho in Fedorec & Marcer, 1996; AuxilRef 5) within a coherent (bio)substrate might well provide solution to this problem (Bohm, 1980; Peruš, 1997a,b,c; AuxilRef 5). Thus, coherent neuronal oscillations realize merely a subset of perceptual binding phenomena, but relation with conscious experience remains unclear.

Ad 4. There are two branches of the quantum hypothesis:

4.1. Conscious experience, or (more precisely) at least its physical correlates, are ontologically of quantum nature, albeit not exclusively. This means that conscious experience is essentially related to quantum phenomena, but might not be reduced to them. It might have additional, unknown features. (Reasons for this hypothesis are described in, e.g., Lockwood, 1989; Goswami, 1990; AuxilRef 6, 5, 2; Peruš, 1997a,b,c).

4.2. Quantum theory can help in modeling (sub)neuronal complex-systems' parallel-distributed processing by analogical reasoning. So, it can help us epistemologically, indirectly in understanding (sub)neuronal correlates of consciousness (e.g., Barahona da Fonseca et al., 1999; AuxilRef 3, 4, 7, 9).

Pribram (1991) advocates both possibilities as either–or: Conscious experience is either related to quantum (4.1) or to quantum-like (4.2) parallel-distributed processing, or both. For me, 4.2 is true for sure[4], and 4.1 is very likely, but not definitely proved, except by taking parapsychology seriously. Here I mean especially the reports on non-local transpersonal and transcendental experiences. Although many individual reports are not (entirely) reliable, the enormous amount of parapsychological reports is not to be rejected *a priori*. On the other hand, for instance, a single clear telepathy-like case (not to mention more exotic phenomena) is enough to reject the theory of exclusively classical-physical (e.g., exclusively skull-limited neural-net) nature of consciousness.

It has been definitely theoretically (Bohm & Hiley, 1993; Gould, 1989) and experimentally (Aspect et al., 1982) demonstrated that the (sub)quantum world is essentially *non-locally inter-connected* and *entangled* (AuxilRef 5). *If* this might enable, e.g., proto-telepathic manifestations in altered states of consciousness, this would help quantum views. However, if it will turn out definitely that conscious experience has quantum roots (Stapp, 1993; AuxilRef 6) and that it might have in exceptional cases also transpersonal manifestations, this does not yet mean that the problem of consciousness is solved. (Sub)quantum unity (coherence) or re-unification can (at least in principle) provide *binding* at the ultimate level, but cannot explain qualia.

This is my opinion in spite of the fact that Penrose and Hameroff (1998, p. 125) considered qualia as fundamental as the most microscopic[5] *networks of quantum spins*: "Qualia might be particular patterns in fundamental space-time geometry [...] encoded in Planck-scale spin networks." To say again, this might well turn out to be true, but it does not decipher the enigma of the essential nature of qualitative appearances within phenomenal consciousness.

Ad 5. I think, hypotheses from the positions of philosophical *idealism* (i.e., non-materialism) or *dualism* (mind distinguished from matter: e.g., Popper & Eccles, 1977)

[4] Consider numerous analogies between quantum systems and other brain's complex systems listed in Peruš (2000b; 2001a,b). Cf., Wang (1998), Dubois (2000), Bonnell & Papini (1997), Zak & Williams (1998), Ezhov & Ventura (2000), Ezhov (2000), Marcer (1999), Marcer & Schempp (1997, 1998), Nobili (1985), Pessa & Vitiello (1999), Snider et al. (1999).

[5] Planck-scale dimension is 10^{-35} m. Spins are the most fundamental "binary" variables attributed to particles.

neglect too much the natural phenomena, especially the numerous physio-psychological signatures or brain–mind parallels, inter-dependence, interactions, complements.

3.2. Special Cases and Problems

Eccles Theory

Eccles, however, is not a trivial member of neither category 5 nor 4, but of a mixed category. This Nobel laureate certainly did not neglect his own field – physiology. He emphasized the current explanatory gap between material and non-material processes. As I am a monist, I think that dualist Eccles maybe makes this gap too "ontological", but one cannot exclude possibility that he might be right. Eccles' idea that *mental events cause neural events by influencing the probability fields in quantum mechanics* (Eccles, 1986) is potentially very perspective and too far-reaching to get large support soon (details in the next subsections). Possible experimental evidence still lies mainly in the field of parapsychology (e.g., so-called Princeton anomaly research led by R. Jahn), but not exclusively (e.g., Pop-Jordanova & Pop-Jordanov, 2002). As often in philosophical questions, words (categories like monism / dualism, interaction, control) may be the main reason for misunderstandings…

Quantum Effects in Synapse

Eccles (1986, 1993) pioneered the idea that conscious mind, using attention, could influence the probability of discrete (quantal) release (*exocytosis*) of vesicles full with neurotransmitter-molecules at the hexagonal-paracrystalline presynaptic vesicular grid.[6] Conscious mind would impose effect on probabilistic quantum processes (e.g., the wave-function collapse) underlying the probabilistic exocytosis in synapses (Beck & Eccles, 1992; AuxilRef 2). So, conscious process would selectively modulate, through quantum fields, the essential ingredients of memory-storage and associative processes – synaptic efficiacies.[7] To be precise, quantum influences should trigger electronic rearrangements resulting in movement of hydrogen-bridges which would effect vesicle-release from the presynaptic hexagonal grid (Hameroff, 1994).

Mind—Brain Relation

Eccles (1993, p. 10) writes: "Mental experiences, such as feelings, may not be vague nebulous happenings, but may be microgranular and precisely organized in their immense variety so as to bring about accurate description of the type of feeling. […] Each of these mental units [Eccles calls them *psychons*, note by M.P.] is reciprocally linked in some unitary manner to a *dendron* [a bundle of apical dendrites with their branches, note by M.P.] […]"

It is not enough clear (to me) what Eccles means with psychons. However, his parallelism of *dendrons* and psychons might be meant somewhat similar to the parallelism of configurations of neuronal activities together with *dendritic* processes (or polarizations around them) on one hand and the *attractors* of system's dynamics on the other hand (details

[6] Originally, Eccles (1986) writes: "On the biological side, attention is focused on the paracrystalline presynaptic vesicular grids as the targets for non-material mental events. On the physical side, attention is focused on the probabilistic fields of quantum mechanics which carry neither mass nor energy, but which nevertheless can exert effective action at microsites."

[7] Experimental modulation of neurotransmitter-function by quantum fields see in Rein (in Pribram, 1993).

on attractor dynamics: Peruš, 2001b, 2000a). In such a case, a psychon would correspond to an *attractor* of processing in a dendron. An attractor is *emergent* or, rather, *supervenient* (cf., Kim, Chalmers) on a neurophysiological configuration.[8] Inter-level influences (between configurations and attractors) are (alternatingly) bi-directional (i.e., top-down *and* bottom-up).

Pribram on Conscious Experience

The Holonomic Brain Theory by Pribram (1991) distinguishes the surface-structure of memory and conscious experience, which correlates with processing in neural systems, and the deep structure which correlates with synapto-dendritic field dynamics (Pribram, 1997b). Pribram (in Hameroff et al., 1996, ch. 10) writes that the *electrochemical synapto-dendritic states are coordinate with the states of conscious experience* as well as with the results of (conscious) learning including memory-storage. Varieties of conscious states are produced by various organizations of the polarization field within the same synapto-dendritic substrate (Pribram, 1971).

Apart from different philosophical interpretations of conscious experience, there are some notable parallels between views of Pribram and Eccles (cf., Pribram, 1993, with Eccles' keynote), which are in any case richer than between these two pioneers (which both emphasize dendritic processing and allow quantum influence) and the present main-stream neuroscience (focusing on axonal signaling).

Conscious experience is attenuated when actions cohere with their guides and so actions become automatic (Pribram in Hameroff et al., 1996). Similar attenuation happens when perceptions match with expectations, or when no novelty is perceived, respectively. In such cases, neural circuits do the job without involving any conscious processing (which has vanished as much as dendritic *processing* has vanished). Unconscious processing in neural circuits is sufficient also for reflex-actions (where dendrites seem merely to transmit signals without processing them inside the dendritic-web field). For a reflex-move of a hand, neural circuits guide the muscles more directly and thus more quickly than dendritic nets — because neurons are closer in size and power to muscle-cells than dendrites are. So, the conscious processing in dendritic nets is excluded during reflex-actions — in order to cause no dendritic-processing delays, to be as fast as possible.

Pribram (1998, Summary) writes: "Weiskrantz's, Stoerig's and Cowey's distinction between reflexive, phenomenal and conscious by accessible processing[9] becomes [...] a distinction between automatic, referential (semantic) and episodic executive processing. Automatic, reflexive, processing occurs whenever a neural circuit has become thoroughly established with a minimal synapto-dendritic delay. Referential, phenomenal, processing is semi-automatic but easily accessible to monitoring when shifts among reference frames[10] are initiated within circuits, either "spontaneously" or by some more organized sensory or central input.

---

[8] Eccles has probably considered psychons as more autonomous entities than merely emergent or supervenient.

[9] Note by M.P.: This divison is presented in: P. Stoerig & A. Cowey (1995): Behav. Brain Res. 71, 147-156. They conclude that phenomenality is prerequisite for the so-called conscious accessibility (a term for conscious representations available for verbal reports, judgements, planning, guiding of action, and the like) to the present or previous visual processing (i.e., for "usable" visual introspection) which needs higher cortical functions.

[10] Note by M.P.: Pribram (1991) defines a frame as the immediate local context influencing an input-perception.

Executive monitoring comes about when frontolimbic processes become involved in processing delays allowing shifts among reference frames in the systems of posterior convexity."

Dendrites and Consciousness

Pribram (1998) emphasizes that conscious experience depends on delay-causing processing in modifiable dendritic patches between synaptic arrival signals and axonal departure signals. Shifts of reference-frames and monitoring of such shifts occur during this dendritic processing. Namely, dendritic processing introduces delay in overall neuronal input-to-output processing (resulting in the axonal output-signal) and this "gives opportunity" for the consciously-monitored and controlled shifts. Delays are unavoidable, if signals get distributed over the dendritic trees and their junctions and if they are extensively processed there. So, delays are a sign of extensive dendritic processing accompanied by structural subcellular changes (which are induced by transit of signals and thus increase delays). Reflex-actions, as it is necessary, avoid (dendritic) delays as long as they are executed by neuronal-circuit processing only. Dendrites probably just transmit signals in such cases. (Pribram, 1998, 1991)

Delays also "give time" for communication with other brain-areas and their cooperation in processing. Other systems, especially the frontal cortex, seem even to facilitate the delays (Pribram, 1998, after Fuster) in order to intervene and consequently, to appropriately change the dendritic and global structure of the circuits. The rapidity of changes in awareness would so be proportional to the delay, i.e. to the duration of interference-processing in synapto-dendritic polarization- or quantum fields (Pribram, 1991, 1997b). Peruš & Dey (2000) and Peruš (2001a) interpret such interference-processing as the phase-Hebbian memory-storage, but strongly top-down modulated, or shaped by other processing layers, respectively. This has been successfully computer-simulated by Peruš et al. (2005).

Conscious monitoring processes are: attention which monitors sensory input, intention which monitors action and volition, and thought which shapes memory (recall) (Pribram, 1998). In all three cases, Eccles' idea that mind influences the quantum probability fields, specifically (I guess) the probability coefficients c_k in the quantum superposition $\Psi = \sum_k c_k \psi_k$ (cf., Peruš, 2000b, Peruš & Dey, 2000), could in principle work. Stapp in Velmans & Schneider (2007, ch. 23) discusses this possibility.

To summarize Pribram, processing delays, dendritic processing and consciousness are coincident, and seem to be thus actively correlated. Processing in circuits of neurons which by-passes consciousness, is possible in the cases of automatic behavior, subliminal perception, priming and unconscious learning. On the other hand, reference-frames, which configure sub-neuronal processes, become conscious (a precondition for introspection) during dendritic-processing delay. Pribram's idea of coincidence of dendritic processing and conscious processing is in agreement with Eccles. They might disagree merely in the ontological status of this coincidence and in explicating details about it (cf., Pribram, 1997a).

Quantum Basis?

As long as mental processes are not conscious, they might be (to a large extent) realizable by classical physics and neural networks. When they become conscious, deeper processes like quantum wholeness (*entanglement*) (Aczel, 2003) may start to operate significantly. Pribram

(1991) supposes that dendritic field dynamics, significantly quantum-influenced[11], are needed for conscious experience.

Peruš (2001a) develops the hypothesis, which expands Pribram's view to the quantum domain, that purely-quantum processes (e.g., wave-function "collapse") also are *necessarily* involved in conscious recognition and other processes. It is based, additionally to the numerous unrepeatable hypothetical indications of parapsychological phenomena, on mathematical-physical observations and considerations: Peruš & Dey (2000) and Peruš (2000b, 2001b) describe where quantum dynamics are similar to neural-net dynamics allowing information processing, and Peruš (1997b, 2000a) discusses *why* the quantum idea. This hypothesis is not yet experimentally provable or testable *more than* parapsychology can provide, but Peruš & Dey (2000) and Peruš (2000b, 2001b) allow establishing such a hypothesis which has many relatives by other authors (Woolf & Hameroff, 2001; Hameroff et al., 1998, 1994; Jibu et al., 1995, 1996, 1997).

Multi-Level Binding

The mainstream view versus the Pribram (1991) view differ in neuronal versus sub-neuronal binding suggestions. Main-stream cognitive neuroscience (Kosslyn & Andersen, 1992) has problems with *binding*, i.e. connecting all the specialized, partial perceptions of "feature detectors" and similar sorts of cardinal neurons into an unified perception (e.g., how a stable phenomenal image emerges from a "mosaic of neurons").

Roelfsema (1998) proposed two possibilities for assembly-labeling and binding: first, the *synchrony of neuronal discharges*, and second, the *enhanced firing-rates* (evoked by *attention* [12]) of the neurons encoding features which belong to a common object and should be bound together. Precise timing of action potentials could also be needed as a label of such features. Responses of lower visual centers, which are triggered through an eye that participates in conscious visual perception, are different from responses which are evoked through an eye with suppressed vision, only in the higher rate of synchrony (Roelfsema, 1998). This stronger synchrony might cause higher activities in related subsequent processing areas.

According to Pribram, the binding problem, and the problem of needing extremely many cardinal neurons for all the special features, are artificial problems which emerge from wrong premises. The neuronal coherence is probably, at least indirectly, bi-directionally related to coherence-phenomena at subcellular (e.g., dendritic) and quantum levels. The ultimate origin of inter-connectedness and coherence of (neuro)physical states is doubtlessly at the (sub)quantum level.[13] The neuronal and dendritic coherence seems to be like a higher-order stimuli-induced *re*establishment of the fundamental (sub)quantum coherence or its fractal-like replica.

So, the "binding solutions" of Roelfsema (1998), Baars (1997), and of other scientists who do not want to involve quantum physics, are just auxiliary *re*binding suggestions, since (potentially conscious) states have always been bound quantumly and can *re*bind quantumly if *re*arrangements of (conscious) states are needed, I believe. Quantum binding (at least sometimes manifested as the so-called quantum entanglement, i.e. inseparability) is always more perfect than (sub)neuronal binding.

---

[11] Cf., Bob & Faber (1999), Jibu et al. (1995–1997), Nishimori & Nonomura (1996), e-journal Neuroquantology.

[12] Eccles' hypothetical influence of mind onto neural transmission in synapses could in principle act here.

[13] Sub-quantum dynamics (in "vacuum" or "holomovement") are defined in Bohm (1980). See also AuxilRef 5.

So, binding is, like brain processes in general, probably a multi-level phenomenon related to multi-level (quasi)coherence (e.g., neural and sub-cellular or quantum wave dynamics appear more or less in rhythm, oscillations of some sort in one level trigger oscillations of another sort in another level) (cf., Haken's synergetics). Coupled oscillatory processes can be enough perturbation-adaptive/compensatory and "error-corrective".

Awareness

Consciousness, or better to say conscious experience, is representational; awareness is *re*-representational: Conscious experience is about a content (it is, by a philosophical word, intentional); awareness is (more or less) about consciousness (of that content). The division is not strict, of course, since representational levels may often "inter-penetrate" or even "merge" (Gennaro, 1995).

Awareness is needed for many specific changes in learning and behavior. Namely, in experiments by McIntosh et al. (1999), such changes[14] occurred only in aware subjects. Awareness was correlated mainly with the (left) prefrontal cortex, but only if its activity was integrated with many other (higher) distributed, large-scale networks.

The parallel-distributed processing background or correlates of awareness are very probably *self-interaction or self-interference processes* of collective states (configurations or assemblies that constitute patterns–attractors) in numerous bionetworks (details in: Peruš, 2001b, 2000a). In short, a content of our awareness bases in the "virtual action" of a specific attractor upon or within itself, or within its own basin of attraction, respectively.

Qualia Unexplained

None of the present theories explains qualia, although there is an increasing number of scientists who take them as essential and irreducible. The "quantitative skeleton of qualia" (AuxilRef 1, 9, 10) is increasingly traced with PET-scans etc., but the qualitative features (AuxilRef 11, 12) still remain encoded in subjects' adjectives used in verbal reports.

4. Visual Conscious Experience

Visual Pathway

Processing of the visual content proceeds along the pathway *eye—LGN—V1—V2&higher—ITC*. LGN is the lateral geniculate nucleus; V1 is the primary visual cortex called striate cortex; V2, V3, V4,... are the extrastriate cortices; ITC is the inferior temporal cortex. After the retinal and LGN net pre-processing, the left and right images are fused into the "cyclopean" representation in V1, enabling stereopsis (three-dimensional vision) and depth perception. Details on further image-processing, mainly in V1, and object perception, mainly in V2, up to further associative processing and global comprehension, mainly in ITC, see in Peruš (2001a) and Peruš et al. (2005).

[14] Subjects became aware or not that a tone predicted a visual event and another tone did not.

Neural Correlates of Conscious Vision

Experiments by Logothetis (1999) and many others show that we are (almost) always unaware of (the great majority of) neural processes subserving visual conscious experience. It was evident from fMRI-images that monkeys had active visual areas although they were anaesthetized. Logothetis (1999) concludes that only a small percentage of the whole community of neurons seems to represent the direct "neural correlates" of visual consciousness, and such neurons are probably distributed all over the visual pathway, with some concentration in higher areas like inferior temporal cortex (ITC).

Clinical Data on Selective Loss

Conscious experience appears to normal subjects to be integrated, unified, holistic and stable, but this is so only "in the final product" or "on the level of envelope" of a heterogeneous multi-level process. Its internal structure becomes evident after selective loss of performance in lesions:

In visual agnosia, conscious object-perception functions well, but patients find no meaning (or purpose) of the object with recognized form and color (Baars, 1997). Prosopagnosia, where a patient cannot recognize faces of known-to-him persons, for instance, is shown (Young & de Haan in Davies & Humphreys, 1993, p. 64) to be a failure of being conscious about the recognized rather than a loss of the recognition mechanism.

In color-agnosia, patients describe their world as "black-white, gray, dirty, erased".

Reticular formation is necessary for wakefulness; intralaminar nuclei are necessary for *waking* consciousness, but cortex is not (not for wakefulness!). Cortex *shapes the contents* of conscious experience (Baars, 1997). Damage to visual association-cortex does not lead to any overall loss of consciousness, but usually to a specific, partial loss (e.g., failure to consciously recognize known faces). Some patients can be blind to specific parts of the external visual field only (e.g., to the left half-space); the other part they can see.

Blindsight

So-called blind-sight is such a case, presumably a result of lesion in V1 and almost always limited to one hemisphere. Such a damaged V1 gives no (conscious) visual responses, and although the extrastriate cortex is active, the patient is cortically blind in the contralateral visual half-field.[15] Blindsight patients can visually discriminate objects in the blind half-field, although without phenomenal visual experience. That is, they are not conscious of seeing anything as if they would not see at all, but if they are "forced" to guess, they give correct location and description of presented objects or patterns. Some of blindsight patients "have conscious residual vision in response to fast moving stimuli or sudden changes in light flux,

[15] From this, it was concluded that "blindsight does not depend on functional islands of tissue preserved within the deafferented striate cortex" – in the abstract of: P. Stoerig et al. (1998): Neuroreport 9, 121-125.

characterized by a contentless kind of awareness, a feeling of something happening, albeit not normal seeing".[16]

Blindsight may reveal something about the role of V1 in *conscious* image processing. Koch (in Hameroff et al., 1996) argues that we are *not vividly aware of* the majority of processing in V1 (e.g., not of the eye-dominance in imaging, and not of the high spatial frequencies which V1 responds to). Koch and Crick (in *ibid.*) reasoned that "we are visually aware of certain [results of [17]] neural activity in V4 that could be triggered by activity in V1." It appears that V1 is crucial for vision including conscious vision[18], but our visual awareness resulting from neural correlates of visual processes starts with higher extrastriate areas (V4, ITC?) etc.

Visual Consciousness

Mainly all the characteristics of consciousness and awareness are valid specifically for their visual branches if the content is a pictorial representation (e.g., a "real", externally-induced image in imaging, a "virtual", internally-induced image in imagery and in so-called lucid dreaming). Conscious vision is the processing where we are conscious of its, usually colorful, pictorial content – image.

Visual awareness is awareness using visual resources. It is strengthened, often, by seeing and recognizing one's own image in the mirror. Since visual attention is being attentive to visual stimuli, all characteristics of attention are valid for its visual branch also, except the physiological pathways of attentional influence onto vision which are much specific / limited to visual cortex (in contrast to, say, auditory attention).

5. Processes Accompanying Consciousness

Systems of Visual Attention

Attention (Styles, 2005) helps in object identification by support of V4 and ITC (inferior temporal cortex), but these areas are not alone. Mesulam (1981) proposes that the limbic (cingulate) cortex adapts and directs the attentional focus according to subject's interests. The reticular formation arouses the whole cortex (Luria, 1973; et al.).

According to the "searchlight" hypothesis by Crick (1984), selecting and moving-around of the restricted region of the visual field while subject attends is regulated by the reticular

---

[16] From the abstract of: L. Weiskrantz, J. Barbur, A. Sahraie (1995): Proc. Natl. Acad. Sci. USA 92, 6122-6126.

[17] Note added by M.P., as also Pribram would say. Namely, it is not clear what do Koch and Crick mean by "awareness of neural activity". The note is our interpretation – after personal communication with Prof. Pribram.

[18] Cognitive psychologist Baars (1997, p. 67) writes: "V1 is the only region whose loss abolishes our ability to consciously see objects, events, people, dramatic gestures, delicate textures, the bold shapes of a Picasso painting. But the cells in V1 respond only to a sort of pointillist level of visual perception, like the dots of light on a television screen. Thus it seems that area V1 is needed for such higher-level experiences, even though it does not contain higher-level elements! [... ...] Damage to V1 'blocks out' all higher-level visual consciousness, while damage to object perception [areas] causes only a local loss." (Emphasis his.) See also chs. 43–45 in Velmans & Schneider (2007).

nucleus of the thalamus.[19] Vidyasagar (1999) writes that attentional spotlight has origin in the dorsal magno-cellular pathway (toward posterior parietal cortex), but acts in serially searching the visual field for targets in the ventral (or temporal, i.e. toward ITC) parvo-cellular pathway (so, first "where", then "what").

Attention Windows

Interesting experiments showed that ITC neurons have very large receptive fields, covering often the most of the visual field, in anaesthetized animals.[20] In alert animals, on the other hand, receptive fields are restricted to the attended window (Moran & Desimone, 1985) which is constantly changing, moving around, adapting to the situation and shifting spatial scale in focus. Moran & Desimone (1985) also showed that when there are two stimuli in the cell's receptive field, those cell (in V4 or ITC) responds much more to the attended stimulus, even if it was ineffective, than to the unattended stimulus, even if that one was effective. Let us note that effectiveness is a property of stimulus, but attention a property of the (conscious) visual system.

So, indeed, attention is a real process, based on *top-down inhibitory influence* which focuses the information flow to important features. For instance, evidence that attention activates the visual nucleus reticularis of the thalamus from V1 using cortico-geniculate top-down pathways is reported by Montero (2000). Bickle et al. (1999) simulated effects of these back-projections including the resulting lateral inhibitory influence of the *nucleus reticularis* (NR) onto LGN. Computational simulation of the LGN–V1–NR circuit, using the so-called Interactive Activation and Competition Neural Net by Grossberg, gave plausible results: Activity in V1-columns with less stimulated neurons decays relative to the most stimulated striate cortical column. Beside voluntary bases, attention has internal automatic bases (Stillings et al., 1995, ch. 12). The visual system somehow "gets attracted" by the stimulus itself.

Attention manifests mainly in V4 cells and more widely in ITC cells, but usually not in V1 cells which do not give such results in the Moran & Desimone (1985) experiment as just described. However, the top-down attentional influence is realized, among others, by feedbacks from ITC and V4 onto V1 [21] which affect changes in position, size or shape, and scale of the window, "trying systematically" to increase the information content. There is one exceptional case where attentional spotlight is effective as early as in V1: when several competing objects (not just one) are presented in the visual field simultaneously (Vidyasagar, 1999).[22]

[19] Pulvinar should be responsible for engaging attention, but parietal cortex for disengaging it, after Olshausen & Koch in Arbib (1995) who cite a work by M.I. Posner & S.E. Petersen (1990): Annu. Rev. Neurosci. 13, 25-42.

[20] R. Desimone & C.G. Gross (1979): Brain Res. 178, 363-380.

[21] Lower-order cortical areas may be modulated by higher order-areas. So, V1 neurons are indirectly influenced by attention. V2, getting feedbacks from higher visual areas, acts with strong feedback further on V1 and even LGN (Aine, Supek & George, 1995 – after: M. Mignard & J.G. Malpeli (1991): Science 251, 1249-1250).

[22] This is reported also in: S. Luck et al. (1997): J. Neurophysiol. 77, 124-142. In V1, "complex attentional window is consistent with object-based attention, but is inconsistent with a simple attentional spotlight," is concluded after fMRI experiments by: D. Somers et al. (1999): Proc. Natl. Acad. Sci. USA 96, 1663-1668.

Figure/Ground Segmentation

Each perceived pattern usually has the figure-part (with figures, i.e. object-forms) and the background-part. The representation in ITC is probably the figure, extracted by an *image segmentation* process (i.e., extracting the figure from its uninteresting background) which is supported by appropriate attention-window positioning. Segmentation of a figure, e.g. a face, is accompanied by object-centered coordinates. Sompolinsky & Tsodyks (1994) invented a network of oscillators which can selectively segment stored objects.[23] Belonging to the same object is encoded by the *coherence of the phases* of individual oscillators. Different objects are "*bound* by" different phases. Without the phase information, all objects of the image would be equivalent for the net, and segmentation would not be realizable. So, also selective attention would not operate, because the whole image, encoded by a single attractor, would remain undifferentiated. Namely, objects of the image would be undividedly trapped together in the attraction-basin of the common pattern–attractor.[24]

Top-down filtering, or even ignoring the unwanted information, probably has origins in conscious process, or its motivational, volitional and ego-centric components, respectively. Conscious I and volition are supposed to have neural correlates in the prefrontal cortex (cf., Kosslyn & Andersen, 1992, V.)

Pictorial Experience

Visual patterns–attractors, i.e. carriers of mental images[25], move from one physiological and virtual level to another, because of fractal-like self-similarity and "mirror-like reflectivity" of networks and their holography-like processing at various levels, and because of universality of attractors irrespectively of the substrate. When an image-attractor is projected to the quantum level, quantum coherence or/and the Bose-Einstein condensation could provide the *unified* conscious experience of the image, we propose. The quantum associative net (basics in: Peruš & Dey, 2000; for vision: Peruš, 2001a; Peruš et al., 2005) (potentially) realizes *associative processing of images, associative storage, (conscious) image recognition (and binding) simultaneously*. This statement is true without words in brackets for our computer simulations, but with words in brackets it may be valid for brain (I hope).

Imagery

There is (much) evidence that *visual imagery* shares representation and neural correlates with visual perception (e.g., Baars, 1997; Changeux, 1986), and that waking and dreaming

---

[23] Memory-storage is realized by a learning procedure which I call phase-Hebbian. It combines the Hebb correlation, or convolution, with phase-coupling for associative memory. (Peruš & Dey, 2000; Peruš, 2001a,b)

[24] A distributed pattern of activities of an assembly of neural units acts as an attractor if it represents an image with a meaning. Attractor means here that other neuronal configurations, which are within the basin of attraction (of the pattern acting as attractor), transform into the attracting pattern. See Peruš (2000a, 2001b) for details.

[25] A pattern of neural-units' activities which acts as an attractor represents a gestalt — it is the neural substrate of the mental image or gestalt. The "mentality" itself emerges from the "virtual influence" of the attractor's potential well and from consciousness (which one cannot explain, but just see its relation to the "virtual action").

visual imagery (Kosslyn, 1994) also share underlying processes. On the other hand, visual imagery might be autonomous with respect to other memory and reasoning processes.[26] It is no wonder that imagery shares perceptual areas since it bases on remembered perceptions. Even visual illusions may appear in imagery (Kosslyn, 1988). Imagery is nevertheless divided into two kinds of processes — those using stored shapes and those using separately stored relations between shapes while arranging an image and its derivatives (Kosslyn, 1988). Tootell et al. (1998) ask whether the representation, which is during imagination reconstructed from memory in higher visual areas (especially V2, V3, also V4), is projected back, via existing point-to-point topographic back-projections, to V1 where "it appears on mental screen".

There is enough evidence for existence of *mental images*. If subjects were solving a task where they had to identify objects, depicted on one side, with their rotated views, depicted on the other side, they used an amount of time proportional to the angle of difference in views (Shepard's experiment). This shows that they "mentally rotated" the depicted objects to match pictures of both sides. Kosslyn's subjects had to recall a map. For that, they used an amount of time proportional to distances between the locations on the map which they had to recall. Namely, they "mentally traveled" from one to another recalled location, as on the real map before. The mental image is not clear and has somewhat better resolution only under introspective attention. It is stored in an abstract, condensed form. We have impression of an unified image, but this is much misleading. As Dennett said, "we cannot recall how many stripes a tiger had." Eidetic people are a remarkable exception. Subjects are usually able to make a "mental zoom" to trace some details in their mental image.

Perky demanded that subjects visualize a tomato on a screen, but have eyes fixated to a light-point. Then he displayed a tomato-image on the screen. Subjects have still been convinced for a while that the tomato is imagined – they confused their mental image with the real image. Segal and Fusella made subjects to visualize a tree on a screen where a weak light-point with slowly increasing intensity was projected – until they saw the light-point. They needed more time to finally see the light-point than when they had to imagine bell-ringing. In contrast to the bell-sound case, imagining the tree caused a direct competition of the tree-image with the light-image. This is possible only when the *same* perception-center is activated or there is at least much neural relation. So, imaging and perception have to *share* the neural substrate for vision or be significantly connected. (Changeux, 1986)

6. Color Qualia?

In this section, the color science will be presented. However, we cannot afford entering here the rich descriptive psychology and phenomenology of color.

[26] This is the result by: M.J. Farah (1984): Cognition 18, 245-272. It is interesting to add that, according to blood flow experiments, thinking as purely-abstract mental process never activated association areas involved in perceptual processes of various types. Only other different task-dependent cortical areas, e.g. superior prefrontal cortex, were used – as reported in: P.E. Roland & L. Friberg (1985): J. Neurophysiol. 53, 1219-1243.

Wavelength- and Color-Responsive Neurons

Zeki (after Lockwood, 1989) has found three types of neurons in V4 and V1:[27]

1 *wavelength-selective neurons* respond to any color-area which reflects enough light of that wavelength;
2 *color-coded neurons* respond preferentially when a colored patch would *look to us* to be, say, red (remarkable!);
3 *wavelength-selective opponent neurons* have a base-level of firing which increases in response to light of one wavelength and decreases in response to light of a complementary wavelength. Their hypothetical relatives, "color-coded opponent neurons", could take part in producing colored after-images — if a patch *looks* red, then it produces a green after-image.

Color as Secondary Quality

Extension or form, for example, are in philosophy called primary qualities of an object, because they are "more objective" (i.e., depend on object's own characteristics) than secondary qualities. Color, for example, is called secondary quality, since it is relatively "more subjective" (i.e., depends on mentality). Here we shall not enter the philosophical debate how much is color a (co)product of characteristics of the object or illumination and how much a (co)product of conscious experience (Peruš, 2000a; AuxilRef 11, 12). By the way, dogs and cats are color-blind (Trstenjak, 1996; Baars, 1997). In other words, colors do not exist for them apart of different rates of grayness. Indeed, colors are subjective (i.e., species-specific or even subject-specific). Secondary qualities like color are not so much something in the objects themselves, but rather "potential powers" to trigger various sensations and perceptions in our brain and conscious experience — based on the objects' primary qualities like extension, form, or wavelength of the object-reflected light.

Color is phenomenally (i.e., reported as) equal, if the responses of cones are equal, to a corresponding superposition (i.e., weighted sum) of the principal colors. This is a consequence of sensational equivalence like, for example, in the case of monochromatic yellow with wavelength 580 nm versus a mixture of yellows with 540 nm and 620 nm. The binding of color is so perfect that the original colors "poured together" cannot be phenomenally traced any more (in contrast to auditory tone-mixing). Perception of the optically-mixed color is thus completely unified or unitary (Trstenjak, 1996).

Basic Characteristics of Color

Subject's impression of color consists of experience of *brightness* (intensity of the effect by light imposed on cones), *hue* (the "folk-color" – proportion of activities of the "red", "blue" and "green" cones), and *saturation* (how "rich, clear" is a hue, how much it is

[27] Partially such or similar neurons, especially item 3, could be found already in retina.

unspoiled by grayness[28]). Saturation is determined by the degree to which all three cones are stimulated to the same degree by the object and its background.[29]

Trstenjak (1996) phenomenologically describes hue as the "classical color", brightness as the "similarity of color to whiteness", and saturation as the "rate of clearness, distinctness, (unique richness) of a non-gray color" or as the "rate of difference from the grayness". About brightness Trstenjak (1996) says that it is not simply a result of the intensity of light-effect, but is also related to "intrusiveness" of color. The rate of intrusiveness is not an absolute function of illumination, but it is rather a relative function of illumination-difference between the observed color and its immediate external context or background. This direct dependence on background distinguishes intrusiveness from brightness which is conditioned by constancy laws.

Theory of Color Qualia?

Color qualia incorporate aspects which remain, in their very phenomenological essence, non-quantifiable. So, no real theory is possible. However, colors also have quantifiable backgrounds in perceptual processing and its stimuli, i.e. in unique wavelengths. This allows a schematic theory of relations (comparisons, contrasts) and combinations (constructive and destructive mixing) of proto-qualia like "elementary" colors (MacLennan in Železnikar & Peruš, 1998). Visual conscious experience is a global superposition of more elementary experiences. Flashing of a red light for 20 ms, followed immediately by 20 ms of green light gives impression of a transient yellow light corresponding to the mixture of red and green (Koch, 1996, after Efron).

7. Neuropsychology of Color

This specialized section demonstrates how secondary perception of color is intermixed with primary perceptions, like edges and contours, which together (partially)-consciously construct visual experience. Holistic approach is indispensable.

Origins of Color

Color phenomena can be divided into those dependent on (Trstenjak, 1996):

1. coloring matter or dye-stuff;
2. illumination and its context or circumstances;
3. psychical integration or completion with other perceptions (the so-called subjectivity).

[28] Similarly, DeValois & Jacobs (1968) define saturation of a light as the measure how much the light is chromatic in contrast to achromatic one.

[29] These "definitions" were based mainly on color-correlates in the activity of retinal cone-cells. We can later trace deeper physiological correlates of color in LGN- and cortical (assemblies of) cells.

Essentials of neurobiology of color follow, but an advanced description of color-vision phenomena can be found in R.L. & K.K. DeValois (1990).

Three-Cones System

The wavelength of light only determines the *probability* that the photon will be absorbed during an all-or-none conformational change of the pigment in the outer segment of a cone-photoreceptor, but the electrical response of the cone's inner segment is not affected by the light directly. In spite of the fact that a cone reacts (if it does) always with the same electrical signal regardless of the wavelength, the number of photons absorbed by the cone does vary with wavelength. A cone responds equally to a light of any wavelength as long as the light-intensity compensates for the cone's absorption-rate at that wavelength. Cones have different absorption-probabilities, or sensitivity-maximums, respectively. So, in spite of the fact that different cones are maximally sensitive to different light-wavelengths, *responses of the three types of cones need to be combined, in their relative strengths, to recognize the color.* Individual cones are not sufficient for detection of an arbitrary light-wavelength (Kandel et al., 1991, ch. 31). Each of the three classes of cones responds maximally to its own wavelength corresponding to phenomenal blue, red and green.

(Double) Opponency

Many ganglion neurons of retina, LGN- and cortical neurons are excited by one class of cones and inhibited by another. Such *opponent cells* encode the hue so that they subtract the output of one of the three cone-types from the output of one of the others – in various combinations. These opponent interactions of the red-, green- and blue-preferring cones are responsible for the phenomenon of color-mixing: e.g., an object that is both red and green appears yellow because the red and green neutralize each-other. Apart of the simple *color contrast* between opponent colors, there is a more complicated ("double") opponent phenomenon called *simultaneous color contrast*: e.g., a gray object with a green background gets a tinge of red.

Color is processed in retina and LGN by so-called *color-opponent cells.*[30] Some other cells react to brightness of the center compared to its surround, and do not contribute to color detection. Color information is further processed in cortical blobs by *double-opponent cells* [31], found in V1 (orientation-selective ones) and area 18 (*complex* ones), which combine inputs from *single*-opponent cells.[32] Saturation of a color is encoded neurally by the proportion of

[30] E.g., response of such a neuron (in V1), determined actually by its receptive field (!), is increased to red and decreased to green. The other pair of complementary response is yellow/blue. Color-opponent cells are thus of the following types: 1. red excitatory & green inhibitory (+R–G); 2. green excitatory & red inhibitory (+G–R); 3. yellow excitatory & blue inhibitory (+Y–B); 4. Blue excitatory & yellow inhibitory (+B–Y).

[31] The receptive field of such a neuron (in V4 or sometimes V1) has a central excitatory region with inhibitory flanks responding in opposite manner to a color and to its opponent color (ibid.).

[32] See Figures 31-5, 31-6 & 31-8 in Kandel (1991) for illustrations of various concentric receptive fields where their internal circles excite / inhibit stimuli-components of a specific basis-color (e.g., red), but their surrounds (vice versa!) inhibit / excite stimuli of the color which is opponent (complementary) to the first one (e.g., green is opponent to red, and yellow to blue).

the total response by the color-opponent neurons rather than by the non-opponent (achromatic) neurons (De Valois & Jacobs, 1968).

Light-wavelengths which trigger excitation in a (single)-opponent neuron, denoted +Y–B because of having an excitatory (+) center of the receptive field in response to yellow (Y) light and an inhibitory (–) surround for suppressing the "yellow" center "with blue (B) centripetal influence", are those which *appear yellow* to a subject. De Valois & Jacobs (1968) continue that colors[33] correspond to the points of maximal response of the corresponding color-opponent cells rather than to the absorption-peaks of the cone-photopigments.

Color Constancy

The brain is able to recognize color in various circumstances like the ambient light, because it compares not only the cones directly affected by the object-stimulus, but also all other cones. This neutralizes to some extent the contextual effect of illumination (which can not be neglected), therefore the color of an object anyway appears roughly the same in spite of different illuminations of the environment-space. This is called *color constancy*. It is top-down constrained, similarly as object perception is. Like object-constancy is the next phase of perception after image processing, color-constancy follows color-imaging (Pribram, 1991). For the effect of color constancy also, red/green, yellow/blue and black/white double-opponent processes are needed.

Color experience diminishes in Ganzfeld experiments (i.e., using a pattern-less, uniform color-field). This demonstrates the importance of object–background comparison also in color perception. Experiments suggest that object's color is detected, or recognized, respectively, by comparing all the objects in the scene and its background (Kandel et al., 1991, ch. 31).

Trstenjak (1996) observes that color contrast and color constancy are antagonistic effects, although interdependent and finally integrated. Color-contrast mechanism, which is mainly early-physiological (retinal), tends to "pour together" the colors of illumination and of the illuminated object into an unified / uniform visual perception, so that the object looses its intrinsic color. Color-constancy mechanism, which is conditioned mainly by memory and cognition, tends to separate effects of illumination from the intrinsic color of the illuminated object in order to produce perception of the object in its "real", more-or-less "permanent" color, regardless of the illumination and shading conditions.

Color Perception of Edges

The boundaries of color-fields usually fit the boundaries of object-forms, therefore color (a consciously co-shaped percept!) helps in edge detection.[34] Within the boundary of an object, colors cancel each-other out by action of color-opponent neurons; across the boundary, however, they support each-other's diversity by action of double-opponent

[33] I.e., the color-experience – which is, however, not confined to any of the mentioned cell-groups, I would say.

[34] Pribram (1991, p. 104) notes: "For changes [in color] to be perceived, the color must remain invariant within boundaries [between figure and ground]. But color constancy is dependent on the entire extent of the perceived space. There are parallels in the changes of organization of this processing space for color and for shape from retina to striate-peristriate and prestriate systems." (Cf., Grossberg)

neurons. Apart of the cases of "competing" color-patches with blurring "edges" *within* the form of an object, color perception enhances contrast of the object versus background. Special *double-opponent neurons* are responsible for *contrast-enhancing by color-differences* at the edge of an object. The internal and the external color are compared synergistically, i.e. perceptual responses of contrasting colors enhance each-other across the boundary because of double-opponent neuronal action.

Color Helps in Contour-Integration

Mullen et al. (2000) write about the role of color in form-perception. Color-detecting system as well as luminance-detecting system are both able to trace edges. Color- and luminance-edges are not coincident in shadows, for example, but are mainly coincident at object's boundaries. Mullen et al. (2000) concluded that optimal color-based contour-integration is only slightly worse than the luminance-based one. Contour's curvature was first established, in their experiment, to be independent from contrast. They compared red/green, yellow/blue and luminance post-receptoral contour-integration systems while linking simple oriented Gabor wavelets into a contour-gestalt. It was concluded that all three systems use a common contour-integration process which is also sensitive to color-contrast and phase of the stimuli.

The last two paragraphs uncover top-down effects of color-qualia, or at least color processing, onto object perception.

8. Looking Back and Ahead

Current (insufficient!) knowledge on the visual component of conscious experience has been presented – together with accompanying processes, like attention, figure/ground segmentation, imagery, which are under essential *top-down influence*. Some emphasis was given on the *quantum* approach to modeling system-dynamic background of conscious visual experience which is in the process of combining (also by this author – book in press) with other multidisciplinary studies and hypothetical theories of conscious experience and their *neural* correlates.

9. Conclusion

The conclusion of this descriptive meta-model based on present wealth of experiments and models is that the topic has to be researched *holistically*, including *all* (collective) neuronal and sub-neuronal levels down to the (sub)quantum levels.

The dilemmas of the nature of qualia and of the origin of top-down influence (whether it comes from prefrontal or other "higher" lobes, or from an autonomous consciousness) remain unsolved. Therefore, the whole conglomerate problem of conscious perception and experience remains more or less in the area of philosophical speculation. This requires intensive collaboration of scientific disciplines and philosophy.

The last but not least remark that I share with Pribram (1998, Epilogue): "The so-called hard problem[35] is common to all epistemological endeavor. Scientific and other types of knowledge always *begin* with our conscious experience. We can no more ontologically "know" a quark or a table than we ontologically "know" a color or a pain. We relate our experience to one another and match these communications to our own experience. [...]"

Reductionism or epiphenomenalism are thus *in*sufficient (Peruš, 1998a,b), but this is unfortunately still a view of scientific minority. One has to consider visual consciousness in widest contexts, like multi-modal or social co-learning, or sensory-motor contingency (i.e., the phenomenal world as an out-side or "out-projected" memory: O'Regan & Noë, 2001).

In sum, *everything* (up to our "whole world") *that we experience happens within our consciousness*, at least indirectly (e.g., radioactivity). We will never know anything reliable on the presumed objective world beside its *appearances* (which our mind co-creates!).

10. Ethical and Ecological Consideration

Consciousness and vision are both separately and together of central importance in science, philosophy and ecological pragmatics (in psychological, economical, political and environment-protectional sense). Imagery is (mis)used in ideology and religion (having the so-called visions), business, interpersonal relations, education, psychotherapy (visualization), etc. Recall artistic effects, optical illusions and other "active vision" phenomena. Our consumer society is mainly visually motivated. Images-of-achievement (preceded by propaganda images) direct our behavior (Pribram, 1991) and trigger our will (or vice-versa). Computer vision is (mis)used in control, security and military applications.

Let me note here that almost the whole paper is valid also for many (higher) animals (usually probably-wrongly classified as "merely instinctive automata"), *except mainly* in their limited wealth of cognitive and memory-*structure* capabilities (and partially different qualia).

Perceived visual characteristics (like size) are correlated with cognitive status (like importance), e.g., children draw important (for them) persons bigger, famous places have been drawn bigger on medieval geographic maps. Indeed, cognition has profound effect on vision, and vice-versa. And consciousness shapes them both, and vice-versa. Motivation, will, emotion, social realm, ethics, etc., are involved into the broad multi-lateral process.

Conscious vision is our link to beauty, (dis)harmony, pollution, to the bulk of existence, and so indirectly also to anticipation and responsibility. *Our* objective existence is an intersubjective compromise of existential aspects of *ours*. And the *purely*-objective existence of a stone (or of "reality", "truth") is unknowable to subjects. It is ethical to be aware of this ecology.

Acknowledgments

My sincere thanks for precious advice to Professor Karl H. Pribram, and for hospitality of his and of Mrs. Katherine Neville. Many thanks for discussions and/or support also to

[35] I.e., how to explain qualia, which is beyond capabilities of contemporary science — note by M.P.

Professors Andrej O. Župančič, Andrej Ule and Janek Musek, to Dr. Mila Božič, Mr. Andrej Detela, and many others.

REFERENCES

Aczel, A.D. (2003): *Entanglement.* New York: Plume/Penguin.

Aine, C.J., S. Supek & J.S. George (1995): Temporal dynamics of visual-evoked neuromagnetic sources: Effects of stimulus parameters and selective attention. *Internat. J. Neuroscience* 80, 79-104.

Arbib, M.A. (Ed.) (2002): *The Handbook of Brain Theory and Neural Networks.* Cambridge (MA): MIT Press.

Aspect, A., P. Dalibard & G. Rogier (1982): Experimental test of Bell's inequalities using time-varying analyzers. *Physical Review Letters* 49, 1804-1807 (and also: A. Aspect, P. Grangier & G. Rogier: *Phys. Rev. Lett.* 47 (1981) 460- &. 49 (1982) 91-.)

ASSC (1998): Neural correlates of consciousness; ASSC conference abstracts. Bremen: Hanse-Wissenschaftskolleg.

Baars, B.J. (1997): *In the Theater of Consciousness.* New York: Oxford Univ. Press.

Barahona da Fonseca, J., I. Barahona da Fonseca, C. Paz Araujo & J. Simoes da Fonseca (1999): A quantum theoretical approach to information processing in neural networks. *In:* D. Dubois (Ed.): *AIP Conference Proceedings*, vol. 517: *Computing Anticipatory Systems – CASYS'99* in Liege. Melville (NY): American Institute of Physics, pp. 330-344. (*Also:* Cognition: characteristic waveform correlates. CASYS 2000, Liege: full-length preprint & abstract book.)

Beck, F. & J.C. Eccles (1992): Quantum aspects of brain activity and the role of consciousness. *Proceedings of the National Academy of Sciences of USA.* 89, 11357-11361.

Bickle, J., M. Bernstein, M. Heatley, C. Worley & S. Stiehl (1999): A functional hypothesis for LGN-V1-TRN connectivities suggested by computer simulation. *J. Computational Neuroscience.* 6, 251-261.

Bob, P. & J. Faber (1999): Quantum information in brain neural nets and EEG. *Neural Network World.* 9, 365-372.

Bohm, D. (1980): *Wholeness and Implicate Order.* London: Routledge & Paul Kegan.

Bohm, D. & B. Hiley (1993): *The Undivided Universe (An ontological interpretation of quantum theory).* London: Routledge.

Bonnell, G. & G. Papini (1997): Quantum neural network. *Internat. J. Theoretical Physics.* 36, 2855-2875.

Changeux, J.-P. (1986): *L'homme neuronal.* Beograd: Nolit (in Serb. transl.).

Crick, F. (1984): Function of the thalamic reticular complex: The searchlight hypothesis. *Proceedings of the National Academy of Sciences of USA.* 81, 4586-4590.

Damasio, A.R. (1999): *The Feeling of What Happens.* New York: Harcourt Brace.

Davies, M. & G.W. Humphreys (Eds.) (1993): *Consciousness.* Oxford: Blackwell.

De Valois, R.L. & G.H. Jacobs (1968): Primate color vision. *Science.* 162, 533-540.

De Valois, R.L. & K.K. De Valois (1990): *Spatial Vision.* New York: Oxford Univ. Press.

Dubois, D.M. (Ed.) (2000): Proceedings of CASYS'99. *Internat. J. Computing Anticipatory Systems*. 5, 6, 7. Liege: CHAOS. Especially, in vol. 7: Proceedings of the Symposium "Quantum Neural Information Processing: New Technology? New Biology?" (espec. papers by Marcer, Sutherland, Farre, Mitchell, Dubois; also Citko, Luksza & Sienko).

Eccles, J.C. (1986): Do mental events cause neural events analogously to the probability fields in quantum mechanics? *Proceedings of the Royal Society of London. B* 227, 411-428.

Eccles, J.C. (1993): Evolution of complexity in the brain with the emergence of consciousness. Keynote in Pribram (1993).

Ezhov, A.A. (2000): Spurious memory, single-class and quantum neural networks. *Proceedings of the Int. Conf. on Computational Intelligence and Neuroscience. 2000* (part of *Proceed. JIC on Information Sci. 2000*, Atlantic City, NJ), pp. 635-638, and other refs. in Proceed.

Ezhov, A.A. & D. Ventura (2000): Quantum neural networks. Ch. 11 in: N. Kasabov (Ed.): *Future Directions for Intelligent Systems and Information Sciences* (Series "Studies in Fuzziness and Soft Computing", vol. 45). Heidelberg: Physica-Verlag (Springer), pp. 213-235.

Fedorec, A.M. & P. Marcer (Eds.) (1996): *Living Computers.* (symposium proceedings). Dartford: Greenwich Univ. Press.

Flanagan, O. (1992): *Consciousness Reconsidered.* Cambridge (MA): MIT Press.

Frith, C., R. Perry & E. Lumer (1999): The neural correlates of conscious experience: an experimental framework. *Trends in Cognitive Sciences.* 3, 105-114.

Gennaro, R.J. (1995): *Consciousness and Self-Consciousness*. Amsterdam/Philadelphia: John Benjamins.

Goswami, A. (1990): Consciousness in quantum physics and the mind-body problem. *J. Mind & Behavior.* 11, 75-96.

Gould, L.I. (1989): Nonlocal conserved quantities, balance laws and equations of motion. *Internat J. Theoretical Physics.* 28, 335-364.

Gray, C.M., P. König, A.K. Engel & W. Singer (1989): Oscillatory responses in cat visual cortex exhibit inter-columnar synchronization which reflects global stimulus properties. *Nature.* 338, 334-337.

Gray, C.M., P. König, A.K. Engel & W. Singer (1990): Synchronization of oscillatory responses in visual cortex: A plausible mechanism for scene segmentation. *In:* H. Haken & M. Stadler (Eds.): *Synergetics of Cognition.* Berlin: Springer, p. 82-.

Hameroff, S.R. (1994): Quantum coherence in microtubules: a neural basis for emergent consciousness? *J. Consciousness Studies.* 1, 91-118.

Hameroff, S.R. (1998): "Funda-Mentality": Is the conscious mind subtly linked to a basic level of the Universe? *Trends in Cognitive Science.* 2, 119-127.

Hameroff, S.R., A.W. Kaszniak & A.C. Scott (1996): *Towards a Science of Consciousness: Tucson I.* Cambridge (MA): MIT Press. (Especially: K. Pribram: The varieties of conscious experience: Biological roots and social usages (ch. 10) & C. Koch: Towards a neuronal substrate of visual consciousness).

Jasper, H.H., L. Descarries, V. Castellucci & S. Rossignol (Eds.) (1998): *Consciousness: At the Frontiers of Neuroscience (Advances in Neurology vol. 77).* Philadelphia: Lippincott-Raven (espec. ch. 17).

Jibu, M., K.H. Pribram & K. Yasue (1996): From conscious experience to memory storage and retrieval: The role of quantum brain dynamics and boson condensation of evanescent photons. *Internat. J. Modern Physics.* 10, 1735-1754.

Jibu, M. & K. Yasue (1995): *Quantum Brain Dynamics and Consciousness.* Amsterdam / Philadelphia: John Benjamins.

Jibu, M. & K. Yasue (1997): Quantum field theory of evanescent photons in brain as quantum theory of consciousness. *Informatica.* 21, 471-490.

Kandel, E.R., J.H. Schwartz & T.M. Jessel (1991): *Principles of Neural Science.* London (UK): Prentice Hall Internat., 3rd ed.

Koch, C. (1997): Computation and the single neuron. *Nature.* 385, 207-210.

Kosslyn, S.M. (1988): Aspects of a cognitive neuroscience of mental imagery. *Science.* 240, 1621-1626.

Kosslyn, S.M. (1994): *Image and Brain.* MIT Press: Cambridge (MA).

Kosslyn, S.M. & R.A. Andersen (Eds.) (1992): *Frontiers in Cognitive Neuroscience.* Cambridge (MA): MIT Press.

Lockwood, M. (1989): *Mind, Brain and the Quantum.* Oxford: Blackwell.

Logothetis, N.K. (1999): Vision: A window on consciousness. *Scientific American,* November 1999, 45-51.

Luria, A.R. (1973): *Fundamentals of Neuropsychology.* Moscow: Izdateljstvo Moskovskogo Univ. (in Russian).

Marcel, A.J. & E. Bisiach (Eds.) (1988): *Consciousness in Contemporary Science.* Oxford: Clarendon Press.

Marcer, P. (Ed.) (1999): Proceedings of the Symposium 9: Quantum Neural Information Processing: New Technology, New Biology? *In:* D. Dubois (Ed.): *Abstract book of CASYS'99 (Computing Anticipatory Systems).* Liege: CHAOS asbl.

Marcer, P. & W. Schempp (1997): A model of neuron working by quantum holography. *Informatica.* 21, 517-532.

Marcer, P. & W. Schempp (1998): The brain as a conscious system. *Internat. J. General Systems.* 27, 231-248.

McIntosh, A.R., M.N. Rajah & N.J. Lobaugh (1999): Interactions of prefrontal cortex in relation to awareness in sensory learning. *Science.* 284, 1531-1533.

Mesulam, M.-M. (1981): A cortical network for directed attention and unilateral neglect. *Annals of Neurology.* 10, 309-325.

Montero, V.M. (2000): Attentional activation of the visual thalamic reticular nucleus depends on 'top-down' inputs from the primary visual cortex via corticogeniculate pathways. *Brain Research.* 864, 95-104.

Moran, J. & R. Desimone (1985): Selective attention gates visual processing in the extrastriate cortex. *Science.* 229, 782-784.

Mullen, K.T., W.H.A. Beaudot & W.H. McIlhagga (2000): Contour integration in color vision: a common process for the blue-yellow, red-green and luminance mechanisms? *Vision Research.* 40, 639-655.

Newman, J. (1997): Toward a general theory of the neural correlates of consciousness. *J. Consciousness Studies.* 4, 47-66 (part I), 100-121 (part II).

Nishimori, H. & Y. Nonomura (1996): Quantum effects in neural networks. *J. Physical Society of Japan.* 65, 3780-3796.

Nobili, R. (1985): Schrödinger wave holography and brain cortex. *Physical Review. A* 32, 3618-3626.

Oakley, D.A. (Ed.) (1985): *Brain and Mind*. London / New York: Methuen.

O'Regan, K.J. & A. Noë (2001): A sensorimotor account of vision and visual consciousenss. *Behavioral & Brain Sciences.* 24, 939-1031.

Peruš, M. (1997a): Mind: neural computing plus quantum consciousness. *In:* M. Gams, M. Paprzycki & X. Wu (Eds.): *Mind Versus Computer.* Amsterdam: IOS Press & Ohmsha, pp. 156-170.

Peruš, M. (1997b): System-processual backgrounds of consciousness. *Informatica.* 21, 491-506.

Peruš, M. (1997c): Neuro-quantum coherence and consciousness. *Noetic J.* 1 (no. 1), 108-113.

Peruš, M. (1998a): Conscious representations, intentionality, judgements, (self)awareness and qualia. *Informatica.* 22, 95-102.

Peruš, M. (1998b): Consciousness: network-dynamic, informational, and phenomenal aspects. *Noetic J.* 1 (no. 2), 183-197.

Peruš, M. (2000a): *BioNetworks, Cognition and Consciousness*. Ljubljana: DZS (in Slovene). [Rus. transl. by A. Ippa: *Vse v odnom, odno vo vsem (Matematičeskie modeli associativnjih neironnjih setei)*. St. Petersburg: KARO, 2000.]

Peruš, M. (2000b): Neural networks as a basis for quantum associative networks. *Neural Network World.* 10 (no. 6), 1001-1013.

Peruš, M. & S.K. Dey (2000): Quantum systems can realize content-addressable associative memory. *Applied Mathematics Letters.* 13 (no. 8), 31-36.

Peruš, M. (2001a): Image processing and becoming conscious of its result. *Informatica.* 25, 575-592.

Peruš, M. (2001b): Multi-level synergetic computation in brain. *Nonlinear Phenomena in Complex Systems.* 4, 157-193.

Peruš, M., H. Bischof & C.K. Loo (2005): Bio-computational model of object recognition: Quantum Hebbian processing with neurally shaped Gabor wavelets. *BioSystems.* 82, 116-126.

Pessa, E. & G. Vitiello (1999): Quantum dissipation and neural net dynamics. *Bioelectrochemistry & Bioenergetics.* 48, 339-342.

Pop-Jordanova, N. & J. Pop-Jordanov (2002): Psychophysiological comorbidity and computerized biofeedback. *Internat. J. Artificial Organs.* 25, 429-433. [And: Brain-wave dynamics related to cognitive tasks and neurofeedback information flow. 7th Exp. Chaos Conf., San Diego, CA, 2002.]

Popper, K.R. & J.C. Eccles (1977): *The Self and Its Brain*. Berlin: Springer.

Pribram, K.H. (1971): *Languages of the Brain (Experimental Paradoxes and Principles in Neuropsychology).* Orig.: Englewood Cliffs (NJ): Prentice-Hall; 5th publ.: New York: Brandon House.

Pribram, K.H. (1991): *Brain and Perception (Holonomy and Structure in Figural Processing).* Hillsdale (NJ): Lawrence Erlbaum Associates.

Pribram, K.H. (Ed.) (1993): *Rethinking Neural Networks: Quantum Fields and Biological Data.* Hillsdale (NJ): Lawrence Erlbaum Associates.

Pribram, K.H. (1997a): What is mind that the brain may order it? *Proceedings of Symposia in Applied Mathematics.* 52, 301-329 (Vol. 2: Proceed. of the Norbert Wiener Centenary Congress, 1994, eds. V. Mandrekar & P.R. Masani; Providence: Am. Math. Soc.).

Pribram, K.H. (1997b): The deep and surface structure of memory and conscious learning: Toward a 21st century model. *In*: R.L. Solso (Ed.): *Mind and Brain Sciences in the 21st Century*. Cambridge (MA): MIT Press, pp. 127-156.

Pribram, K.H. (1998): Brain and the composition of conscious experience. *J. Consciousness Studies.* 6 (no. 5), 19-42.

Pribram, K.H. (2004): Consciousness Reassessed. *Mind & Matter.* 2, 7-35.

Rakić, Lj., G. Kostopoulos, D. Raković & Dj. Koruga (Eds.) (1997): *Brain and Consciousness — Proceedings of the First ECPD Int. Symposium* (vol. I) & *Workshop* (vol. II) *on Scientific Bases of Consciousness '97*. Belgrade: ECPD.

Roelfsema, P.R. (1998): Solutions for the binding problem. *Zeitschr. für Naturforschung* 53c, 691-715.

Roth, G. (2003): *Fühlen, Denken, Handeln*. Frankfurt/Main: Suhrkamp.

Snider, G., A. Orlov, I. Amlani, X. Zuo, G. Bernstein, C. Lent, J. Merz & W. Porod (1999): Quantum-dot cellular automata: Review and recent experiments. *J. Applied Physics.* 85, 4283-4285.

Sompolinsky, H. & M. Tsodyks (1994): Segmentation by a network of oscillators with stored memories. *Neural Computation.* 6, 642-657.

Stapp, H.P. (1993): *Mind, Brain and Quantum Mechanics*. Berlin: Springer.

Stillings, N.A., S.E. Weisler, C.H. Chase, M.H. Feinstein, J.L. Garfield & E.L. Rissland (1995): *Cognitive Science*. Cambridge (MA): MIT Press. Ch. 12: Vision.

Styles, E.A. (2005): *Attention, Perception and Memory*. Hove (E. Sussex): Psychology Press.

Tootell, R.B.H., N.K. Hadjikhani, J.D. Mendola, S. Marrett & A.M. Dale (1998): From retinotopy to recognition: fMRI in human visual cortex. *Trends in Cognitive Sciences.* 2, 174-183.

Trstenjak, A. (1996): *Psychology of Colors*. Ljubljana: Inštitut A. Trstenjaka (in Slovene).

Velmans, M. & S. Schneider (Eds.) (2007): *The Blackwell Companion to Consciousenss*. Malden (MA): Blackwell.

Vidyasagar, T.R. (1999): A neuronal model of attentional spotlight: parietal guiding the temporal. *Brain Research Review.* 30, 66-76.

Vitiello, G. (2001): *My Double Unveiled*. Amsterdam: John Benjamins.

Wang, P.P., et al. (Eds.) (1998): *Proceedings of the 4th Joint Conference on Information Sciences '98*. Research Triangle Park (NC, USA): Assoc. Intellig. Machinery. *Volume II*: Proceedings of the 3rd Internat. Conf. on Computatational Intelligence & Neuroscience (Ed. G. Georgiou); sections on neuro-quantum information processing: pp. 167-224.

Woolf, N. & S. Hameroff (2001): A quantum approach to visual consciousness. *Trends in Cognitive Science.* 5, 472-478.

Zak, M. & C.P. Williams (1998): Quantum neural nets. *Internat. J. Theoretical Physics.* 37, 651-684.

Železnikar, A.P. & M. Peruš (Eds.) (1998): Consciousness in Science and Philosophy '98 (Math. Dept., Eastern Illinois Univ., Charleston, IL) – abstracts. *Informatica.* 22, 373-403.

Auxiliary References

[1] Neural correlates of conscious experience: G. Globus: J. Cognit. Neurosci. 4 (1992) 299. B.L. Lancaster: J. Consc. Stud. 4 (1997) 122-142. B. Libet in: Experimental and Theoretical Studies of Consciousness; John Wiley & Sons, Chichester, 1993 (pp. 123). D.W. Orme-Johnson, C.T. Haynes: Neuroscience. 13 (1981) 211.

[2] Quantum features of neurophysiological processes: P.C. Bressloff: Phys. Rev. A (1992) 7549-. A. Freitas da Rocha et al.: Progr. in Neurobiol. 64 (2001) 555-573. H. Fröhlich: Int. J. Quantum Chem. 2 (1968) 841-649. E.H. Walker: Noetic J. 1 (1997) 100-107.

[3] Quantum neural information processing: Amlani et al.: Appl. Phys. Lett. 72 (1998) 2179-2181. E.C. Behrman, J. Niemel, J.E. Steck, S.R. Skinner: A quantum dot neural network; Proceed. Workshop on Physics of Computation, 1996, pp. 22-24 (full-length paper submitted to IEEE Transac. Neural Net.). S.C. Kak: Advances Imaging & Electron Phys. 94 (1995) 259-313. S.C. Kak in: Proc. Intl. Conf. Info. Sci., North Carolina, 1997 (pp. 141-144). S.C. Kak: in K.H. Pribram, J. King (Eds.): Learning as Self-Organization; Lawrence Erlbaum A., Mahwah (NJ), 1996; ch. 7 (and also ch. 21). S.C. Kak: Tech. rep. ECE/LSE 92-13; 94-42/1994. T. Menneer, A.Narayanan: Quantum-inspired neural nets; Tech. rep. R329, Univ. of Exeter (Comp. Sci. Dept.), 1995/6 (& Tucson II Conf.). J.E. Steck, S.R. Skinner, A.A. Cruz-Cabrera, M. Yang, E.C. Behrman: Field computation for ANN hardware: Examples in nonlinear optical materials; in Proceed. IJC Info. Sci. 2000, Atlantic City (NJ) (there see also: Gould, MacLennan, Menneer & Narayanan, Ventura). V. Tryba, K. Goser: A modified algorithm for self-organizing maps based on Schrödinger equation; in Proceed. Int. Workshop ANN '91. D. Ventura, T. Martinez: in Proc. ICANNGA '97, Norwich (pp. 482-485). D. Ventura, T. Martinez: in Proc. JIC NN '98, Anchorage (pp. 509-513).

[4] Artificial quantum neurodynamics: A.A. Ezhov, G.P. Berman: Introduction to Quantum Neural Technologies; Rinton, Princeton (NJ), 2003. R.L. Dawes: Parametric avalanche stochastic filter; Martingale Res. Corp. (Allen, TX) final tech. report, contract no. N60921-91-C-0071, for NSWC in White Oak Lab., 1993. R.L. Dawes, H.L. Bodine, E.v.K. Hill: Automatic detection and classification...; Martingale Res. Corp. final report of Phase I SBIR Project, contract no. III-9361410, for NSF, 1994.

[5] Sub-quantum backgrounds of consciousness: non-locality, implicate order: P.C.W. Davies, J.R. Brown (Eds.): The Ghost in the Atom; Cambridge Univ. Press, Cambridge, 1986. B.S. DeWitt, H. Graham (Eds.): The Many-Worlds Interpretation of Quantum Mechanics; Princeton Univ. Press, Princeton, 1973. G.I. Shipov: A Theory of Physical Vacuum (espec. chs. 2.4, 2.5); Moscow, 1997. D.Z. Albert: Sci. Amer. (May 1994) 34. R.L. Amoroso: Informatica 19 (1995) 585-590. D. Bohm: Foundat. Phys. 17 (1987) 667. D. Bohm, B.J. Hiley: Foundat. Phys. 11 (1981) 529. C. Dewdney, P.R. Holland, A. Kyprianidis, J.P. Vigier: Nature 336 (1988) 536. D. Deutsch: Phys. Rev. D 44 (1991) 3197-3217. F.A.M. Frescura: Foundat. Phys. 18 (1988) 777-808. F.A.M. Frescura, B.J. Hiley: Foundat. Phys. 10 (1980) 705. F.A.M. Frescura, B.J. Hiley: Foundat. Phys. 10 (1980) 7. L.I. Gould: Arkhimedes 2 (1993) 144-157. J.S. Hagelin: Mod. Sci. & Vedic Sci. 1 (1987) 29-87. B. Hiley: Annal. Fondation L. de Broglie 5 (1980) 75. B. Hiley, N. Monk: Mod. Phys. Lett. A 8 (1993) 3625. B. Hiley, M. Peruš: J. Critics Sci. 174 (1995) 33-49 (in Slov.) & Dynam. Psych. (1997) www (Engl. orig.). B. Hiley: in Saunders, Brown (Eds.): The Philosophy of the Vacuum; Oxford Univ. Press, 1991. B.D.

Josephson, F. Pallikari-Viras: Foundat. Phys. 21 (1991) 197. P. Kwiat, H. Weinfurter, A. Zeilinger: Sci. Amer. (Nov. 1996) 52-58. I. Pitowsky: Foundat. Phys. 21 (1991) 343. G. Taubes: Science 285 (1999) 512-517.

[6] Quantum backgrounds of consciousness: M. Kafatos, R. Nadeau: The Conscious Universe; Springer, New York, 1990. R. Penrose: The Emperor's New Mind (Concerning Computers, Minds, and Laws of Physics); Oxford Univ. Press, London, 1989. R. Penrose: Shadows of the Mind (A Search for the Missing Science of Consciousness); Oxford Univ. Press, Oxford, 1994. J. Polkinghorne: The Quantum World; Princeton Univ. Press, Princeton (NJ), 1984. H.P. Stapp: Mind, Matter and Quantum Mechanics; Springer, berlin, 1993. A. Stern: The Quantum Brain (Theory and Implications); North Holland / Elsevier, Amsterdam, 1994. G. Blommestijn: in Proceed. Conf. Consc.- Tucson II, 1996 (abstract & full-length preprint). J.S. Hagelin: Mod. Sci. & Vedic Sci. 3 (1989) 3-72. K. Ludwig: Psyche 2(16) (1995) www. P. Marcer: Proceed. 14th Int. Congress Cybernetics, Namur, 1995 (pp. 429-434 & 435-440). P. Marcer: World Futures: J. Gen. Evol. 44 (1995) 149-159. M. Peruš: Sci. Tribune (1997) www. D. Raković: Informatica 21 (1997) 507-517 E.J. Squires: Synthese 97 (1993) 109-123. A. Scott: J. Consc. Stud. 3 (1996) 484-491. H.P. Stapp: Foundat. Phys. 21 (1991) 1451-1477. H.P. Stapp: Psyche 2(5) (1995) www.

[7] Informational and cybernetic aspects of cognition and consciousness: F.J. Varela: Principles of Biological Autonomy; North Holland, New York, 1979. M. Gams, C. Bavec (Eds.): Proceed. Int. Multi-Conf. Info. Soc. '98 & '99; Ljubljana, IJS. D. Dubois (Ed.): CASYS'99 Book of (Extended) Abstracts; CHAOS asbl, Liege (espec. Symp. 9 & Pribram; Sienko; Kirvelis; Dubois; Hoekstra & Rouw; Araujo; Tsirigotis). A.J. Bell: Phil. Transac. Roy. Soc. London B 354 (1999) 2013-2020. D. Bojadžiev: Informatica 19 (1995) 627-634. G.A. Christos: Noetic J. 1 (1998) 149-161. G. Farre: Informatica 21 (1997) 533-540. D. Gernert: BioSys. 54 (2000) 165-180. C. Green, G. Gillet: Philos. Psych. 8 (1995) 333-340. M.W. Ho: J. Consc. Stud. 3 (1996) 231-244. J. Horne: Informatica 21 (1997) 675-682. J. Horne: Informatica 24 (2000) 275-279. S.C. Kak: J. Intell. Sys. 6 (1996) 133-144. M. Peruš: Anthropos 1-2 (1996) 84-89 (in Slov.) M. Peruš: World Futures: J. Gen. Evol. 51 (1997) 95-110. J.D. Valentine: Psychoenerget. 4 (1982) 257-274. A.P. Železnikar: Cybernetica 40 (1997) 261-296.

[8] Cognitive aspects: E. & W. Baumgartner et al. (Eds.): Phenomenology and Cognitive Science; Röll, Würzburg, 1997. D.C. Dennett: Content and Consciousness; Routledge & Kegan Paul, London, 1969. W.G. Lycan (Ed.): Mind and Cognition (A Reader); Blackwell, Oxford / Cambridge (MA), 1992. H.L. Pick, P. van den Broek, D.C. Knill (Eds.): Cognition; APA, Washington (DC), 1992 (espec. ch. by E.J. Gibson).

[9] Consciousness – implementation; complex-system background processing: A.D. Linkevich: Informatica 21 (1997) 435-464. M. Peruš: J. Critics Sci. 174 (1995) 11-22 (in Slov.). M. Peruš: Horizons Psych. 4 (1996) 73-84 (in Slov.).

[10] Consciousness – general: T. Nagel (Ed.): Experimental and Theoretical Studies of Consciousness; John Wiley & Sons, Chichester, 1993; especially: M. Kinsbourne (pp. 45), B. Libet (pp. 123). D. Raković, Đ. Koruga (Eds.): Consciousness; ECPD, Belgrade, 1996. J.R. Smythies, J. Beloff (Eds.): The Case for Dualism; Univ. Press of Virginia, Charlottesville, 1989. Towards a Science of Consciousness -"Tucson II"; Consciousness Research Abstracts, JCS, 1996. J.F. Kihlstrom: Cognit. & Consc. 2 (1993) 334. J.R. Searle: Cognit. & Consc. 2 (1993) 310. Trends Cog. Sci. 4 (2000) 372-

382. C. Tart: J. Consc. Stud. 4 (1997) 71-92. A. Ule: J. Critics Sci. 176 (1995) 117-130. A. Ule: Informatica 21 (1997) 683-688. A.P. Železnikar: Informatica 20 (1996) 475-484.

[11] Consciousness – phenomenal aspects: F. Brentano: Psychology from an Empirical Standpoint; Routledge & Kegan Paul, London, 1973 (German orig. 1874). D.W. Hamlyn: Perception and Sensation; Routledge, London, 1961. N. Nelkin: Consciousness and the Origins of Thought; Cambridge Univ. Press, Cambridge, 1996. S. Blackmore: J. Mental Imagery 11 (1987) 53. A.J. Deikman: J. Consc. Stud. 3 (1996) 350-356. M. Draganescu: Noetic J. 1 (1997) 28-33. H. Hendriks-Jansen: Informatica 21 (1997) 389-404. J.J. Kupperman: Philos. Psych. 8 (1995) 341-351. M. Peruš: Poligrafi 7-8 (1997) 115-124 (in Slov.).

[12] Qualia problem: W.P. Banks: Consc. & Cognit. 5 (1996) 368-380. D.J. Chalmers: Sci. Amer. (Dec. 1995) 62-68. T.L. Hubbard: Consc. & Cognit. 5 (1996) 327-358. P. Hut, R.N. Shepard: J. Consc. Stud. 3 (1996) 313-329.

In: Philosophical Insights about Modern Science
Eds: Eva Žerovnik, Olga Markič and Andrej Ule
ISBN: 978-1-60741-373-8

Chapter 13

NATURAL AND MACHINE LEARNING, INTELLIGENCE AND CONSCIOUSNESS

Igor Kononenko*
Faculty of Computer and Information Science, University of Ljubljana, Ljubljana, Slovenia

SYNOPSIS

If you understand others you are intelligent.
If you understand yourself you are enlightened.
— Lao Tse

In the first part of this chapter, which is more "scientific", we discuss learning and its relation to natural and artificial intelligence, and we relate learning and intelligence to (phenomenal) consciousness. Practising, imitating the teacher, and repeated trial and error is called *learning*. The process of transformation due to learning is called *knowledge acquisition*. Learning by a living system is called *natural learning*; if the learner is a machine – a computer – it is called *machine learning*. Learning is a precondition for intelligent behavior. *Intelligence* can be defined as the ability to adapt to the environment and to solve problems. *Artificial intelligence* research deals with the development of systems that act more or less intelligently and are able to solve relatively hard problems. With respect to the *complexity of the learning process* we differentiate several *learning types* and learners use different *search strategies* which are also used in machine learning. Important aspects for understanding the abilities of artificial intelligence are the impact of learning on intelligence, the speed of problem solving, the principal limitations of algorithms (stemming from the theory of computability), and the imitation of intelligent behavior. We expect an intelligent system to be (at least to some extent) intelligent in all the areas which are characteristic for human problem solving.

However, most speculations about artificial intelligence do not take into account yet another level: *consciousness*. Consciousness seems to be fundamentally related to the following notions: life, intelligence, and free will. The world (the *real*-ity – the analogy in mathematics are real numbers, which are in great majority *irrational*) is most probably

* Trzaska 25, 1000 Ljubljana, Slovenia; e-mail:igor.kononenko@fri.uni-lj.si

non-describable by any of the symbolic formalisms which we are able to use with our rational mind (that can tackle only the *rational* part of reality – the analogy in mathematics are rational numbers), and the same limitation holds for computers.

In the second part, which is less scientific and expresses the author's own viewpoint, we discuss the relation between objectivity (in the sense of measurability and describability), and (phenomenal) subjectivity (in the sense of the 1st person, direct experience) which seem to be the relation between intelligent and conscious behavior and is in turn the relation between objective science and subjective spirituality (in the sense of spiritual, phenomenal experience). Phenomenal consciousness is highly subjective (in the sense of the 1st person experience), while science - by definition - is struggling to be objective. Scientists mostly use the rational mind (intellect) in order to *indirectly* and objectively study matter and to derive *knowledge*. Mystics on the other hand mostly use the intuitive mind (inner sense, heart) in order to *directly* and subjectively experience consciousness and to attain *wisdom*. Both extremes, objective science and experiental (phenomenal) spirituality search for the truth. They are complementary to each other and we need both.

INTRODUCTION

Intellect separates, locates and compares details by searching mutual contrasts;
Wisdom unites and joins apparent opposites into one uniform harmony.
— Sri Aurobindo

We can define *learning* with the following general situation: the learner (a system) has to perform a certain task. Initially the performance is poor or even impossible. With practice the performance gradually becomes better. "Better performance" can be more accurate, cheaper, faster, etc. Practicing (imitating the teacher or repeated trial and error) is called *learning*. Herbert A. Simon (1983) one of the pioneers of Artificial Intelligence research, has defined learning in a similar way:

"Learning denotes changes in the system that are adaptive in the sense that they enable the system to do the same task or tasks drawn from the same population more efficiently and more effectively the next time."

The system has *learned* to perform the task, if it can repeat the task equally well without relearning. The process of transformation, due to learning, that enables the system – learner – to repeat the task equally well is called *knowledge acquisition*. We define *knowledge* as an interpretation of the information contained in data. Knowledge can be given in advance (for example pre-coded or in living organisms inherited), or is the result of learning. It can be wrong, correct but useless, incomplete, etc. In practice only useful knowledge, i.e. knowledge that enables the system to perform better, is interesting.

Learning takes place in almost all living beings. *Natural learning* is learning by a living system and *machine learning* is learning by a machine – a computer. The development of machine learning methods (Kononenko and Kukar, 2007) aims, besides better understanding of natural learning and intelligence, to enable the problem solving that requires specific knowledge. Often the target knowledge is unknown to humans or is used by a limited number of human experts. By using machine learning algorithms we can often efficiently generate such a knowledge. The derived knowledge can then be used to solve new problems.

Even the evolution can be regarded as learning: with natural selection, genetic crossover, and mutation the evolution creates better and better systems, capable to adapt to different environments. The principle of evolution can also be used in machine learning (with so called *genetic* or *evolutionary algorithms*) as a search mechanism that searches the hypothesis space (Goldberg, 1989).

Learning, knowledge and *intelligence* are closely related. There is no universally accepted definition of intelligence, yet it can be roughly defined as follows:

> Intelligence is the ability to adapt to the environment and to solve problems.

Adaptation itself implies learning and in order to solve problems one obviously needs knowledge (and the ability to use it).

Artificial intelligence (AI) research deals with the development of systems that are able to solve relatively hard problems and/or in general act more or less intelligently. The developed methods often imitate the human problem solving. AI areas are machine learning, knowledge representation, theorem proving and automatic reasoning, logic programming, qualitative modeling, the development of expert systems, natural language understanding, game playing, the development of artificial senses, cognitive modeling, and robotics (Bratko, 2000). A long term goal of machine learning research (which currently seems unreachable) is to create an artificial system that could achieve or even outperform the human intelligence.

In all the areas of AI learning plays an essential role (Kononenko and Kukar, 2007). By using machine learning techniques, the systems can learn and improve in perception, reasoning and theorem proving, language understanding, game playing, and heuristic problem solving. Logic programming is also highly related to inductive logic programming – a subfield of machine learning – that aims to develop logic programs from examples of the target relation. In qualitative modeling the machine learning algorithms are used to generate descriptions of complex models. Machine learning is often used to generate the knowledge base of an expert system. Intelligent robots need learning in order to improve their problem solving. Nowadays, without learning algorithms, cognitive modeling is practically impossible.

Self awareness, differentiation of self from others, awareness of your own problems, tasks and your own (ethical and moral) responsibilities – all these are related with *consciousness*, however, what consciousness is by itself is much harder to define. Nowadays scientists from various fields study various aspects of consciousness: psychologists, psychiatrists, neurophysiologists, physicists, biologists, chemists and biochemists, computer scientists, philosophers etc. On annual international scientific conferences in Tucson, Arizona hundreds of scientists all over the world each year try to clarify at least some aspects of consciousness (Lorimer, 1998). Through years it became clear that no one really knows how to define consciousness. In recent years they invite to the conference also people that study and practice various spiritual techniques and meditation. It seems that phenomenal consciousness is highly *subjective* while science by definition is struggling to be *objective* (Wallace, 2000).

In the following we are going to analyse more deeply various aspects of human and machine learning, its impact on intelligence and in the second part the relation between intelligence and consciousness.

LEARNING IN HUMANS

The ultimate goal of machine learning is knowledge.
The ultimate goal of human learning is wisdom.

We learn throughout whole our lifetime. Newborn learns to look, to listen, to distinguish the voices and faces of mother and father from other people, etc. Babies learn the connection between sight and touch, the meaning of words, and various moving skills like grasping of objects and creeping. Children learn to walk, to climb, to speak, to swim etc. Pupils learn to read and write, to calculate, they train in logic inference, abstract reasoning, in understanding and speaking foreign languages etc. In high-schools learners assimilate new, general descriptive knowledge, improve in abstract reasoning, learn general heuristics for general problem solving etc. Students restructure their knowledge, assimilate new specialized descriptive knowledge, learn special heuristics for specific problem solving, learn to explore and are trained in hypothesis testing. At work we learn our profession practically the whole life through. With practice we improve our performance and broaden our knowledge.

Our knowledge is changing, broadening and improving practically every day. Besides humans, also animals learn. The ability to learn depends on the evolutive stage of species. Investigation and interpretation of natural learning is the domain of *the psychology of learning* (Borger and Seaborne, 1966; Anderson, 1995) and *the educational psychology* (Driscoll, 1994). The former investigates and analyses the principles and abilities of learning. On the other hand, the latter investigates the methods of human learning and education and aims at improving the results of educational process. Educational psychology considers various human attributes like attention, tiredness, forgetfulness, and motivation, to be of crucial importance for a successful educational process. Therefore education carefully takes into account the relation between teacher and learner, and suggests various motivation and rewarding strategies. They are of course of great importance for human learning, however, they are much less important or even completely unimportant for the (contemporary) machine learning.

The basic question, addressed by the psychology of learning, is the distinction between *innate* and *learned knowledge* (this question is important also for machine learning researchers: what kind of background knowledge – program – is necessary and/or sufficient for efficient and successful machine learning). Many skills of animals are innate and not learned. For example, grain pecking is innate to chickens and practice has no influence on its success (Borger and Seaborne, 1966). The age, however, is important due to growth of the body into its adult form (on which learning has no influence). This ripening process can have periods when learning is possible and/or necessary for the subsequent development (Anderson, 1995).

Instincts are innate skills. Also with humans instincts play a significant role. Children have innate inclination towards learning and learn spontaneously. Innate ripening process is necessary for certain forms of learning. Learning itself is crucial for human existence, while with many animal species this is not the case. The higher the evolutive stage of species, the more important role has learning. Higher final level of the learning capability of species implies slower learning in childhood. For example, primates spend much more time for perceptual learning than simpler species (Anderson, 1995).

Newborns have certain innate recognition abilities such as, for example, the ability to distinguish faces from other objects. Most of perception is yet to be learned, especially seeing and hearing. When a blind-born person gains eyesight after an operation, the sight disturbs him and he can't use it. Only after training he starts to recognize faces, figures etc. The ability to use eyesight is even worse than with newborns. This fact can be explained with the loss of innate abilities during growing up.

A living organism has certain innate abilities that determine its learning abilities and also its intelligence. For example, a worm has 13 neurons and a bee has about 900 neurons. Higher animals learn to solve simple tasks as quickly as humans do. On the other hand, solving harder problems can be learned only by most evolved species. For example, a gorilla was successfully trained to understand more than two thousand words which is typical vocabulary of a pupil. Human's "hardware" is more advanced and contains about 10^{10}-10^{11} neurons and about 10^{14}-10^{15} synapses.

Human Memory

Without the ability to memorize learning is impossible. This can be illustrated with patients that due to certain brain damage lost the ability to store new experiences in the long term memory (Sacks, 1985). Such a person is even not able to remember the experience from a few minutes ago. If the experience is repeated the patient considers it as a novel one. Such people behave in the same way all life through. Their knowledge does not change through time, for them the time "stopped" in the moment of their brain damage. Their reactions are the same again and again. They are like computer programs - for a given input they return the same answer.

Neurons in the brain are interconnected with synapses which transmit the impulses between neurons. Besides receiving impulses from some neurons and transmitting them to other neurons, each neuron participates in various activities. There is a lot of neurophysiological evidence that the following activities are the basis of memory: creating new connections between neurons, changing the connection strengths on the synapses, and protein construction. The second basic principle - changing the connection strengths – was discovered by Hebb (1949) and is used in artificial neural networks, which is one of machine learning subareas (Haykin, 1998). Artificial neurons are able to sum up the input impulses and, in case that a certain threshold is exceeded, send the impulses forward to other neurons. If biological neurons were very simple processors, similar to artificial neurons, and if the only memory mechanism was the connection strength, then the brain would have enough capacity to memorize everything one can experience in one hundred years (Kohonen, 1984). We may say that the abilities of the human memory are indeed unlimited. Nowadays, researchers agree that a person can memorize just every experience from the lifetime.

However, recollecting of memorized facts is not always easy and sometimes seems impossible. On the other hand, in special states of consciousness, such as deep relaxation or under hypnosis some people are able to recollect details from their experiences that happened many years ago (even from early childhood). People with photographic memory can recollect all the details from scenes they saw many years ago. For example, they are able to recollect from their childhood a picture with some text and they are able to read from that picture (although, at the time they saw the picture, they were not able to read yet).

One special example is a Russian journalist (Russell, 1979). He was able to recollect every detail that happened to him in his life. When asked, for example, to describe, what he was doing seven years ago on a particular date, he thought for a while and then asked: "At what time of the day do you want to know?" On one occasion he had read a string of several hundred meaningless syllables and recollected it without a mistake. Several years later he recollected it without mistakes.

Due to its associative nature we say that the brain is an associative memory and is content-addressable. That is why we are better at *recognition* than at recollection or *reconstruction* of known concepts, images, melodies etc. Some types of artificial neural networks (for example Hopfield's artificial neural networks (Hopfield, 1982; 1984)) are also content-addressable and exhibit in some aspects similar behavior to that of the brain (Rumelhart and McClelland, 1986). The most difficult task for ordinary people is to recollect a data which cannot be related to another known data. In fact, the address of that data in the memory is missing.

We quickly "forget" the details of a concept, a picture, or an event, however, the basic outline and the idea are kept in memory. Such a rationalization is a basis for *generalization* where the memory covers a series of similar concepts that differ in details. Machine learning algorithms are based on similar principles.

Types of Learning

Most frequently, the types of learning are classified according to the learning complexity, the learning material, and the learning strategy. With respect to the complexity of the learning process we differentiate the following learning types (Borger and Seaborne, 1966; Anderson, 1995): imprinting, conditioning and associating, probabilistic learning, memorizing, learning by trial and error, imitation, and learning by understanding and insight. The latter one is the most pretentious kind of learning. It requires memorizing, abstract (symbolic) thinking, logical reasoning, and causal integration which leads to problem understanding. The insight comes suddenly, when the learner discovers the solution by integrating the relationships in a given problem situation. To some extent, higher developed species are able to learn by understanding and insight, however, by far the most qualified for that kind of learning are humans.

As regards the learning material we differentiate the following learning types: sensorial learning, learning of motor skills, and semantic (textual) learning. In all kinds of learning there appears a *spontaneous generalization*. In sensorial learning the learning material is generalized to similar sensations. For example, if we learn to react to the sound of a given frequency, we will react to all similar sounds – the closer is the frequency to the learned one, the stronger is the reaction. During learning of motor skills with one part of the body the skill will generalize also to other parts of the body. For example, the learned reactions with right hand will be transferred to some extent also to the left hand and even, to lesser extent, to legs.

In semantic learning the learned reaction generalizes to phonetically and semantically similar words (synonyms). Again, the more similar is the word, the stronger is the reaction.

FROM INTELLIGENCE TO CONSCIOUSNESS

When you remove all thoughts, what remains is pure consciousness.

— Ramana Maharshi

Nowadays, most of the researchers agree that without learning there is no intelligence. However, learning alone is not enough. In order to be able to learn, a system has to have some capacities, such as memory capacity, ability to process data, ability to perceive (input and output) etc. These abilities per se do not suffice if they are not appropriately integrated or if they lack an appropriate learning algorithms. Besides, for efficient learning one needs also some initial knowledge – background knowledge (which is inherited in living systems). By learning the abilities of the system increase, therefore intelligence also increases. Opinions of various scientists and philosophers are not united whether hardware, background knowledge, and learning suffice for (artificial) intelligence. Defenders of the opinion that natural intelligence is the only possible intelligence (for early discussions see for example (Taube, 1961; Dreyfus, 1972)) disagree with their opponents who claim that it is possible to create an intelligent machine (Gams et al, 1997; Penrose, 1989; Searle, 1992; Sloman, 1992).

Types of Intelligence

We have to consider various types of intelligence (abilities): numerical, textual, semantical, pictorial, spatial, motor, memorial, perceptive, inductive, deductive etc. Lately, even emotional intelligence became widely recognized (Goleman, 1997). Some authors describe more than one hundred types of human intelligence. Therefore the systems cannot be strictly ordered with respect to the amount of intelligence. A system (a human or a machine) can be better in some types of intelligence and worse in others and vice versa. When speaking about artificial intelligence we do not expect from an intelligent system to be extremely capable in only one narrow aspect of intelligence, such as for example the speed and the amount of memory, the speed of computation or the speed of searching the space or (almost optimal) game playing – nowadays computers in each of these aspects already have very advanced capabilities. We expect from an intelligent system to be (at least to some extent) intelligent in all areas which are characteristic of human problem solving. It seems that we need an integration of all different types of intelligence into a single sensible whole (a kind of supervisory system) so that during problem solving it is possible to switch appropriately between different types of intelligence. However, most of the speculations about artificial intelligence do not take into account yet another level: *consciousness.* Although we are not able to formally define consciousness, it may be good candidate for the supervisory system.

Consciousness as a Mode of Being

Many different aspects of consciousness are nowadays extensively studied by many researchers from various fields. There is a lot of literature available. For our purposes we shall only briefly state some basic properties of consciousness and later we shall focus on phenomenal consciousness (i.e. the 1st person experiential consciousness or consciousness as a mode of being) which seems to be relevant for the limits of artificial intelligence.

Due to the subjective character of consciousness, it is hard to define it objectively. Currently we have no widely acceptable definition of consciousness. We can speak of self-awareness, however, it only describes a certain state of consciousness. With humans we can differentiate several *states of consciousness*. One possible classification is as follows:

- Wakefulness: a usual state when the mental content is present and we are self-aware.
- Dreaming: when we are sleeping, dreams are our mental content but we are not self-aware.
- Dreamless sleep: there is no mental content and we are not self-aware (therefore, such a state could be classified even as without consiousness). Note that we can have also the opposite situation: dreaming during wakefulness.
- Meditation: there is no mental content and we are self-aware.

The boundaries between the waking state, dreaming and dreamless sleep are fuzzy. While dreaming one can be self-aware (so called lucid dreaming). Besides, we do not know if a dreamless sleep exists at all. When we do not remember any dreams we cannot be sure that in fact there was no dream.

The human consciousness can be further divided into several *levels*. One possible classification is as follows:

1. pure consciousness (without mental content, which corresponds to deep meditation): about pure consciousness we have only subjective descriptions of sages that are able to intentionally achieve such a state (Lorimer, 1998);
2. super-consciousness (corresponds to special abilities, such as direct vision – vision without physical eyes, clairvoyance, and telepathy) and the altered state of consciousness. Such states of consciousness are still a matter of debate, although there exist numerous scientific papers and books that study these phenomena, for example (Jahn et al., 1997; Lorimer, 1998; Korotkov, 2004; Korotkov et al., 2002);
3. normal consciousness (waking state – mental content depends on our attention): this seems an obvious state of consciousness as everybody experience it;
4. unconsciousness (corresponds to all mental processes, which we are not aware of, but in principle we could become aware of them with the appropriate focus of attention): most widely accepted psychological models include unconsciousness but only rarely you can find super-consciousness;
5. no consciousness (which most probably corresponds only to the dead body).

Some quantum physicists relate consciousness with the collapse of the wave-function which is used to describe the probability distribution of all possible states of the observed system (for example a set of particles). When the measurement takes place the wave function collapses and from all possible states one particular state appears as real – the result of the measurement (Capra, 1983). The great mathematician John von Neumann, who provided a rigorous mathematical foundation of quantum mechanics, believed that only the human consciousness can collapse the wave function. The eminent Nobel prize-winning physicist Eugene Wigner writes: *"It follows that the quantum description of objects is influenced by impressions entering my consciousness ... It follows that the conscious being must have a different role in quantum mechanics than the inanimate measuring device."* The famous physicist John Wheeler has taken this one step further. According to him the entire universe can emerge into true physical existence only via observation of the consciousness!

Therefore, the quantum principle of non-determinism of state "until the measurement" could actually mean "until the measurement, performed by a conscious being". By this principle we can speculate that the reality is not determined until a conscious observer measures it. If confirmed, this hypothesis could clarify many currently unexplained phenomena, such as telepathy, precognition, tele-kinetics and clairvoyance. Note that this is currently only a speculation and is far from being a grounded hypothesis.

However, many researchers assume (note that this is only an assumption), that consciousness appeared in a certain stage of evolution and is a result of complex (hard-wired) system, such as human brain.

Computability

In order to analyse the capabilities and limitations of artificial intelligence, we refer to the results from theoretical computer science. The theory of computability (Hopcroft and Ullman, 1979; Manna, 1974) reveals that only a small part of all problems, which can be formally posed, can be algorithmically solved. The number of all possible algorithms is countable infinity $\aleph_0$ which is equal to the power of the set of all natural numbers: $\|N\| = \aleph_0$. However, the number of all problems is uncountable infinity $\aleph_1$ (which is equal to the power of the set of real numbers $\|R\| = \aleph_1$). Uncountable infinity corresponds to the power of the powerset of a countably infinite set: $\|2^{\aleph_0}\| = \aleph_1$. Therefore, the number of all possible problems is so huge, that almost none of them is algorithmically solvable.

Science uses the following formal symbolic languages for describing (modeling) reality: mathematical logic, programming languages, recursive functions, and formal grammars. All these formalisms have equivalent expressive power and they all have equivalent limitations (Manna, 1974): they can partially describe the phenomena in the discrete world (discrete functions), and practically a negligible part of the continuous world (continuous functions). Therefore, if the world is indeed continuous (note that this is only a hypothesis), then most probably it is undescribable by any of mentioned formalisms. And we (science) are able to use only those formalisms with our rational mind. This implicates that any knowledge that can be reached by science, described in books or by teachers, cannot be ultimate. It is always only an approximation of the reality.

The number of all rational numbers (fractions) is the same as the number of all natural numbers: $\|N\| = \|Q\| = \aleph_0$. We can assign to each rational number a unique natural number and vice versa. The set of rational numbers corresponds to the *discrete world*, while the set of all real numbers corresponds to *continuous world*. The names here are indeed suggestive. *Rational* numbers are analogous to the rational, discrete world, reachable by our *rational* mind, while *real* numbers are analogous to the *reality* which (by above assumption) is much (much, much, much!) more rich and is (if the assumption holds) in principle unreachable to rational mind!

Important Steps towards ArtiFIcial Intelligence

In all years from the very beginning of electronic computers we cannot notice any crucial progress towards the ultimate goal of creating an intelligent machine by using machine learning algorithms. Anyway, we can mention some important steps, such as discovery of new concepts with Automatic Mathematician, successes of machine learning in playing complex games, modeling the cognitive processes with artificial neural networks and generating new knowledge from data.

- *Automatic Mathematician (AM)* – an interesting program for discovering new concepts in mathematics (Lenat, 1983). The system could model a remarkable selection of cognitive activities as search in which the program is guided by a large collection of heuristic rules. Initially AM was given the definitions of 115 simple set-theoretic concepts and a collection of 243 heuristics for guidance. AM managed to discover the basic mathematical operations such as addition, multiplication, and their inverses. Multiplication was discovered in different ways: as multiple additions, Cartesian product, and the length of the list where each element was replaced with another list. Following this, AM discovered the concept of divisibility and noticed that some numbers have a number of divisors. One of AM's built-in heuristics tells it to explore extreme cases. Its first attempt was to find all the numbers with '0' divisors, but found none, then one divisor (found 1) and two divisors. Before creating a new concept of prime numbers, it listed all the numbers with three divisors, for example 4. The reason that the prime number concept was noted is due to AM's "Look for something interesting" heuristic. Other interesting heuristics were "Matching is interesting" and "For interesting operators, define also inverse operators". AM achieved similar successes in planar geometry.
- *Great successes of computers in complex games,* such as checkers, backgammon, and chess. One of the most impressive game playing applications of machine learning is Tesauro's TD-Gammon for the game of backgammon (Tesauro, 1995). TD-Gammon required little backgammon knowledge and yet learned to play extremely well, close to the level of the world's strongest grandmasters. The learning algorithm in TD-Gammon was a straightforward combination of the reinforcement learning (Sutton and Barto, 1998) and nonlinear function approximation using a multi-layer neural network trained by the backpropagation algorithm (Rumelhart and McClelland, 1986). TD-Gammon was developed in several phases. Each phase was improving on the structure and the size of the neural network from the previous one. Tesauro was able to play his programs in a significant number of games against world-class human players. Based on these results and analyses by backgammon grandmasters, TD-Gammon appeared to be at the playing strength of the best human players in the

world. In the long run, TD-Gammon may play even better than grandmasters, because of the possible human player's weariness and preferences for certain strategies. TD-Gammon has also changed the way the best human players play the game. It learned to play certain opening positions differently than was common among the best human players. Based on TD-Gammon's success and further analysis, the best human players now play certain positions as TD-Gammon did.

- *Artificial neural networks for modeling the cognitive processes in the brain.* Algorithms for artificial neural networks (ANNs) mimic biological neural networks by abstracting neural functions to a simple element (neuron), only able to summarize its input and normalize its output. Neurons are interconnected into arbitrarily complex artificial neural networks (Haykin, 1998). The task of the learning algorithm is to determine weights on connections between neurons (used for calculating weighted sums in each neuron) in order to optimize the given criterion function. For modeling the cognitive processes the most suitable are variants of associative memories, originally developed by Hopfield (1982; 1984). Many cognitive processes of the brain can be emulated by ANN.

  Similarities with brain are very interesting (Rumelhart and McClelland, 1986). The brain is content-addressable, as is the case also with ANN. Mental processes use the brain as a whole. It is not possible to locally separate the memory function from processors, as we can do with classical digital computers. The memory in cortex is distributed. As already mentioned, in neurophysiology it is known that synaptic connections contribute significantly to human memory as was discovered by Hebb (1949), and ANN use synaptic weights as their memory. In the brain as well as in ANN the performance gradually decreases with the number of destroyed neurons. There is no single neuron of crucial importance whose damage would collapse the whole system. In the cortex there is no single part on which all the other parts depend. Also, no part of the brain is irreplaceable. For example, with children born without one of the hemispheres, the other hemisphere takes over the functions of the missing hemisphere and children grow up normally. Human perception is to a certain extent invariant to height, translation, and rotation. The brain is able to automatically generate a reduced representation. Different cortex regions have specialized for different sensorial signals while preserving the topology of spatial signals. The same principles can be simulated with ANN. There exists much neurophysiological evidence that brain neurons use self-organization which can be modeled by a certain kind of ANN (Kohonen, 1984). In the brain as well as in ANN there is no "hardware" in the strict sense of the word, nor is there any software in the usual sense of the word. The brain and ANN are not appropriate media for recursive processing (and recursive statements are hard to understand by untrained humans).

  ANN models do not strictly follow the analogy with the brain. The majority of biological neurons have either excitatory or inhibitory synapses. On the other hand, an ANN can have both types of synapses with single neuron. Besides, as opposed to ANN, the size of the signal in a biological neuron is given with the impulse frequency. Another major difference is that currently known ANN models do not use global communication between the neurons, while in the brain the global communication is not only present but seems to be of crucial importance.

- *Several successes of machine learning in generating new and beneficial knowledge from data* in various domains, such as production processes and medical diagnostic (Kononenko and Kukar, 2007). In most production processes there are small deviations from the normal course of events. The important deviations are those that cannot be anticipated by the controller (or the controlling system) because of partially incomplete knowledge of the underlying production process. Machine learning methods can, from the periodical observations (measurements) of the production process, synthesize a process model that can be used for forecasting in the future.
  Based on the records of the patients who had been treated in the same hospital for the same (or similar) disease, machine learning methods can be used to induce knowledge that can be used for diagnosing new patients. The induced knowledge may be used as an explanation for given diagnoses, and may provide insight into the diagnostic problem. Tools based on induced knowledge are also used to assist medical students and inexperienced physicians (Kononenko, 2001).

Note, however that the principal limitations for programming languages and other formalisms, that stem from the computability theory, hold also for any machine learning algorithm, no matter how advanced and complex it is. Very strict limitations are posed by the theory of learnability (Gold, 1967; Osherson et al., 1986). The latter is derived from the computability theory – the machine learner is necessarily an algorithm. As it may be expected, all the limitations for computability hold also for learnability.

Potentials for ArtiFIcial Intelligence

Research of AI methods tries to develop systems that behave intelligently and are able to solve relatively hard problems. The developed methods are often based on imitating the human problem solving. As a long-term goal we are interested whether computer intelligence (capability) can indeed achieve or even exceed the human intelligence. Important aspects for understanding the abilities of artificial intelligence are the impact of learning on intelligence, the speed of problem solving, the principal limitations of algorithms, and the imitation of intelligent behavior.

Learning has great impact on intelligence. By learning the capabilities of the system increase, therefore also its intelligence increases. Human intelligence is dynamic and is changing throughout the whole life, mostly increasing. However, when determining the amount of intelligence one has to take into account numerous different types of intelligences.

More intelligent systems are faster. Adaptation to the environment and problem solving are better (more efficient) if they are faster. Therefore, intelligence is highly related to speed and time. All tests of intelligence are timed as are also all examinations. Therefore, we can conclude, in that sense, that faster computers are more intelligent than slower ones, that parallel processing is more intelligent than serial one, etc.

Algorithms have mathematically determined limitations. If humans were equivalent (degradable) to a computer algorithm then all the limitations posed by the computability theory would hold also for humans – this would have strong impact on the abilities of human intelligence. If, however, we assume that humans are stronger "machines" than (digital) computers (for example continuous and not discrete machines) then the human activity is algorithmically undescribable. The consequence of this assumption is that it is impossible to algorithmically derive an artificial intelligent system which would completely reproduce the human behavior.

Imitating of intelligent behavior is becoming possible with the improved technology. Nowadays the technology of movies, multimedia, computers, robots, and virtual reality is very convincing and suggests that it is possible to imitate just everything and induce the sensation of reality. Therefore, if we omit the consciousness, by huge amount of memory and by storing the solutions to all possible situations a machine can in principle be intelligent enough to induce the sensation of artificial intelligence. If we add also extraordinary processing abilities (super parallelism with super-fast processors), algorithms for efficient search of huge spaces and machine learning algorithms that would enable online improvements of algorithms and heuristics, then such a machine could rightly be named "intelligent" – it could outperform the humans in many if not in all "practical" tasks. Of course, such a machine still lacks consciousness.

ArtiFIcial Consciousness?

We can determine whether a system has certain learning capabilities and a certain level of intelligence. On the other hand, the phenomenal (experiental, 1st person) consciousness is much different – it is necessarily related to subjective experience and any objective observer has no means to verify it. It is objectively possible (but, of course, nontrivial) to determine the ability to learn, the amount of acquired knowledge, the ability to intelligently adapt to the environment and solve problems. Numerous tests of intelligence are able to measure only specific types of intelligence (and the results are typically only partially reliable). Opposed to intelligence, in principle it is not possible to verify the phenomenal consciousness of the system. Whether a (biological or artificial) system is conscious or not is known only to the system itself, but only in the case that it is conscious. An observer from the outside has no means to verify it. One can speak about phenomenal consciousness only if she herself is conscious and if she assumes that systems, similar to her, are also conscious. Note that any conscious system can be imitated with an unconscious system to arbitrary (but always incomplete) resemblance. It follows that any objective observer can be fooled.

We continue with speculations about some interesting viewpoints. A system can in principle be more or less intelligent but without consciousness. Such system is a robot (or in extreme cases a human zombie). On the other hand, a system can be conscious but much less intelligent. Such are less intelligent people, animals etc. Therefore, we may strictly differentiate between the system's abilities – its intelligence – and the system's consciousness, which is necessarily subjective. From this point of view, the (phenomenal) consciousness (as a state of being) seems to be fundamentally related to the following notions: *life*, *intelligence*, and *free will*.

One attractive hypothesis is that *life implies consciousness* (and also vice versa). Humans are conscious, dogs and cats are conscious (which are typical claims of pat owners), and even ameba may be conscious (at least to some extent). Nowadays science is still not able to explain the origin of life. By "materialistic" assumption life appeared by chance which is highly improbable. The other, more probable possibility, is that it is a result of the complexity and self-organization of matter. Current theory of evolution has many supporters and also many opponents and there exist a lot of arguments for and against (Meyer et al., 2007). Another theory states that life came out of the space – aminoacids were found on the meteors. But then we have to ask, where and how were those aminoacids created. On the other hand, by vitalistic assumption, the life was created by a higher force – a universal consciousness. None of these hypotheses can be completely proved or disproved by current scientific knowledge. If life indeed implies consciousness, a very interesting question can be posed: would artificial life with highly developed artificial intelligence acquire consciousness, i.e. be conscious? We cannot say much about it as we do not know if artificial life is possible at all.

One interesting speculation is about the levels of consciousness, namely that more complex systems enable more intelligent behavior. One could argue that *more intelligence enables higher level of consciousness*. Although consciousness is not objectively verifiable nor measurable we can speculate that with greater capabilities, i.e. greater intelligence, the higher level of consciousness can be achieved. We can assume that less developed species are less conscious than more developed ones. However, you can have obvious counterexamples. Let us have a super intelligent system (for example a highly intelligent man) and remove consciousness (such as brain washing or simple blindness with his or her own ego), you can obtain a highly intelligent system (for example a fanatic or an extremely avaricious man for money or power) that is not conscious of his or her actions. If we paraphrase: a child (in the sense of lack of consciousness) is playing with a nuclear bomb. The consequences can be catastrophic. On the other hand, you may have an uneducated person that is not able to solve rather simple intellectual problems, however, he or she can be surprisingly wise. It is known from various traditions that wisdom in not the consequence of knowledge but rather of high level of consciousness (Kononenko, 2007).

Another interesting speculation is that *consciousness implicates free will*. If a system only reacts to outside stimuli then its responses are determined in advance and therefore could be unconscious. A conscious system can by itself, without any outside cause or stimulus, decide for an action (and this is not a reaction). This means that it has free will. Scientists and philosophers still argue whether free will exists at all. Anyway, it seems sensible to assume that consciousness as a state of being implies also free will. Note that the latter speculation together with the second last (which assumes that less developed species are less conscious than more developed ones) implies that the less developed species have also a certain kind of free will.

Conclusions of Part I

The appropriate relationship between human and machine is complementarity and broadening and not imitation.

— Mortimer Taube

Cognitive modeling is intended to investigate and explain cognitive processes in the human brain (and maybe in other parts of the body as well). Humans always wanted to know themselves better. We still want to create systems that would exceed human abilities.

Nowadays it is impossible to explain cognition without learning. Most of recent cognitive models include various learning algorithms.

Human learning may be considered very slow (although we do not know any more efficient biological system): twenty years of education is necessary for a Master of Science to start to learn her profession at work. Only after the next fifteen years is enough experience gathered and finally we have an expert in a narrow problem domain. Such experts are not able to simply transfer their knowledge and experience to their younger colleagues. Besides, an expert has all human weaknesses: he or she can be forgetful, regularly tired, his/her effectiveness depends on current mood, motivation, weather, and the state of health.

One of the principal purposes of the development of machine learning is automatic generation of knowledge bases for expert systems. An expert system is able to help human experts with their work. In exceptional cases it can even replace human experts, at least temporarily (for example when an expert is absent or ill). For expert systems we require that they are able to explain and argue about their decisions. Only then can a user trust such a system and transfer to it important decisions. For such a system a knowledge base is needed. It can be developed manually, such as by examining the existing literature and/or interviewing available human experts. This is typically time consuming, non-consistent, unreliable, and sometimes even impossible. Another possibility is the automatic generation of the knowledge base from examples of solved problems by machine learning algorithms. It is well known that experts can easily provide examples of solved problems. The descriptions of solved problems can be obtained from archives and from stored records. Descriptions of problems, solved in the past, can be used for machine learning and the automatically derived rules can serve as a knowledge base for an expert system. One has to be careful with automatically derived knowledge bases. It has to be checked and evaluated. Most often the accuracy on unseen cases is evaluated. It is sometimes essential that automatically derived rules are transparent – we need to understand them. We usually need to correct and/or adapt the rules and iteratively repeat the learning phase by appropriate modification of the description of learning examples.

All the advantages that hold for computers hold also for expert systems. They are never tired, available every day 24 hours per day. They are reliable and provide replicable results. The knowledge base of an expert system is trivially transferrable to other computers. Of course, expert systems can consult huge data bases, collected in the past.

Expert systems cannot and at least in the near future will not replace human experts, however, the most important issue is complementarity and broadening (as suggested by Mortimer Taube) - they will and already do complement the work of human experts and help them with their decisions. The comparison of human and computer characteristics shows the advantages and disadvantages on both sides. Although computers can with extremely high speed manipulate huge amounts of data with mathematical accuracy, they by far do not reach the wideness of human knowledge and the intelligent and common sense access to the huge human memory. Due to enormous parallelism in the brain, humans are able far more quickly and with a surprising ease to solve certain kinds of problems which nowadays computers are not able to solve, or the solution would be much too slow. A major advantage of humans is that they are flexible. They dynamically adapt, upgrade and improve their knowledge. This is also the reason why we need to develop advanced machine learning systems. With appropriate learning algorithms we need to make our computers more flexible and adaptable to new situations. Even algorithms themselves will be in the future dynamic and will gradually change through learning.

On the other hand, we have to be aware that the behavior of such systems may be unpredictable. Therefore, we have to include the mechanisms for verification and control over such systems. Some researchers predict that intelligent systems will have to lie to the user in order to achieve optimal performance in critical situations (Michie and Johnston, 1984). For example, when immediate action is needed and there is no time to argue with the user, a system may decide to lie to the user to quickly achive the desired results. Such a system would be hard to control. Therefore, there always has to exist the possibility of the complete control over the system.

Part II. Objective Science and Subjective Spirituality Are Complementary

Science without faith is lame, religion without science is blind.

— Albert Einstein

In this part, which expresses the author's own viewpoint, we discuss the relation between objectivity (in the sense of measurability and describability), and (phenomenal) subjectivity (in the sense of the 1st person, direct experience). This relation seems to implicate that the relation between intelligent and conscious behavior is similar to the relation between objective *science* and subjective *spirituality* (in the sense of spiritual, phenomenal conscious experience). Phenomenal spiritual conscious experience is highly subjective (in the sense of the 1st person experience), while science - by definition - is struggling to be objective. We start by emphasizing the extreme positions and later we claim that extremes need to be balanced.

Science (from extreme, mostly "materialistic" position) tends to be *objective* and is therefore limited to *rational (logical) mind*. Spirituality (in the extreme sense of the 1st person, experiential spirituality) tends to be *subjective* and primarily uses *intuitive mind (heart)*. Both, extreme objective science and extreme experiential spirituality search for the truth, but mostly use completely different tools and mostly interpret their results on different grounds. In agreement with the above statement by Albert Einstein (1940) we argue that science and spirituality are complementary to each other and that we need both (note that the Einstein's statement can be interpreted that spirituality without science is not spirituality but religion). See table 1 for main contrasts between the two extremes.

Science models empirical data. It derives a model (hypothesis, theory) which *describes measurements* and, if the model describes the data accurately, reliably and repeatedly, it is eventually accepted to be a (natural) law. If new measurements (which can be more accurate or measured under different conditions) deviate from the current knowledge, the laws are changed/widened in order to correspond also to new measurements. Objective science is guided by objective principles. It admits only the rational mind, which is limited with symbolic representation (computability, learnability) although, of course, scientists during creative research use also, most probably undescribable intuition. Objective science is interested in HOW the nature operates. It is not concerned in WHY the universe exists and what is the purpose of life. Due to ignorance of the latter two questions many scientists unfoundedly assume that the universe and the life appeared *by chance.* This assumption can lead to a conclusion that there is no deeper meaning of existence, which however is disputable. There may be individuals who find meaning in the sole (objective) fact of the high complexity, diversity and "beauty" of life.

On the other hand, (extreme experiential) spirituality is mainly concerned with the *purpose of life*. In all traditions, in all religions and spiritual movements, from the east and the west, we find the same basic issues: the purpose of life goes beyond the material world; everything that exists is one, originates from the same source and serves the same purpose; the truth is non-describable and unreachable to logical mind, it is necessary that everyone tries to feel it by him or herself by subjective (1st person) experience; the purpose of life is learning, the goal is to overcome the limitations of ego, to subjectively recognize the truth and to attain the wisdom; spiritual life is based on cultivation of spiritual virtues, such as unconditional love, compassion, faith, humility, patience, tolerance, simplicity, spontaneity, modesty, courage, sincerity, forgiveness etc. However, due to non-describability of truth, we have to consider the latter descriptions only as mental approximations of merely subjectively experienced purpose of life. Experiential (1st person) spirituality is necessarily subjective and uses intuitive mind, inner sense (heart). Various relaxation methods, meditation and spiritual ceremonials tend to calm down the rational mind, to eliminate thoughts. When all thoughts are eliminated, the *direct sense of reality* is enabled and the one's consciousness is widened.

Table 1. The Main Contrasts between the Extreme Objective Science and Experiential (1st Person) Spirituality

EXTREME OBJECTIVE SCIENCE	EXPERIENTIAL SPIRITUALITY
intellect	*inner sense, heart*
objectivity	subjectivity
measurable, describable	non-measurable, non-describable
describing reality	conscious sensing, awareness of reality
logical, rational mind	intuitive mind
HOW? describing	WHY? searching purpose
studies matter	experiences consciousness
life appeared by chance	life is chance
doubt, verification	faith, gnosis
logic, experiments, statistics	relaxation, meditation, ceremonials
analysis, differentiates	synthesis, joins
reductionistic, parts	holistic, whole
discrete, *rational world (analogy from math: Q)*	continuous, *real world (analogy: R)*, irrational, transcendence
objective, indirect experience	subjective, direct experience
theory, approximation of reality	practice, reality itself
separation,*space-time dimension*	all is one, *spectral dimension*
causality, thinking of past and future	no causality, now!
knowledge	wisdom

In the original sense of the word, "philosophy" uses both, science and spirituality, objective and subjective experience, in order to achieve the balance and *harmony between rational and intuitive mind* (between head and heart). True philosophy deals with both questions: how the universe operates and why universe exists and what is the meaning of life. Great philosophers and sages from all cultures remind us that we need both, rational and intuitive mind. As Dalai Lama has stated several times in his inspiring talks: "*We need education and the sense of ethics – these two have to go together.*"

From even higher perspective, one can speculate that true spirituality encompasses science – spirituality can encompass science as one form of human life and work. The analogy from mathematics was described above (in subsection on computability). The rational numbers are a tiny subset of real numbers – the analogy suggests that the rational world is a tiny part of the real world: we can speculate that the majority of the world is irrational. If we continue in this spirit, we may suggest the hypothesis that objectivity is a tiny part (a special case) of experiential subjectivity and that intellect is a tiny part of consciousness.

Intelligence is the capability that artificial systems are gaining and in the future they will continue to gain more and more of it. On the other hand, consciousness has deeper meaning and purpose. It is necessarily connected with *ethics of life*. Intelligence without heart is unconscious intelligence, able to demolish and destroy the environment and itself. Artificial (and natural) intelligence is a *tool* which can be beneficially used or abused. The responsibility remains on our consciousness and conscience.

Conclusions of Part II

> Knowledge is important, however, much more important is its beneficial use. This depends on human mind and heart.
>
> —Dalai Lama

It seems that the basic question which everybody has to answer by him or herself is whether life has a purpose. The basic assumption that life appeared by chance (note, however, that there is no scientific evidence whatsoever for neither such an assumption nor for the opposite one) many people take for granted. This, however may be a very dangerous assumption. If life appeared by chance, one may conclude (but not necessarily does) that there is no deeper meaning of life – the only purpose is survival. This may result in fear and competition. The obvious next step is struggling for PROFIT (that enables better survival), which results in greediness, exploiting of nature, violence etc (however, there are also other possibilities which ego may select, such as ease and laziness). There is no room for ecology – why should I bother myself with ecology, if I only want to get more for myself? (Of course, sooner or later ego is forced to take into account ecology as there is no other possibility to survive). Within this scenario there is no room at all for ethics – there is nonsense to act ethically if my purpose is to serve only my own profit.

On the other hand, (not necessarily bound to scientific or spiritual orientation) one can assume that life has a deeper meaning – in fact it could be a chance to learn and to develop VIRTUES, the basic of which is humility and modesty. This leads to harmony with nature, ecological concerns, nonviolent acts of peace, giving, sharing and finally to recognition that all is one – and this is the only basis of true and sincere ethics. Everything else is just a trained/inforced morality which was always and will always be relative and dependent on "who wrote the rules". True ethics, however, comes from the very Source, knows no fear and is always there to advocate the Truth.

Maybe there will be times when the apparent gap between science and spirituality will be closed. Maybe the same Truth will become evident from both angles, objective and (experiential) subjective, once. Perhaps there is no contradiction in creation or evolution, both could be either started from Oness or they could lead again to One.

References

J.A. Anderson. *Learning and Memory, 3rd Extended Edition.* John Wiley and Sons, New York, 1995.

R. Borger and A.M. Seaborne. *The Psychology of Learning*. Harmonds worth: Penguin Books, 1966.

I. Bratko. *Prolog Programming for Artificial Intelligence, 3rd Edition.* Addison-Wesley, 2000.

F. Capra. *The Tao of Physics*. Flamingo, GB, 1983.

H. L. Dreyfus. *What Computers Can't Do*. New York: Harper & Row, 1972.

M.P. Driscoll. *Psychology of Learning for Instruction*. Allyn and Bacon, London, 1994.

A. Einstein. Science and religion. *Nature*, 146:605–607, 11 1940.

M. Gams, M. Paprzycki, and X. Wu (eds.). *Mind Versus Computer*. IOS Press, Amsterdam, 1997.

E. M. Gold. Language identification in the limit. *Information and Control*, 10(5):447–474, 1967.

D. E. Goldberg. *Genetic Algorithms in Search, Optimization and Machine Learning*. Addison-Wesley, 1989.

D. Goleman. *Emotional Intelligence: Why It Can Matter More Than IQ*. Bantam, 1997.

S. Haykin. *Neural Networks: A Comprehensive Foundation*. Prentice Hall, 2nd edition, 1998.

D. O. Hebb. *The Organization of Behavior*. Wiley, New York, 1949.

J. E. Hopcroft and J. D. Ullman. *Introduction to Automata Theory, Languages, and Computation*. Addison-Wesley, 1979.

J. J. Hopfield. Neural networks and physical systems with emergent collective computational abilities. *National Academy of Sciences*, 79:2554–2558, 1982.

J. J. Hopfield. Neurons with graded response have collective computational properties like those of two-state neurons. *National Academy of Sciences*, 81:4586–4590, 1984.

R.G. Jahn, B.J. Dunne, R.D. Nelson, Y.H. Dobyns and G.J. Bradish: Correlations of random binary sequences with pre-stated operator intention: A review of a 12-Year Program, *Journal of Scientific Exploration*, 11(3)345-367, 1997.

T. Kohonen. *Self-Organization and Associative Memory*. Berlin: Springer-Verlag, 1984.

I. Kononenko. Machine learning for medical diagnosis: History, state of the art
and perspective, Invited paper, *Artificial Intelligence in Medicine*, 23(1)89-109, 2001.

I. Kononenko. Objective science and subjective spirituality are complementary. In: I. Kononenko (ed.). *Proceedings of Measuring Energy Fields : International Scientific Conference*, (Kamnik, Tunjice, Slovenia, 13-14 October 2007). Kamnik, Slovenia: Zavod zdravilni gaj, 2007, pp. 117-124.

I. Kononenko and M. Kukar. *Machine Learning and Data Mining: Introduction to Principles and Algorithms.* Chichester, UK: Horwood Publ., 2007.

K. Korotkov (ed.). *Measuring energy fields : state of the science, volume 1.* Fair Lawn: Backbone, 2004.

K.Korotkov, P.Bundzen, V.Bronnikov, L.Lognikova (2002) Bioelectrography correlates of direct vision phenomenon. *Proc. 6.International scientific congress on GDV Bioelectrography*, St.Petersburg, July 13-14, pp.47-50.

D. Lorimer (ed.). *The Spirit of Science*. Floris Books, Edinburgh, 1998.

Z. Manna. *Mathematical Theory of Computation*. McGraw-Hill, 1974.

S.C. Meyer, S. Minnich, J. Moneymaker, P.A. Nlson, R. Seelke. *Explore Evolution: The Arguments for and against Neo-Darwinism.* Melourne/London: Hill House Publ., 2007.

D. Michie and R. Johnston. *The Creative Computer: Machine Intelligence and Human Knowledge*. New York: Viking., 1984.

D. N. Osherson, M. Stob, and S. Weinstein. *Systems That Learn.* Bradford Book, The MIT Press, 1986.

R. Penrose. *The Emperor's New Mind.* Oxford University Press, 1989.

D.E. Rumelhart and J.L. McClelland (eds.). *Parallel Distributed Processing, Foundations*, Volume 1. Cambridge: MIT Press, 1986.

P. Russell. *The Brain Book: Know Your Own Mind and How to Use It*. Routledge and Kegan Paul, London & Hawthorne, New York, 1979.

O. Sacks. *The Man Who Mistook His Wife for a Hat and other Clinical Tales*. Harper & Row Pub., 1985.

J.R. Searle. *The rediscovery of the Mind.* MIT Press, Cambridge, MA, 1992.

H. Simon. Why should machines learn. In: R. Michalski, J. G. Carbonell, and T. M. Mitchell, editors. *Machine Learning, An Artificial Intelligence Approach.* Tioga, 1983, pp. 25-37.

A. Sloman. The emperor's real mind: review of roger penrose's the emperor's new mind: Concerning computers, minds and laws of physics. *Artificial Intelligence*, 56: 355–396, 1992.

R. S. Sutton and A. G. Barto. *Reinforcement Learning: An Introduction*. MIT Press, 1998.

M. Taube. *Computers and Common Sense: The Myth of Thinking Machines*. New York: Columbia University Press, 1961.

G. Tesauro. Temporal difference learning and td-gammon. *Communications of the ACM*, 38(3), 1995.

B.A. Wallace. *The Taboo of Subjectivity: Towards a New Science of Consciousness*. Oxford University Press, 2000.

In: Philosophical Insights about Modern Science
Eds: Eva Žerovnik, Olga Markič and Andrej Ule
ISBN: 978-1-60741-373-8

Chapter 14

PHILOSOPHICAL REFLECTIONS ON THE HISTORY AND FUTURE OF SCIENCE AND SPIRITUALITY

Thomas C. Daffern*
International Institute of Peace Studies and Global Philosophy,
Llanerfyl, Powys, Wales, United Kingdom

SYNOPSIS

As a firm believer in non-violence as the only way to solve political and religious conflicts, the author has spent some 25 years working not only on the detailed history of specific conflicts and problems but also on a wider overall theoretical approach to provide a methodological perspective which can help heal and solve the current global crisis we are living through. It is from the background of this body of work, therefore, that the current remarks are addressed. The chosen structure of this paper is therefore to present firstly a synthesis of studies concerning the relationship between science and spirituality in different world civilizations, secondly to propose a metatheory (transpersonal history) within which the specific insights of each civilisational enterprise can be both affirmed yet also integrated , and thirdly to propose a scheme whereby these collected insights can help shed light on some of the more pressing and crucial of our global dilemmas facing us at the present time. The author submits that the current volume, examining the future and contemporary scientific advances in discrete fields of knowledge, needs to be situated in a longer time framework, which it is the purpose of this essay to provide.

This paper is a brief resume of the fruits of research into the overlap between science and spirituality by the author in an intercultural context. Specifically, its purpose is to propose a methodological revolution in our epistemological approach to knowledge, that can help further our collective task of achieving inter-cultural peace and conflict prevention. The author has systematically pursued a study of comparative philosophical and spiritual systems on the planet, going into great detail in his study of the different

* Address: International Institute of Peace Studies and Global Philosophy, Llanerfyl, Powys, Wales, UK, SY21 0ER; e-mail: *iipsgp@educationaid.net*;
URL: www.educationaid.net/www.lulu.com/iipsgp/ www.holisticchannel.org.uk

intellectual and philosophical lineages of the worlds major and minor civilizations. The Indian and the Chinese civilisations, Greek and Roman as the Classical European civilization, Egyptian, Sumerian, Babylonian and Persian, Jewish-Hebrew and Christian cultures are all dealt with, examining in each case something of the details of the relationships between scientific achievement and spiritual insights operative in each context. The study concludes that the notion of science has a deep and complex intercultural history, and that the mono-syllabic pursuit of science as a "Western" invention is a false idea which needs challenging – that science, in the sense of exact, detailed, rigorous and methodologically precise ways of knowing. It is precisely through the rediscovery of the richness and cultural depth of the scientific enterprise, that a way can be found to divert the resources of the world's intellectual elites from military research and development into more peaceful and beneficial scientific research fields.

INTRODUCTION

The history of science is a flourishing industry, as are the histories of philosophy, and the history of religion. Too often however these three areas of thought are treated as if they are separate projects. Reviewed through the lens of an intercultural approach, however, it is apparent that the boundaries between fields of knowledge are permeable, and vary from culture to culture depending on the norms and values and priorities of its own epistemological evaluations. With the world's scientific and intellectual communities facing complex problems, partly of their own making, such as war and violence, continuing economic and social instability, environmental threats and challenges, not least the dilemmas of global warming and resource depletion, notably energy sources, this volume comes at an opportune moment. Too seldom do specialists in scientific knowledge talk to one another across the detailed vocabularies of their own specialisms. Too often experts retreat into a nest of similarly trained semantic specialists, who merely exchange technicalities with one another, and rarely stray outside their own particular furrow. Such detailed specialist work is important and vital to the advancement of knowledge. But there are certain times, certain moments of kairos, when a broader view is called for. The dawning of the third millennium is one such cusp, it would seem. For several hundred years now, since the Baconian revolution in empirical method and the re-invention of scientific empiricism in the 17th century, there has been steady progress made in all fields of scientific knowledge, from mathematics, to medicine, to technology, to chemistry and biology and physics – such that if Bacon returned among us he would no doubt be flabbergasted at the progress in knowledge of the past 4 hundred years. At the same time he would be sad how little progress had been made in some fields of knowledge, particularly in terms, perhaps, of man's spiritual and divine understanding, which mattered greatly to Bacon, and also in regards to his knowledge of the irenic arts. It is the purpose of this present essay to propose several projects at once – firstly, to enrich discourse in relation to the unfolding fields of specific sciences, by placing the developments of scientific progress within the greater context of their more ancient histories. Secondly, by proposing a more useful enterprise for the human minds, for the collective intelligences and wits, that instead of plotting against one another in nationalist blocs or religious ensembles, we should rather collaborate together and share, across the barriers of our respective expertise and specialist discourses, and fumble back towards a common conversation, a common enterprise of knowing, that is at once, personal and collective, leading to the advancement of collective learning which is our human birthright as members of one species, Homo Sapiens Sapiens. After the arms races, and the cold wars, and the long dismal litany of warfare and military enterprises which litter the history of our globe, the author contends that it is time once again to resaddle the original horses of the scientific enterprise, and make tracks in the right direction, for the benefit and advancement of mankind. Need spirituality and empiricism be opposites in this enterprise, like some pantomime horse whose secret members are at war with each other and move in variant directions? Meta-Baconian in method, the purpose of this current essay is to show that this needs not be the case, but rather, science and spirituality can and must work hand in hand, answering to different needs, different persuasions of the human condition. In 1623, when Bacon published in Latin *De Dignitate et Augmentis Scientarum*, he felt it too early and dangerous a time to expound his review of the religious and theological sciences of his day. Bacon's work was an overview of the state of knowledge in all its branches – a kind of encyclopedia of all departments of human knowledge reviewed, analyzed, and corrected. In each division of learning, be it history, or geography, or poetry, or literature, or natural philosophy, Bacon made a survey of what institutions and methods were being employed in his time to advance learning therein, and secondly, made concrete suggestions as to how to improve and increase the knowledge quotient, so to speak, within each discrete field of knowing. Nothing quite this ambitious had been attempted since Aristotle in the history of thought, although intellectual historians have since discovered that Muslim encyclopaedists in the Ottoman Empire, Chinese thinkers and India philosophers, came close – but the difference in Bacon's case was his proposal of scientific empiricism as the guiding methodology for the overall advancement of learning. The question which confronts us now, nearly 400 years later, is how successful this empiricist turn of Bacon, followed subsequently by mainstream European scientific thinkers and practitioners, has actually been in practice. What has been accomplished? What has been achieved? Which branches of learning have flourished most in the 400 years since the *De Dignitate et Augmentis* was published? Which have been neglected? In which fields of knowledge have there been great gaps and little progress made? Which totally new areas of thought have been opened up, unknown or undreamed of by Bacon and his contemporaries (for such is the very nature of the advancement of scientific thought, that as one distant far horizon of hills is surmounted, another visage of previously unknown peaks swarms into view) – at once both exhilarating and exhausting those who range freely in the fields of knowledge as general thinkers, surveyors of the overview, in short, philosophers. This present volume, therefore, can be thought of an attempt of a first survey, a preliminary report on the advancement of learning in some small selected areas of modern scientific research. It is the purpose of this conclusive (Part B) essay to ask some wider questions of the nature of the scientific entries as a whole by placing the work in the widest possible intercultural and historical framework – asking perhaps some of the same questions which Bacon would have asked, and perhaps also some which he would not have dared to ask publicly in those dangerous days, when religious divisions often brought a dagger to the throat, or an axe to the neck for too much freethinking. Sadly, 400 years later, our religious wars have not gone away – but merely become more technologically sophisticated, and frighteningly more destructive in terms of potential harm to mankind at large. The task before us is an urgent one – to reframe the original scientific mission back to its fundamentally irenic and beneficent purposes, and to encourage intercultural and transcultural discourse, rather than narrow nationalisms or tribalisms, or the promulgation of scientific knowledge in the service of some party interest or national fraction. Without further ado, therefore, let us begin this task, and see what light it throws on the wider enterprise.

1. Indian Science and Spirituality

In the Indian subcontinent since ancient times, the clash between science and spirituality has been a non-starter. The ancient Vedic sages of India would have found amusing the notion that there can be such a think as "scientia" (knowledge) divorced from ethics or metaphysics. Indeed it was a hallmark of Indian philosophical thought from the very beginning that true knowledge cannot be divorced from ethics and morality, and from what we can call "spiritual awareness". Indeed, Indian thinkers became adept early on in achieving advancing states of altered consciousness, in which their mental faculties expanded to achieve heights of awareness that enabled calamity of thought and consciousness to allow new vistas of speculative thought to unfold. The Upanishads, at the end of the long sequence of the Vedic writings, remain a testament the Indian conquest, so to speak, of inner space.

The same motif was operative in the other great philosophical systems of India, especially Buddhism and Jainism, which both focused on aspects of this metaphysical and consciousness based spirituality, developing a kind of spiritual empiricism which set liberation from suffering and ignorance as its soterioloigcal target. Buddha and Mahavira between them, in slightly different but complementary ways, pioneered for all mankind paths to enlightenment beyond labels or particualarisms. To the Indian mind must also go the glory of realising that the innermost secrets of nature and supernature cannot be reached by storm; that true science requires grace and humility on the part of its practitioners, and that non-violence (ahimsa) must be a sine qua non for the innermost delving of the mysteries of existence. Such profound metaphysical insights as the Indian mind achieved, in the person of thinkers as diverse as Shankara, Ramanuja, Nagarjuna and Patanjali, took place against a backdrop of a colorful pantheistic polytheism in which the daily presence of the divine was anthropomorphized in a countless panoply of deities, dakinis and Bodhisattvas.

But amidst such a rich and fertile spirituality, science in the formal sense also flourished, for example in medicine, where the ayurvedic traditions of healing developed to a level of sophistication equal to any other medial system on the planet, and also in mathematics, where the Indian invention of zero as a discrete mathematical quantity in its own right, made advanced mathematics possible, as well as in the invention of the decimal numbering system later adopted by Islamic scientists and through them by Christian and European mathematicians.

Small surprise then that in the 19th century many advanced Western scientists joined the Theosophical Society, which was set up by the Russian metaphysician H.P Blavatsky with the advice and support of numerous Indian thinkers, some public and some "occult" (the Mahatmas), and which took as one of its objects the advancement of the reconciliation of science and spirituality, and as another, the advancement of a new humanity free of race and class prejudice. Nor is it surprising that Mahatma Gandhi should have received his first major support in his career of the "science of applied ahimsa" (or an "experiment with truth") from the Theosophical Society of South Africa, nor that Annie Besant and other theosophists (A.O.

Hume) should have helped found the Congress Party, nor that these actors working in harmony eventually gave India her independence.

Nor is it surprising that it is current Indian thinkers who are continuing to pioneer the new synthesis between science and spirituality made possible through advanced quantum mechanics and the new physics, in the form of thinkers such as Deepak Chopra, author of numerous work (such as How to Know God, The Science of Quantum Healing etc.). The list of Indian thinkers from the 20th century alone who have written about the links and reconciliation that is necessary between science and spirituality would be enormous.[1]

Nor must we overlook the contribution of India's Sikh religion, which achieved a remarkable synthesis between Hindu and Islamic thought, and which continues to inspire and enlighten innumerable Sikhs and their associated traditions, not just in India but throughout the world. Many great modern scientists have been of Sikh origin, perhaps because of the insistence on the part of the 10 gurus, starting with Guru Nanak, that spirituality should be focused not on other worldly pursuits, but on integrating and grounding spiritual insight into the everyday market place of the ordinary world.

Suffice it to say however, in summation, that Indian work of relevance to the synthesis of science and spirituality includes the following: the chakra theory of human psycho-spiritual physiology; work in quantum theory and the relation of physical matter to consciousness; medical healing systems based on ayurvedic traditions; research on the nature of intention and the power of mantras and invocation for manifestation, indicating the traditional mind-body split of 19th century atomistic science needs completely rethinking; research in paranormal faculties and potential in human beings, including out of body experiences and near death experiences, indicating that mind can exist outside of the normal body-brain continuum; research into past life regressions, and the scientific evidence for reincarnation, indicating that human minds and spiritual intentions date from longer time perspectives than hitherto realized, and that we may well carry the summation of our spiritual achievements in one lifetime on into future life experiences. In addition, work in India has been advanced into the ecological life of plants, and the consciousness of animals, indicating that we humans share a common life force with all sentient beings, thus providing a metaphysical basis to vegetarianism and ahimsa in environmental practices. Finally, in the science of altruism and agapology (the science of love) Indian researchers have long pioneered the understanding that it is in the fullness of the flowering of human relationships and in the drama of love, that sustainable wisdom and enlightenment can be found, and thus in modern day psycho-physiology, the study of hormones and hormonal exchanges, and subtle energy exchanges, seems to support what the tantric traditions of Indian spirituality have been saying all along – that life is a dance of lila, of bliss in action, and that our destiny as human beings is to awaken our fullest possible potential, as peace lovers, and dwell amidst this great harmony, and actualize it within ourselves, skimming between the transcendent and the immanent like a pebble skipping across a lake's surface, reflecting for a few brief moments the glory of the shimmering sunshine….

[1] Many of them, such as Iyer, Radhakrishnan, Aurobindo, Mukerjee etc are documented in the author's recently published biographical encyclopedia A-Z of transpersonal theorists, historians, psychologists and philosophers 1945-2001 (2006)

2. Chinese Science and Spirituality

In Chinese civilisation we have another vast history of the successful synthesizing of the best of scientific wisdom with the very highest possible esoteric metaphysical speculations and practices. In the indigenous Taoist traditions of China, going right back to the semi-mythical era of the Yellow Emperor and the other Taoist immortals, we have the idea lodged securely in the Chinese philosophical mind, that esoteric wisdom is not only accessible to the sage who cultivates spiritual knowledge in humility and harmony, but also that such knowledge must and can be put at the service of the common people. In the Tao Te Ching, we have the image of Tao as water, taking the lowest possible way to reach the sea, equated with the spiritual insights of the sage – there to serve everybody, anonymously, without asking for reward. In the self effacing service of the sage towards society, we have in a sense a precursor of Bacon's ideal of the scientist who cultivates knowledge only to help advance humanity's common lot, or indeed in the best traditions observed by Marxist inspired scientists, working for the common good.

As the great historian of Chinese Science and Civilisation, Dr. Joseph Needham has explored over a lifetime of scholarship, the list of Chinas's contributions to the global marketplace of scientific and technological ideas and inventions would fill several volumes in their own right, including: paper, gunpowder, advanced ceramic and pottery techniques, printing, silk cultivation, advanced metallurgical procedures, alchemical techniques and practices, advanced medical knowledge and the development of acupuncture and the knowledge of meridians and subtle physiology– the full details are listed in Needham's magisterial opus.[2] Like Indian civilisation, Chinese thought has long been fascinated by the links between the subtle and invisible realms and the manifest visible world. The function of the sage was to observe through the signs of reality in manifestation, according to the cyclical

[2] The full list of the work as it exists in print at present is as follows: VOL. I. Introductory Orientations. Joseph Needham, with the research assistance of Wang Ling (1954) VOL. II. History of Scientific Thought. Joseph Needham, with the research assistance of Wang Ling (1956) VOL. III. Mathematics and the Sciences of the Heavens and Earth. Joseph Needham, with the research assistance of Wang Ling (1959) VOL. IV. Physics and Physical Technology. Pt. 1. Physics. Joseph Needham, with the research assistance of Wang Ling, and the special co-operation of Kenneth Robinson (1962) Pt. 2. Mechanical Engineering. Joseph Needham, with the collaboration of Wang Ling (1965) Pt. 3. Civil Engineering and Nautics. Joseph Needham, with the collaboration of Wang Ling and Lu Gwei-djen (1971) VOL. V. Chemistry and Chemical Technology Pt. 1. Paper and Printing. Tsien Tsuen-Hsuin (1985) Pt. 2. Spagyrical Discovery and Invention: Magisteries of Gold and Immortality. Joseph Needham, with the collaboration of Lu Gwei-djen (1974) Pt. 3. Spagyrical Discovery and Invention: Historical Survey, from Cinnabar Elixirs to Synthetic Insulin. Joseph Needham, with the collaboration of Ho Ping-Yu [Ho Peng-Yoke] and Lu Gwei-djen (1976) Pt. 4. Spagyrical Discovery and Invention: Apparatus and Theory. Joseph Needham, with the collaboration of Lu Gwei-djen, and a contribution by Nathan Sivin (1980) Pt. 5. Spagyrical Discovery and Invention: Physiological Alchemy. Joseph Needham, with the collaboration of Lu Gwei-djen (1983) Pt. 6. Military Technology: Missiles and Sieges. Joseph Needham, Robin D.S. Yates, with the collaboration of Krzysztof Gawlikowski, Edward McEwen and Wang Ling (1994) Pt. 7. Military Technology: The Gunpowder Epic. Joseph Needham, with the collaboration of Ho Ping-Yu [Ho Peng-Yoke], Lu Gwei-djen and Wang Ling (1987) Pt. 9. Textile Technology: Spinning and Reeling. Dieter Kuhn (1986) Pt. 12. Ceramic Technology . Rose Kerr and Nigel Wood, with additional contributions by Ts'ai Mei-fen and Zhang Fukang (2004) Pt. 13: Mining. Peter Golas (1999) VOL. VI. Biology and Biological Technology Pt. 1. Botany. Joseph Needham, with the collaboration of Lu Gwei-djen, and a special contribution by Huang Hsing-Tsung (1986) Pt. 2. Agriculture. Francesca Bray (1988) Pt. 3. Agroindustries and Forestry. Christian A. Daniels and Nicholas K. Menzies (1996) Pt. 5. Fermentations and Food Science. H.T. Huang (2000) Pt. 6. Medicine. Joseph Needham and Lu Gwei-djen, edited by Nathan Sivin (2000) VOL. VII. The Social Background Pt. 1. Language and Logic. Christoph

ebb and flow, the operations of the hidden Tao, or the spirit which flows through all things, and to help steer and channel the integrated and harmonious interaction of heaven and earth in a continuing cosmic state of harmony. For the Chinese mind, mankind is a harmonic of both heaven and earth, existing at the cusp or intersection of the transcendent and immanent, and the job of the sage or spiritual scientists is to act as a channel in both directions, taking messages to heaven on behalf of humanity, and bringing messages to earth on the part of heaven. Outwardly, such a sage may appear to do very little, but inwardly, through this practice of wu-wei (non-doing) everything is accomplished, and the transcendental unity of heaven and earth and their cyclical interaction sustained and reaffirmed generation to generation. For the Chinese mind therefore, the notion that there could be such a think as "scientia" divorced from ethics and spirituality, would have made very little sense. All knowledge is impregnated with higher dimensions of meaning. Mere astronomy, to the Chinese, existed within the context of astrological divination; mere landscape surveying, although developed to a high art, made sense only in the context of geomancy etc. There were of course problems with Chinese intellectual life however, and often the true sages had to retire into the background of the distant hills and forests as the war lords took control of Chinas's destiny and foci of power. With the tradition of legalism there also grew a skepticism of all spirituality and a denial of the necessity for following the mandate of heaven; the Emperor was seen instead merely as a victorious warlord who could impose his will by force against the needs and wishes of the common people over the learned intelligentsia. From time to time in China's history then we have seen wholesale bloodbaths of the spiritual representatives of learning in the name of some centralising dogmatism imposed by which warlordism happened to reign triumphantly at the time. The fate of Buddhism, in China has therefore similarly followed an up and down path, and whilst in many of the greatest minds of China was seen as compatible with China's philosophical wisdom traditions, was seen by others as a threat. Notwithstanding such vicissitudes, perhaps we can sum up China's final work on the topic of the integration of science and spirituality as a learned humanistic tradition, embodied by Confucius, overshadowed by Lao Tzu, and nourished and fertilized by rich lineages of Buddhist thought, and more recently by the full panoply of Western scientific and philosophical traditions, not least by Marxism , in which China has sought to bring down to earth from the heavenly realms the wisdom of the sages, such that it might once again serve the common people, to their lasting benefit. In the modern era however, China's soul faces a Faustian dilemma: if she embraces to the full the fruits of modern technology and industrialization and the glamour of wealth which its seems to proffer, she might neglect to her own peril the indigenous voices of wisdom and self restraint which her own culture has found in the long run to be the wiser path; if she embraces the profit motif as her sole ruling idea, and seeks to out-profit the other great players in the international game of capitalism, she might run the risk of bringing the whole world down with her as she rises from the dragon's lair, through excessive consumption. Yet if the other wiser voices of China's civilisational traditions can be listened to, she can perhaps pioneer for the world a new form of social progress in which science and spirituality can work hand in hand to satisfy the needs and wishes of all of us in harmony – including our highest wish of all, the wish for mutual enlightenment and spiritual wellbeing.

Harbsmeier (1998) Pt. 2. General Conclusions and Reflections. Joseph Needham , edited by Kenneth Girdwood Robinson, with contributions by Ray Huang, and an introduction by Mark Elvin (2004)

3. Classical European Philosophy, Science and Spirituality

At the heart of Western classical civilisation, lays a paradox – the Orient ! For whereas the term "The West" is really only a modern coinage of cold war times, the true roots of European civilisation were always in Greece and further East, in Anatolia, Crete and Asia. The first philosophers of Europe thus came from the coast between Europe and Asia, along the coasts of Anatolia (Turkey) and were both part sage, and part natural scientist; Thales, the first of them, speculated that the arche (origin, source) was water – all life depends on water for its coming into be.[3] Anaximander disagreed: air, he said, not water is the arche (the source). Anaximenes differed again – there must be an infinite bundlessness which underlies everything in form; the infinite is the arche of the finite. Heraclitus in turn disagreed – fire is the arche, and fire is what dwells at the heart of the logos, the Mind or Reason within and behind the universe as also within us humans. Anaxagoras speculated that Nous or Intellect was the Arche, since the universe indicates that it has been conceived by a creative intelligence of superhuman powers. Pythagoras, known to have traveled to India to hold congress with the sages of that country, both agreed and disagreed with them all – for behind and above all such speculations, he argued, lay the principle of arithmoi, or Number. And the cosmic energy behind everything is expressed best in the simple enumeration of number – worked out in the unfolding of the tetractys, the number count from one to ten, conceived of in the shape of an equilateral triangle, presenting a form of the passage form the transcendent (the one) to the many (the four) which encapsulates all we can know about the ultimate Arche behind all existences. Pythagoras also found in music further evidence of his sense that mathematics underlies the structures of being, and also in geometry. But for Pythagoras, mere science, in the sense of mathematical knowledge or physics, was not worth pursuing except in the wider context of ethical and spiritual knowledge, and it was this wider field, indeed, that he called philosophia – the love of wisdom, as opposed to mere techne, or handicraft skill. For Pythagoras, when he founded his Wisdom School in Crotona, southern Italy, entrance to it's classes was only possible for those students who were prepared to undertake the higher discipline of askesis, submitting their senses and bodies to the disciplines and chastities demanded by the higher muses and the Gods and Goddesses of Wisdom. For Pythagoras, as for the Indian sages, the pursuit of higher wisdom was impossible without the simultaneous cultivating of non-violence, and the true philosopher had to show moral courage and clarity in the face of ignorance and persecution, even unto death, rather than be sucked in to the normal cycles of hatred and revenge and killing; the philosopher's job was to stand for the possibility of transcendent grace over and above the corruptions of ordinary mankind. It was this vision in turn which inspired and enthused both Socrates and Plato, leading to the death of the former, and the life long mission of the later, to establish the Academy, as a place for the anchoring of higher philosophical wisdom at the heart of Athenian and Hellenistic life, a project which endured for some 8 centuries in direct succession, and which endures to this day wherever the world academic is enunciated. From Plato in turn came the work of

[3] Such an idea is not absurd: in 2005 astronomers announced they had discovered that interstellar space is full of water vapour, and not empty at all, to the tune of trillions and trillions of tons of it, thus making it highly likely that water is abundant on planets throughout the trillions of planets that exist in the numerous galaxies of which we have cognizance, and that life must be therefore abundant on many of them… So well done Thales !

Aristotle and the Aristotelian school, which likewise endures in its influence wherever the University system of higher education is found, since it was the recovery of Aristotle's logical, scientific and philosophical writings in the Middle Ages which directly led to the establishment of the Universities of Paris, Oxford and Bologna in the course of the 12th century AD. And which had also earlier inspired the work of Islamic scientists such as Ibn Rushd and the great Jewish philosopher Maimonides. The synthesis of Aristotle therefore was far ranging in its influence as it was far reaching in its cope, for Aristotle sought to take the Platonic ideal, the One (to ten) and delineate its configurations in the everyday world of form and matter; how exactly do bodies form? How do species interrelate? How do different types of plants take their shape and functions? How do the heavens seemingly revolve? How do political states differ from one another in their choreography of power? How do different moods and emotions orbit one another in the revolutions of the soul? How do prophets receive inspiration? How do dreams arise within the soul ? What are the Gods made from and where do they reside? What is energy? What are the different literary forms? Do all tragedies have some similar pattern of unfoldment that the critic can discern ? What is the nature of time itself? How do the forms and styles of speech enable one proposition to be true and another false? How does thought itself proceed either logically or illogically? What are the very inherent structures of Being itself? What are the ten categories under which all that exists can be named and known? Is there a final love which pulls all things toward it through their changes of form as to a great Attractor, and is this love divine?[4]

In all these questions Aristotle was relentless in his tumbling forth with ideas and possibilities and speculations.. Was this science? Was this philosophy? Was this spirituality? To Aristotle the question would have been meaningless – it was all three, and more – for here, with Aristotle, whole sciences were themselves coming into being for the first time, at least in the history of European thought – the sciences of psychology, of biology, of zoology, or politics, of astronomy and cosmology, were themselves first taking form in this work; and so too were the disciplines of literary criticism, of metaphysics and of the history of philosophy. Of course many of the details of what Aristotle wrote have been disproved or augmented – but his epochal significance cannot be underestimated.

Nor was Aristotle the only Hellenistic thinker working on the cusp between science and spirituality – the Stoic school, launched by Zeno of Citium, developed both a metaphysics, an ethics and a naturalistic cosmology; the atomistic school of Democritus (written about by Marx in his doctoral dissertation) also developed through Lucretius and other thinkers, a comprehensive naturalistic explanation of the cosmos as well as an ethics; Neo-Platonism and Neopythagoreanism in the hands of thinkers such as Plotinus, or Iamblichus or Porphyry, developed a complex emanationist[5] metaphysics in which reality as we know it derived from the cosmic outpourings of the divine unity in a series of energetic outflows, much as a stepped waterfall, in a style reminiscent of Hindu yoga philosophy in relation to Advaita

---

[4] This was the assertion of Aristotle in the Metaphysics, where he expounded a theology of love which made subsequent religious thinkers, from Jewish, Christian and Muslim traditions able to harvest his rich insights for their own theistic purposes.

[5] This coinage, comes from Emanate, meaning to flow out, from Latin emanatus, and emanare, meaning to flow out, arise, from e – out, and manare, to flow, from an IE base *mano meaning damp, wet, meaning to come forth, issue, as from a source. This liquid origin of the metaphor is often found in philosophical semantics – water, air, light, often lie at the root of the metaphorical spatio-temporal grids of thought wherewith we think; water particularly, such that Thales' insight into the primacy of water as the Arche of life (bios) seems worthy of a special commendation in the annals of thought.

Vedanta. Alongside these philosophical schools developed Greek and Hellenistic science per se, in the works of mathematician such as Euclid, inventors like Archimedes, geographers like Erasthotenes and Strabo, medical doctors such as Hippocrates, Galen and Celsius, architects such as Vitrivius, Polymaths such as Pliny the Elder. The observant reader will notice that Roman and Latin names have crept into this list almost unnoticed – this being because, whilst the Roman contribution to the advancement of science was enormous, it was largely based on earlier Greek and Hellenistic models of scientific achievement; for many years, the Greek, supplied the teachers to the Roman Republic and then the Empire. Nor did Hellenistic scientific achievements stay only inside the boundaries of Hellas; after Alexander the Great spread his conquests throughout Asia, Greek scientific and spiritual ideas also became part of the lingua franca of philosophical thought throughout the whole Eurasian region as far as India. Likewise in time this Hellenistic tradition extended on till the Christianization of the Roman Empire and beyond, for, as we shall see, Christian thinkers adopted pretty much intact the fruits of Hellenistic science and spirituality both in order to articulate their encounter with the Christian impulse.

The crucial point to enunciate here however is that for Greek and Hellenistic mind, science, knowledge, was pursued and developed within the wider context of philosophy, the love of wisdom, and this latter in most cases involved the conscious cultivation of spiritual wisdom and enlightenment. For Pythagoras and for Plato the point of pursuing philosophy was to achieve lysis, the solution to the problems of life, which was the Greek equivalent to moksha or enlightenment, and which emphasised the workable application of wisdom in the everyday world. The philosopher sage of Hellenistic culture, like his or her Chinese or Indian equivalent, was at once active on transcendental levels of reality, whilst at the same time concerned with the pursuit of justice and righteousness in the affairs of the city state and the local community. Yet one can perhaps point to a gradual deterioration in the ethical quanta available for circulation, as Rome replaced Greece as the centre of gravity; law replaced philosophy as the queen of the sciences, and the finesse and intellectual jousting of barristers replaced the subtle intellectual arguments of Alexandria. Not surprisingly then, when Rome found a new systems to satisfy (partially) its spiritual hunger, it turned to a legalistic interpretation of the wisdom of the east, rather than a Gnostic interpretation, and latched onto the idea of enlightenment by contract (covenant) which it then promulgated by Imperial fiat, ending up with the closure of the philosophical schools that had lasted since remotest antiquity – it was as if the Roman Imperium cast a pall of power and the fear of wisdom (sophiaphobia[6]) which resulted in the eventual split between sapience and science which we in Europe have inherited now in our current era, as bastard children of the pseudo enlightenment. But this story needs telling again properly from the beginning – to do which we must first return to the Judean wilderness and listen to what the prophetic tradition had to say….and before them even, to Egypt and Babylonia and beyond….

[6] Sophiaphobia is a word of the author's coinage, which denotes the fear of wisdom, as a pathological condition underlying much of the institutional and psychological frameworks within which we philosophers have to operate. It equates to an anti-philosophical impulse in the heart of the human condition a fear of too much knowledge, a fear of wisdom – as evidenced in mythological stories and myths worldwide, not least in the expulsion of Adam and Eve from the Garden of Eden for eating the fruit of the tree of knowledge of good and evil. See the author's published work, Sophiaphobia, which explores this dynamic in detail (available at www.lulu.com/iipsgp)

4. Egyptian, Sumerian, Babylonian and Persian Science and Spirituality

Any holistic account of this topic would be incomplete without at least an oblique reference to the ancient wisdom schools of Egypt and Babylonia and Sumeria. Greek thought for one owed an incalcuable amount to the wisdom schools of the Middle East and North Africa. The list of Greek and universal scholars who spent time studying and teaching in Egypt is like a roll call of ancient wisdom traditions, and would include Pythagoras, Plato, Iamblichus, Plotinus, Ammonius Saccas, Archimedes, Erasthotenes, Euclid, Moses, Jesus, Philo, Maimonides, and so on. The Library and Museum of Alexandria became the greatest library and research institute the world had ever known up to that point partly because of the immense prestige that attached to the wisdom of the ancient Egyptians. Hermetic wisdom, as it became known, was really derived from the deity of learning worshipped in Egypt, namely Thoth, the god of scribes and sacred writings. The wisdom literature of Ancient Egypt endured for many thousands of years, and its style and content inform large sections of what was later known as the Bible. The religious myths of Egyptian antiquity, the life and death and rebirth of Osiris, in many ways overshadowed and prefigured the life, death and resurrection of Jesus; in his dismemberment by Seth and his rememberment by Isis we have even a prefigurement for the sacrament of the Christian mass; in the Egyptian archetypal mothering of Horus by Isis we have the iconographic precedent for the later image of the Virgin Mary and the infant Jesus. Scientifically, we know the ancient Egyptians excelled in medicine, and in developing the science of anatomy, partly because of their skill at embalming; they were skilled in geometry and trigonometry; in the use of pigments and colors and fine art and pottery; they were unsurpassed in their architectural and temple building skills, and in Egypt there existed a greater density of sacred temples per square mile than in any other ancient civilisation which has ever been on Earth. Their approach to education and learning, based as it was on the work of the scribe, and the mastery of the science of hieroglyphics, was essentially humanist in tone, and full of toleration and the love of reason. Although polytheistic and pantheistic, worshipping a variety of Neters (Gods) their sages had come to realize that underlying all was an essential energy, an ultimate quantum field from which the various energies of creation manifest at different times and places as appropriate. In their attitude to animals, their worship of the Cat Goddess (Bast) and the Cow goddess (Hathor) they approached the sensibility of the ancient Indian civilisation. The love of life, their eroticism, their appreciation of the female and male aspects of divinity and of existence, were likewise similar to the ancient Indian insights, and so too was their celebration of peace. Osiris, as a world teacher, spread the arts of agriculture, learning and civilization without ever using violence or warfare to do so, rather traveling as far as India teaching and sharing wisdom wherever he went (according to Plutarch's version of the story) but tragically it was whilst he was away that Seth plotted top murder him on his return, having usurped his place. At the time of the later scientific renaissance in Europe, it was all partly triggered by the rediscovery of the Hermetic wisdom in Florence and its translation and inclusion in the pantheon of classical wisdom that the renaissance sage had to master. The notion of Hermetic wisdom entered the vocabulary of learned discourse, and figures such as Pico Della Mirandola in Italy, or Ficino, under the patronage of Cosimo de Medici in Florence, popularized against the notion that al the roads of wisdom led to Egypt; in

Elizabethan England it was Dr John Dee, the Magus of Queen Elizabeth I and coiner of the phrase the British Empire, who popularized the idea of Egyptian esoteric wisdom as underlying the Biblical and Greek corpus. In Renaissance France, Guillaume Postel performed a similar task, declaring furthermore that Christianity, Islam and Judaism all could be traced back to a common Egyptian wisdom source, the uncovering of which would bring about world peace and religious harmony. Not surprisingly, the pioneers of the early scientific enlightenment, who opposed the dogmatism of mediaeval Christendom, such as Giordano Bruno or Francis Bacon, also looked to Egypt as the fount of great learning. But what did this ancient wisdom stand for if not the possibility, indeed the actuality of, the reconciliation of science and spirituality? Why else had generations of Greek sages repaired to the great temple schools of Egypt? Why else had Alexander himself repaired to Siwa for the final clues as to his own divine identity?

But Egypt was not alone – from early on, men and women of wisdom realized how much the world owed to the other ancient Middle Eastern civilizations of Sumeria and Babylonia, and Persia. From the most ancient times of recorded learning, the sages of the earliest times came from Mesopotamia, the land between the rivers. Here Abraham came from; here Adam originated in the Garden of paradise; from near here Zoroaster and the Magi came to share their wisdom throughout the mages, and gave their name to the science of "magic". Factually speaking, we know that Sumerian civilisation first developed urban structures and communities of an advanced type before any other region on earth, and also the first spiritual literature, in the Epic of Gilgamesh and his quest for immortality from the sage Utnapishtiom. We know the Babylonians inherited from the Sumerians their unique style of cuneiform writing, with wedge shapes impressed on wet clay to form ideograms, and developed therefore a complex wisdom literature over many generations, out of which store several of the most famous of biblical stories themselves have their origins. Scientifically, we know the Babylonians were skilled geometers and mathematicians, and experts in urban planning and architecture, metallurgists and workers in precious metals and precious stones to a high degree of artistic skill, and above all expert astronomers who mapped much of the known sky. They also developed the division of the circle in 360 degrees which made the science of geometry as we know it possible. Yet they were also a deeply religious people who worshipped a variety of deities, including Marduk, Ishtar and Shamash, while the scribes and people of learning worshipped Nebo, the God of the scribes and sages. They also developed their medical knowledge to a high degree, and combined surgical and medical interventions with magical and spiritual incantations and exorcisms. To the Ancient Persians, who eventually conquered and absorbed the realms of Mesopotamia into their own empire, was also given a powerful spiritual and religious lineage, namely Zoroastrianism, stemming from the prophet Zoroaster, who lived perhaps as early as about 1700 BC, and in Persia likewise the ethical teachings of Zoroastrianism were combined with respect for the natural elements, which were all regarded as sacred (Earth, Air, Water, Fire) and the ethical duty of mankind to follow the right spiritual path was made clear at all times by the Zoroastrian priesthoods (Magi).

To all these ancient civilizations then, the idea of splitting science off from spirituality would have seemed absurd, much as trying to have fire without flames, or the sun without light; the power[7] that comes from science, as in *knowing how*, was not to be placed in the hands of those unworthy; all these ancient civilisations therefore had closely guarded priesthoods, open to both men and women who had shown their worth, in which the secrets of creation would be revealed over time, through processes of initiation and instruction, and in which both logical and rational aptitude had to be accompanied by moral rectitude and spiritual uprightness. To this day scattered fragments of the thinkers of these ancient civilisations continue to exist and continue to bear witness to the possible reconciliation of science and spirituality. In the case of the Zoroastrians, once expelled from Persia proper they found a welcoming home in India, especially in the Mumbai region, and the Parsees eventually came to play an important role in the spiritual life of India in modern times.

5. Jewish and Hebrew Contributions to Science and Spirituality

Of all the ancient Middle Eastern mystery traditions which have survived into modernity however, and which stands supremely for the reconciliation of scientific and spiritual approaches to knowledge, it is the Jewish or Judaic tradition that lives on most vibrantly into the modern era. And Judaism itself might perhaps be best understood, from the standpoint of the history of religions, as an amalgam and synthesis of the wisdom of the ancient mystery schools of the Middle East which preceded it, for there is much of the Sumerian, the Babylonia, the Egyptian and even the Persian wisdom lineages in the complex mixture that makes up modern Judaism. Even for those who accept revelatory history at face value, it is undeniable that revelatory experiences are layered up, so to speak, on the accumulated wisdom of teachings and teachers that preceded them in innumerably complex and often lost specificities. So Abraham and Isaac and Jacob (Israel) and Joseph and his brothers, and so on, all derived even according to the Biblical stories, some aspects of their wisdom from their neighboring regions and lands. Indeed, it is to the geographical location of Israel that we can look if we are trying to understand why such a numerically small people should have exerted such a vast influence over mankind's spiritual and temporal history. For it is my contention, along with many other modern scholars, that the unique opportunity of the Hebrew and Jewish peoples is that they lived in close proximity to the epicenter of the invention of the notion of an alphabetic script (achieved by the Canaanites or Phoenicians of the region just to their North in modern day Lebanon) and it was with this profound took of wisdom that they were able to shape the sacred texts and scriptures that have come down to modernity as the Tenakh, the sacred Bible of Judaism. In the Tenakh, the emphasis is on wisdom, and the role of the divine creator, again and again it is emphasized that there is an active supernatural and divine force behind creation and behind existence. The world is good and it was created to be that way by its maker, the Lord of All, YHWH, also known as Elohim or Adonai – all of

[7] Knowledge is power, said Bacon – and power, an English-Norman word, comes from Norman French pouvoir, ability, to be able. To be powerful is literally, semantically, etymologically, to be able, and ability, stems from knowing how. Thus Bacon's aphorism is a tautology, at heart. But the power that comes from true knowing, stems from self-knowledge, power over one's lower self, and the ability to know one's deeper essence, or true identity. It is this "know thyself" which is at the heart of the philosophical enterprise, and which is universal to all cultures. Blessed are the pure in heart, said Jesus Christ, for they shall see God. This is the same maxim expressed through a Hebraic theistic framework.

which terms are translated usually in English as God, but which in Hebrew have subtle differences. But the work of reason, of figuring out the nature of reality, is also given to mankind, for the ultimate wisdom we can have as human beings is perhaps that we will never be able to figure it all out – and thus a final closure of scientific knowledge will always forever be unobtainable – and we must return to wonder and praise, for God's ways remain inscrutable and mysterious. Our proper attitude to reality should therefore be one of wonder and awe, and indeed of love and reverence, not only to the invisible and unknowable Creator, but also to our fellow humans, for we, in some mysterious way, are ourselves made *In the Image of God*, and thus bear the imprint of the divine maker. The Prophetic message of the later Hebrew Prophets repeats this same basic message – only if we remain true to our spiritual innermost nature, and reverence the invisible source of all being (the primal arche) can we live in harmony as human beings with one another and attain peace. In the Mosaic mission, the Hebrew slaves escape and take upon themselves the duty of becoming a people, chosen to led not only themselves but also mankind at large away from bondage to false idols and ignorance, and to a promised land based in fairness and justice and plenty. The burden of this historical task has proven at different times almost too great to bear by the ordinary men and women of the Jewish community, particularly as it has often seemed to arouse the wrath of their neighbors, leading ultimately to anti Semitism and the holocaust. The biblical tradition however is only part of the story =- it was understood as the written Torah, to be supplemented by the oral Torah, the oral law enunciated and debated and discussed by countless generations of Jewish sages and Rabbis, from Palestine to Poland, from New York or Baghdad, over many millennia. It is this living lineage of faith that constitutes true Judaism. And what do we find in the Rabbinical sources, such as the Mishnah and the Talmud? We find Judaism as a form, of spiritual humanism, in which usually the most compassionate, the most humane solution, and the most moderate interpretations is teased out from the sometimes uncompromising religious texts which comprise the Tenakh. We find in Rabbinical Judaism and uncompromising commitment to the advancement of learning and knowledge, and the cultivation of literacy and a respect for education, that would lead ultimately to the creation of generations of scientists and religious experts and humanities scholars of the very highest excellence. How else can we account for the inordinately high contribution of the numerically small number of Jews to the intellectual achievements of mankind at large, especially in the sciences in the past 150 years? Another reason might be found in the very deepest spiritual recesses of Judaism, namely in the Kabbalah, which constitutes its mystical heart.

Since Gershom Scholem reclaimed the Kabbalah for legitimate academic inquiry, thousands if not tens of thousands of books and pamphlets have streamed forth about the nature of the Kabbalah. An esoteric teaching system claiming to derive from remote antiquity, it purports to be the oral accompaniment dating from at least the times of Abraham, Jacob and Moses, without which the bible text as we know it remains an impenetrable mass of sometimes ill digested stories, legends and impossible accomplishments. The Kabbalah is effectively the wisdom teachings within the heart of Judaism. The word is related to the Arabic word "kiblah" which is the "direction in which to face" (i.e. Mecca) when praying. The Kabala on the contrary is the direction to face when seeking spiritual wisdom, and is usually interpreted to mean "receiving (instruction) face to face" (i.e. orally) from a teacher. In this way the Kabbalah claims to be the unbroken lineal descent of spiritual wisdom from the most ancient of times, before writing was invented. It was written down however, at least in the 3rd century AD, in an important study called the Sefer Yetzirah (Book of Formation), which is a mystical study of the innermost meaning of the sounds of the Hebrew alphabet, written much as a Hindu sage might interpret ancient Sanskrit mantric qualities. In the Book of Yetzirah the sounds of the alphabet are allotted the different qualities of creativeness or energy, or the different archetypal energies associated with the deity, including maleness and femaleness. By the 12th century a very complex text indeed, known as the Zohar (The Brilliance) had been written down in mediaeval Spain, which was itself an intercultural melting pot of great density, where Jews, Muslims and Christians lived together in relative intellectual harmony (at least among the philosophers) anonymously, but later found out to probably be by one Moses de Leon. The Zohar propounds a complex multilayered world view, almost equivalent in many respects to a kind of Western or Jewish Tantric cosmology, ion which the unknowable Godhead (Ain Soph) establishes creation through a series of energetic stepping downs, or emanations, which produce in turn 10 worlds, of descending vibrational intensity, known as the 10 sefiroth. These are conceived as shells or layers, as in a Russian doll, which hide one layer within each other, but for formal teaching purposes are laid out in a diagram, known as the Tree of Life, or the Tree of the Sefiroth. This shows all ten Sefiroth as the following table demonstrates. Also given are their equivalent to the Indian tantric chakra systems which have become universally used in Hindu, Buddhist, Jain and Sikh spiritual wisdom teachings.

Hebrew name of Sefiroth	English translation	Presumed identity to Indian chakra
Keter	The crown	Sahasrara
Hokhmah	Wisdom	Ajna
Binah	Understanding	Vishudda
Hesed	Loving kindness	Anahata
Gevurah	Sternness / judgment	Anahata
Tiferet	Beauty	Anahata
Nezah	Victory	Manipura
Hod	Vibration	Manipura
Yesod	Foundation	Svadhisthana,
Malkuth	The kingdom	Muladhara

It is part of the author's contention that the oral traditions embodied in the Kabbalah do indeed go back to an ancient prehistoric oral tradition, which constitutes the same source as the Indian metaphysical systems, and which therefore, not surprisingly, contains pretty much the same fundamental cosmological teachings. While orthodox formal Judaism presents sometimes as quite an unyielding religious system, with no place for the characterization of deity, and certainly not for its anthropomorphisation, in contrast to the playful polytheism of ancient Greece or India, in the Kabbalah is found the acknowledge that the One issues forth as Many, and that the Monadic unity takes delight in plurality and difference. The Sefiroth are understood by Cabalists to contain the secrets of the innermost nature of God, since God manifests in these different levels of energy, all the way from the most undifferentiated absolute majesty of the supreme Godhead (Keter, the Crown) down to the variegated divergent splendor of the many species of living and inanimate things we find in the material world (Malkuth, the Kingdom). We human beings also contain all this same divergent range of energies, form the undifferentiated wholeness of absolute Self, through to the many cells and systems of our physical bodies. It is in this wise that the Kabbalah therefore affirms, "Man/Woman is made in the image of God".

Now this system of esoteric Judaism has constituted the heart of living Judaism in European and also Middle Eastern history and tradition. Great Kabalistic centers exist in Safed, in Galilee, associated with the life and teachings of Isaac de Luria and Moses Cordevero in the 16th century, and later in Germany, Poland, France, England and the USA. In the history of the contributions of Judaism to spirituality therefore the history of the Kabbalah is paramount. But so too is this true in the case of the history of the sciences.

Many of the greatest thinkers and scientists in Judaism have been influenced by and had their thought worlds shaped, sometimes unconsciously, by the existence of the Kabbalah as an esoteric teaching system. Indeed, the entire projects known in European intellectual history as the renaissance and then the enlightenment, would, in the author's opinion, had been severed impoverished, had not the discovery by Christians of the existence of the Kabbalah been an essential part of its unfolding. The original academy in Florence sponsored by Cosimo de Medici, where Plato's texts became the centre of a great flowering of intellectual creativity, and where the Corpus Hermeticum of Ancient Egyptian thought was also translated, also acknowledged and included the study of the Kabbalah as the Jewish form of the manifestation of the Prisca Theologica, the eternal primal theology, which the renaissance thinkers believed that all times, all places, and all cultures had upheld (in their view the real meaning behind the word Catholicism).

Among the many renaissance and enlightenment thinkers influenced by the Kabbalah we can but mention here only a few: Lord Herbert of Chirbury, Dr John Dee, Gemistus Plethon, Pico della Mirandola, Ficino, Giordano Bruno, Friar Roger Bacon, Raymond Lull, Bishop William Lloyd, Spinoza, Lord Francis Bacon, Robert Murray, Sir Isaac Newton, Elias Ashmole, Guillaume Postel, Leibniz, Boehme, Hegel, Schelling, Goethe, Marx, Freud, Jung, Rudolf Steiner, Martin Buber, Horkheimer, etc.

As can be seen from this partial and selective list, the history of the Kabbalah and its reception by the European world had an enormous and profound influence on the history of spirituality on the European and hence global mind. One of the ideas which the Kabbalah put forward, for example, was that the traditional God of theism is in fact an obscure anthropomorphized image which often obscures the reality of the actual divine energies themselves, and that to be truly religious and spiritual might sometimes involve the denial of such a chocolate-box image of God. It must also be explained that it was the very idea latent in the Kabbalah that there are secret inner laws behind creation which God has ordained for all things, that underlay the idea in post renaissance European culture, that made the idea of scientific laws possible – the idea that the universe had coherence and inner structure, which it was the job of the scientists to fathom – this idea which became embedded in European and global consciousness in the image of the scientist, began first as the arcane holder of mystic knowledge, indeed as the magician, perhaps even the Faust figure, who could ferret out the secrets of existence (must be by way of a compact with diabolic realm of Lucifer) an image which itself owed much to the idea of the Kabbalist, the secret interpreter of God's scripture and arcane mysteries.

It is not possible in the small confines of this paper to do justice to this theme: but consider for a moment the huge impact of the Jewish mind on the direction of the sciences in the 19th and 20th centuries – name only Marx, Freud, Einstein – each of whom was influenced either directly or indirectly by the Kabbalah – and you will see how large a place this tradition has played in the formation of the very notion of what modernity thinks science, including the social and psychological sciences, consists of. The problem, and it is a problem, however, is twofold:" firstly, that in the course of the 20th century the Jewish people in Europe were subjected to a ferocious onslaught without precedence in world history, and one in which many of the great kabalistic minds and holders of the lineages perished in the flames of the concentration camps, and that this struck a scar at the heart of the global mind which we have still not recovered from. Secondly, there remains a great reticence on the part of European trained intellectuals, including even many Jewish thinkers, to acknowledge the influence of the Kabbalah on their own thought worlds, or to even consider it a topic worthy of research and consideration. Partly it is as if the influence is too pervasive to have even been noticed (much as a fish would deny the existence of water) and also because there has traditionally been a very effective ban on people learning about it, or discussing or even finding out about the existence of the Kabbalah before they were at least 30 years of age, and then found a bona fide teacher who would instruct them personally on a one to one basis. In many parts of the Jewish community this is still the case. In other words, we are dealing here with forbidden knowledge, with esoteric knowledge, that was withheld from the community at large, and access to which had to be proved and earned, within the context of an initiatory system.

My own approach to the matter as a philosopher of history and a specialist in the history of religious and scientific ideas, however, is that we are living in a time of global emergency without precedence in human history, in which weapons of mass destruction have been unleashed, and in which the cultural symbiosis of the wisdom traditions is a matter of manifest survival. I regard it therefore as an intellectual duty on the part of the intelligentsia of mankind that they no longer *conceal their knowledge*, or withhold it, but rather that they share, with one another, the highest insights and wisdom which they have been vouchsafed (through divine grace, or human circumstance, or however understood) to receive. In other words, that the quintessence of the wisdom teachings of each of the great global civilisation and spiritual community must be shared in open heartedness for the common benefit of the illumination of mankind as a whole. To continue concealing, withholding and silencing (all tactics used to protect esoteric knowledge) is too dangerous, for "*a people without understanding comes to grief*" (Hosea 4: 14).[8]

If it is true then that the history of esoteric Judaism in the form of Kabbalah has indeed contributed to the formation of renaissance and enlightenment scientific and rational discourse about reality, and the evolution of scientific theory, then one of the ways we can reconstruct a workable synthesis of intelligent discourse about such matters now, in our times of crisis, is to uncover and research the history of how we got into this mess in the first place? How did we ever think that esoteric knowledge, or spiritual knowledge, and worldly knowledge were separate domains? How did we ever create the artificial divide between

[8] This same impulse prompted Francis Bacon, when sending a complementary copy of his newly published Book De Dignitate et Augmentis Scientarum to the Libraries of the Universities of Oxford and Cambridge, to say the following (in summary): To Cambridge University, he exhorted them "to apply themselves strenuously to the advancement of the sciences, and not to lay up in a napkin the talent entrusted from the ancients, but rather to make a legitimate and dexterous use of the keys of sense, put away the spirit of contradiction and dispute only as if they were disputing with themselves." To Oxford University he urged them to "Not regard the work of the ancients as nothing, nor yet as all; to weigh their own powers modestly, and yet make trial of them. And all would succeed best if instead of turning their arms against each other, they joined their forces to make an impression on the nature of things. This would be sufficient for honour and victory, achievable only through

ethical and moral knowledge and the ability to manipulate through science and technology the outer planes of reality? Within Kabbalistic terms, it is as if our current predicament stems from the failure to perceive the importance of upholding the integrity of the tree of life as a whole, and some civilisations and cultures have focused far too much on the lower sefiroth, while ignoring the upper sefiroth. What is need from the global intelligentsia is a concerted effort to put back together the entire tree of life into a whole again, and to enable the transcendent and the immanent worlds to embrace once again. Ironically, this very coming together of immanent and transcendent is symbolized by the star of David, the 6 point star which comprises the flag of Israel, in which the descending triangle intersects with the ascending triangle (this is also a very ancient tantric symbol often use in Indian sacred art).

Fortunately there is much good news here; many people of the Jewish tradition throughout Israel and further field are indeed at the forefront of exploring how spiritual and scientific aspects of reality can be effectively inter-related. Israelis such as Puharich and Geller have long been at the forefront of research into telepathy and parapsychology, along with many researchers in this field in the former Soviet Union. Many people from the Jewish tradition, who did after all miraculously survive the Holocaust, itself a sign surely of the divine mercy, are leading the global shift in consciousness towards an ecologically sustainable global civilisation, and are pioneering the use new greener technologies etc. If only peace between Israel and the Arab and non-Jewish world[9] could be achieved, then the creative potential of the Israeli people could indeed help with the cultural and scientific development of the whole region of the Middle East, as many of the utopian Zionists of the first generation of pioneers intended. But before that can happen, there has first to be understanding, repentance and forgiveness, on all sides. How this might be possible, requires that we next look at Christian and Islamic teachings on science and spirituality, and see whether there is the possibility of a common ground with Judaism and the other traditions explored above.

6. Christian Philosophy, Science and Spirituality

Within the history of Christianity there are innumerable threads and points to highlight of concerning both the development and contribution of Christian thinkers to the evolution of scientific thought and also to the evolution of spirituality and the religious history (and future) of mankind. For the sake of clarity I will set these out in numbered paragraphs:

a) Jesus was first and foremost a healer-Rabbi, and as such his concern was with the healing of mankind, both individually and collectively – and since healing is something which demands both attention to scientific knowledge and also to spiritual attitude, by definition anyone coming in his wake, any true Christian, by profession, so to speak, must be committed to the reconciliation of scientific and spiritual approaches to reality

cooperative research in the investigation of nature, in which work all men could find all necessary recompense."

[9] To this end the author launched in 2008, through the auspices of the International Institute of Peace Studies and Global Philosophy, the Truth and Reconciliation Commission for the Middle East (TRCME) – about which a book is available at www.lulu.com/iipsgp giving full details

b) Jesus himself obviously had derived from a source or storehouse of esoteric wisdom and spiritual knowledge the wisdom to live out his teaching / healing career, and it must be surmised, without the possibility of so called proof, that this store was from the same ultimate origin as the later Kabbalist thinkers and writers, since so much of his teachings relate to the same fundamental insight
c) St Paul and the other early Christian apostles and disciples, including St Thomas, St John etc. also reference in their writings the existence of an esoteric spiritual wisdom which can be accessed (although with great difficulty, moral earnestness, vigour and intellectual struggle, as well as divine grace) on the part of the sincere seeker of wisdom
d) The Gnostic strand in Christianity, which was more or less equivalent to the kabalistic strand within Judaism, was however on the whole suppressed once Christianity became the Imperial cult of the all powerful Roman Empire, resulting in the suppression or persecution of so called heretics, such as Pelagius, the British Druid Christian, who taught a rival and rational form of Christianity to that promulgated by St Augustine; so too the feminine aspects of the Christian mysteries were suppressed by the male church hierarchy, with the result that the divine feminine wisdom was also excluded from the innermost Christian teachings. This tragedy meant that the real possibility of balance and enlightenment, which the original Christian teachings had promised, were not truly fulfilled and developed. Gnosticism lived on however among the Bogomils in the Balkans and the Cathars in Southern France as well as in pockets in the Celtic Churches
e) The later intellectual renaissance that occurred in Christianity in the 12th century came about as a result of the transmission of Aristotelian and to an extent Platonic texts and the reclamation of ancient Greek scientific and philosophical wisdom generally on the part of a generation of crucial Christian scholars, including Peter Abelard, Michael Scot, Duns Scotus, Albertus Magnus and St. Thomas Aquinas, Grosseteste, Roger Bacon etc.. These scholars were partly inspired, both in love and in rivalry and envy, by the example of Islamic scholarship, which had already evolved to an advanced state of synthesis between Greek wisdom and Islamic revelation. It was this generation of scholars which gave rise to the mediaeval and hence the modern University system, of truly lasting importance, and the whole idea that reality is intelligible. However, when Christian scholarship took the work of Islamic thinkers, they adapted and adopted them to their own Christian biblical contexts, and failed to do justice, arguably, to the Islamic context from which they borrowed. The secularized science as something apart from divine wisdom – and this is something which Islam itself had always warned against. Islamic thinkers, for example, although brilliant at mathematics, would not countenance a secularization of knowledge, or the idea that mere mathematics alone could provide fundamental answers to ethical and metaphysical conundrums. To the Islamic mind, only a moral idiot could think that reality is intelligible without cultivating religious knowledge
f) The later renaissance was accomplished by synthesizing even more reclaimed Greek and Latin classical works and integrating them in an expanding Christian worldview; Christ was seen as an exponent of the ancient wisdom bedded deep within all cultures, all times, all civilisations – the job of the Christian renaissance magus was to explore and expound this universal wisdom found in all places and times and cultures, and then to see how Christ himself had actualized it in his brief life and above all in his resurrection
g) The reformation and counter-reformation were born of the fissures and clashes brought about by the rapid expansion of knowledge, by the discovery of new worlds,

inner and outer, and by the invention of moveable type printing, and the rapid dissemination of books and learning, thus in effect popularizing, secularizing and democratizing knowledge; the idea of science as a socially and spiritually progressive endeavour was launched around the same time, equally by both reformers and counter-reformers. What they both agreed on (Protestants and Catholics) was that knowledge, science, was of inestimable value.

h) The rise of the nation state at around this same time in history meant that the pursuit of science became a matter of national security and national power, both in terms of military and commercial success, and the various Royal Scientific academies and societies were founded as a result. Spiritual knowledge per se was pretty much dropped from their curricula, and indeed was disallowed, as it was not felt to be conducive to the advancement of knowledge-as-power which the budding nation states sought to develop at that time. Spiritual knowledge fell more and more into disrepute and the universities and academies of learned society in Western Europe and North America were increasingly interested in natural science as an answer to solving the needs and dilemmas facing the world community.[10]

i) All this rise of the empire of science, per se, went hand in hand with the expansion of colonialism and imperialism, and technological and scientific might was seen as key to the domination of world trading routes. The British attitude to India in the 19th century was typical: Macaulay denigrated indigenous Indian spiritual knowledge, and argued that the British had arrived bringing science in their wake, to spread enlightenment and to "educate the natives" – something which only science could do. The main University of London college for scientific research was named "Imperial College"

j) In the hands of Newton the scientific enlightenment found someone who seemed to be able to explain reality according to measurable scientific laws, including both the nature of both light and gravity, and after Newton physics as we know it moved inexorably in the direction of attempting to provide a coherent cosmology which could explain all aspects of reality through the working out of physical and mathematical laws. But Newton himself was a deeply committed esoteric philosopher, and a deist Christian, who believed that the spiritual and moral truths of the Bible pointed to a greater spiritual unity underlying reality than mere mathematical laws could do; he wrote a thesis on the transcendental religious unity of mankind which was deemed too dangerous for publication in his lifetime, and which was only published finally in 1952. Newton was also deeply interested in esoteric Christianity – in alchemy, in Rosicruciainism and in freemasonry – all of them important strands in the lost or suppressed wisdom traditions of genuine Christianity, stemming from the ancient suppressed Gnostic wisdom traditions of Christian enlightenment, equivalent to moksha or satori, as practiced in the East. Newton did not dare speak openly of his actual beliefs during his lifetime for fear of censure and suppression.

k) Kant, Schelling, Goethe, Hegel, Kierkegaard - all of them were some kind of esoteric Christians, and attempted in their various philosophies to remain true to the innermost wisdom teachings of Christian civilisation, For them it was an imperative to retain the linkages between spirituality and science; indeed their whole understanding of the work of the philosopher was to retain this linkage and to ensure that the fissure between naturalistic knowledge (science) and spiritual knowledge

[10] See the concluding section below for more details on this point. Basically, the author argues that the original Baconian understanding of the scope of scientific knowledge was understood as encompassing spiritual knowledge as well, but that the times were not fortuitous for its exposition. They are now, however.

could be mended. For this reason Hegel entitled part of his great oeuvre Towards the Phenomenology of Spirit, or An Encyclopedia of the Spiritual Sciences (Geisteswissenschaft). In Britain Coleridge and the Romantic movement of idealistic philosophy attempted the same thing, trying to find a way to remain true to the highest essence of Christian teachings, and yet also open to and affirmative of the insights of scientific thought. It was against the reductionism of natural science, the splitting of everything into separate domains of knowledge, however, that the Romantic Christian philosophers protested; and in the form of Blake, one of their leading spokespersons, they raged against the "dark satanic mills" which had been born of the industrial revolution, itself the bastard child of the scientific revolution.

l) Even Darwin himself had once wanted to train as a Christian minister, and his wife retained her Christian faith right to the end, hoping for the reconversion of her lapsed husband. Also, Darwin, in his belief in the intelligibility of reality, and in his belief in the possibility of uncovering a law behind evolution, was the unconscious inheritor of generations of thinkers and cosmologists before hand the vast majority of whom had believed that such laws of reality are at once spiritual and scientific. He himself remained agnostic about the possibility of building such a bridge between scientific facts and spiritual speculation. But his natural politeness and good nature would not necessarily have ruled out of order all such attempts

m) His other half (or friendly intellectual rival), A.R. Wallace, however, retained far more of his spiritual faith than Darwin, and whereas Darwin seemed to be unable to appreciate the aesthetic side of life, and owned his failure to appreciate literature or poetry or spirituality, for example, Wallace was not only the co-discoverer of the theory of natural selection, but also someone who had witnessed first hand advanced spiritualist phenomena, which were at that time being investigated scientifically by the curious of polite society, which had convinced him beyond doubt that life after death was a fact, and that the human mind had powers of perception and understanding which could not be explained away by reference to the mere physical apparatus that houses the human spirit.

n) Within Catholicism and the Jesuit order in particular, there has been all along a resolute determination not to see the enterprise of science (knowledge) and that of spirituality as being divorced from one another. There has been a consistent refusal to permit the task of the philosopher being envisioned as less than the constant reconciliation and repair of the fractured ness of knowing and being in the world. All the various popes since renaissance times down to now, when this problem of knowledge (scientia) really loomed large, have confirmed and affirmed the need to keep science and spirituality as a seamless web in the service of Christ. Sometimes their zeal in his regard obliged them tragically to oppose the development of science, sometimes even to the extent of persecution, in order to retain this indivisible unity. Nowadays the Vatican organizes large networks of scientists and academics who try to examine the details of scientific research in the light of biblical and Christian wisdom teachings. A former Archbishop of Delhi, the late Archbishop Angelo Fernandes, who played a leading role in the work of Vatican Two, and who was head of the international Catholic Schools Association at one point in his career, showed by example how it is possible to be a modern Catholic committed to a scientific outlook in the 20th century yet at the same time to remain faithful to the core wisdom teachings of Christ. The same could be said of Thomas Merton, or Teilhard de

Chardin, E. F Schumacher, Bernard Lonergan S.J., the late Pope John Paul 2nd, Dom Bede Griffiths, Dom Sylvester Houedard and many others. The difficulties experienced by Hans Kung and Mathew Fox, two great contemporary Catholic theologians, as well as the work of Liberation theologians such as Segundo, Guttierez, however, indicate the difficulty and limits of intellectual freedom within the official church structures, and their suppression by then Cardinal (now Pope) Ratzinger, thus indicating the need for intellectual healing and humility all round, and the solution of the complex political dimensions underlying these issues as a matter of urgency, if the true voice of the Church is to be enabled to be heard contributing to the resolution of our current global predicament in the way being sketched in this essay. The Catholic Church has huge intellectual and spiritual resources within it to contribute to the authentic advancement of spiritual learning – one must pray that they can be released and harnessed to this genuine end of catholicity of wisdom, rather than the cultish and sectarian advancement of one school of Christian learning as opposed to others.[11]

o) Within the Orthodox Christian community there are likewise many currents and trends of hope and service, seeking to place the wisdom of Christianity at the service of mankind, both in Eastern Europe, in Russia and the Greek speaking world; paradoxically it was because the western and eastern churches fell out over the nature of the holy spirit in relation to the trinity that the schism between them came into being in 1054, and the scandal of their division has held ever since. As an ecumenicist is it too much to hope that the Spirit could in turn eventually reunite what has been sundered through human political ambitions and intellectual obscuration. Mark however that this would be a *spiritual* work. What can be affirmed about orthodoxy however is that many great scientists have been themselves of one of the various Orthodox communions, and secondly, that orthodoxy and its various offsprings are widespread in various forms in Asia, including in the scattered churches of the Middle East such as Iraq, or Turkey, or Persia and even as far as southern India. It seems these churches go back to Syriac Christian origins, spread by St Thomas and his school, and intellectually centred on Edessa, and that they were influential in spreading Christian ideas even as far afield as China, where Christian-Buddhist sutras have now been discovered from the 7th Century AD. In the guise of Nestorianism and later Manichean's, spirituality spread like a leaven across Asia, and perhaps cross fertilized with Mahayana Buddhist ideas of universal salvation, tinged with Zoroastrianism left by the older veneers of the Persian Imperial teachings. All of these spiritual currents can be regarded as part of the rich mixture bequeathed to humanity by the Christ impulse, and can be studied objectively by the transpersonal historian, if not by the confessional dogmatists. All of them are part of the legacy of humanity, and its long search and striving for liberation and enlightenment from the darkness of warfare, violence and suffering. Not least among its achievements, was the Nestorian contribution of the alphabet to the Mongolian Empire, which facilitated

[11] Interestingly, in the 1623 edition of De Dignitate et Augmentis de Scientarum, which was a much revised and enlarged Latin translation by Bacon of his own earlier work, The Advancement of Learning, Bacon removed passages which might have been construed as being anti Catholic. Although raised a Protestant, he was too astute a philosopher to not see that party, fraction and contention in religious matters were a key stumbling point in the matter of the advancement of religious knowledge. Woefully, this is still the case in our day.

their literacy; the translation of many Greek texts in to Arabic by the Nestorian community of the Islamic Empires of Baghdad and also in Spain; and finally, by the Oriental Christian influence on the life and times of Muhammad himself and the birth of Islam, a subject to which we must next turn. Indeed, Islam at its best can in many ways be seen as just one of the most successful of the numerous Christian interpretative sects which existed in the strange desert hinterland between the Byzantine and Persian Empires in the 6th and 7th centuries. The desert was awash with mystics, Gnostics, renunciates and Christian monks, each reading their own apocalyptic visions of history into current events, and each praying for wisdom to liberate and transform humanity, and if necessary, for the prophetic fire to be rekindled in the hearts of a stubborn and ignorant mankind. That it had been eventually kindled in the deserts of Mecca and Medina could have been a matter of great rejoicing by Christians worldwide, for after all, the Muslim community acknowledged Jesus as a Prophet and indeed as the Messiah. What went wrong then? How did conflict arise between them? Is peace between them now possible, at both the theoretical level and in the streets of downtown Baghdad and Kabul? If so, how and when, and who can bring this into being? And what role respectively can science and spirituality play in this reconciliation process?

7. Islamic Philosophy, Science and Spirituality

Within the birth of Islam something miraculous and extraordinary happened, as Thomas Carlyle observed among Western commentators for perhaps the first time. The vastness and incalculable wonder of the Creator, of God, of Allah, burst through the bounds of respectable and permitted paths, and revealed matters of highest wisdom to a desert dwelling Arab, Muhammad, (whose name means "praise"). Unlearned, so far as we known in the sciences or in formal philosophy in any shape or form whatsoever, and very probably largely illiterate beyond the rudiments of what was needed for his career as a merchant businessman, Muhammad was largely what one could call an intuitive mystic, someone who worked his way to the divine through his own efforts and through the divine mercy itself. He also received support and encouragement from his own immediate milieu, his wife Khadija, who was some kind of Christianising Arab, and her relatives; he also met Christian monks, one in particular apparently recognized him as a future Prophet. Notwithstanding his lack of formal learning, Muhammad always showed the greatest possible respect to the works of the mind, and to genuine learning and what we would later come to know as scholarship. Numerous hadith are reported from him, including "The ink of the scholar is worth more than the blood of the martyr" and he also placed the injunction on all true Muslims to *"seek knowledge, even as far as China"*. From such an attitude, we can surmise, there grew amongst early Muslims, spread by the immediate companions of Muhammad himself and the first Imams, an attitude of reverence to knowledge and to all sciences. To the Islamic mind, all comes from God, Allah, and therefore knowledge is the first and greatest of gifts God has bequeathed us, his Servants. Our duty as humans is to seek to extend our own knowledge as much as possible, but to do so always in a spirit of reverence and humility, in submission to the absolute stupendous and awesome majesty of the Divine Mystery. For over 1400 years therefore

Muslim scientists and sages have expounded, discovered, researched, and examined the mysteries of existence, both inwardly and outwardly, and contributed enormously to the development of the sciences. Once established in the heartland of the Middle Eastern world, in Baghdad, and Cairo, and as far as Spain, and Northern India, their great scholars and sages had at their disposal the combined teachings and wisdom literature of innumerable civilisations and cultures, including the ancient Greek and the Indian, the Persian and the Babylonia and Sumerian and Phoenician and Egyptian and later the Turkish. All of these threads wove subtly into the tapestry that makes up the history of Islamic thought. For the Muslim mind, the separation of science and spirituality is unthinkable since both come from Allah. The art of the scholar is to trace first one, and now another, as they weave their double thread across the tapestry of human intellectual activity as something divine: in the history of the activities of the mind, there is no place that God is not hiding. This profound insight, which fits in with the radical omnipresence of the divine in the insights of Shankara, and many other great philosophers, enables the scholar to trace the arteries and veins of knowing right back to their source, and to find traces of grace even in the most mundane knowledge. Just as the Hebrew inspired mind conceived as the 10 Sefirot as a way to harmonize the many-ness of the One, and to affirm the beauty of immanence as well as the grandeur of the transcendent, so too Islam developed the notion of the 99 Names of God, each of which encapsulates an aspect of the divine majesty, but together all of which fail to do justice to the real majesty of Allah, the Lord of All Universes.

Of the 99 names of God as understood by Islamic theology, here is a small selection of those names or qualities which specifically seem to have to do with knowing, and with the enterprise of science, thus indicating that scientific knowledge and Islam can indeed go hand in hand in amity. It would be a very strange theology indeed that could affirm on the one hand that God is the source of wisdom, and the creator or endower of reason and wisdom in human beings, and yet to argue that we should not or cannot exercise this faculty we are endowed with! All these divine names shown in what details the Islamic understanding of God endorses the idea that God is on the side of reason, knowing, science and learning.

Alim ,Al - the Knower;
Basir ,Al - the Seer;
Batin, Al - the Hidden;
Fattah ,Al - the Opener;
Hadi ,Al - the Guide;
Hakim,Al - the Wise;
Haqq Al - the Truth;
Hasi ,Al - the Reckoner;
Jabbar,Al - the Repairer;
Jami ,Al - the Collector;
Khabir,Al - the Aware;
Latif ,Al - the Subtle;
Muhsi ,Al - the Counter;
Nur An - the Light;
Quddus ,Al - the Holy;
Raqib Ar - the Watcher;
Rashid Ar - the Director;

Sami As - the Hearer;
Shahid As - the Witness;
Wadud,Al - the Loving
Wajid ,Al - the Finder;
Wasi ,Al - the Comprehensive
Zahir, Az - the Evident

The full and detailed study of these names, and their implications and many layers of meaning, which is given to the spiritual science of Sufism, is beyond the reach of this essay. Suffice it to say however, that sufficient common ground can be found with other global spiritual systems, such as those explored elsewhere in this paper, to guarantee absolutely that genuine Muslim scholars and followers can easily and contentedly live in peace and amity alongside genuine scholars and students of the other great faiths of humanity. There is nothing in Islamic law or teachings that supports the pursuit of political goals by illegitimate violence. Indeed the name of Islam itself means peace, the peace which comes from surrender to the overwhelming power of God. It is God alone who wields such power. On us humans is put the duty of amity, of friendship, of concord, of forgiveness, and of reconciliation.

Fortunately many great scholars and sages in the modern Islamic world have explored these themes, too numerous to mention, such that the intellectual groundwork for an Islamic civilisation at peace with its neighbors is indeed in place, and just needs properly activating and supporting. This is of course a task made doubly more difficult when outside forces invade and despoil and lay waste Islamic countries, such as has happened in Iraq, and Afghanistan, and to parts of Palestine, for example. Given half a chance, the author believes that the Islamic world would genuinely embrace in fellowship and amity the hand of friendship, if only it could be proffered.

But there are also difficulties and obstacles in the way of such concord, and they should not be glossed over. There are other Muslim thinkers who hold to an absolutist position for the overwhelming veracity of Islam over all other religions, and forget that Muhammad himself guaranteed protection and respect to people of the book (i.e. of learning) wherever found. In doing this they go against true Islam. In the xenophobia and Westernophobia practiced by some Muslims, they likewise fail to live up to the full potential of what genuine Islam is all about.

As the Sufis have explained, it is about an attitude of surrender, a state of spiritual grace, a placing oneself in a lineage of light, rather than a set of outer formal worship patterns, or of cultural styles or modes of dress and deportment. Fault lines throughout the world are being caused by the failure both of the Islamic community to live up to its own high calling, and by the failure of the non-Islamic community to properly appreciate the phenomenon of Islam on its doorsteps. Rather than terrorism and counter-terrorism, I would advocate endless patience, dialogue, study, research, prayer, and the searching out of one another's scriptures. Where are the other religious teachings of the planet in their Arabic translations for example? Are they readily available? Where are the Upanishads? The Vedas? The Buddhist Sutras? The Christian Gospels? The Jewish Tenakh and the Rabbinical Commentaries? Where is the Zohar? Where are the writings of Shankara in Arabic, or the Jain or Zen scriptures? These are works for the future, if there is to be a future, in which we can each sit down and compare and study one another's literatures and philosophical and metaphysical systems, and in so doing learn humility and mutual wonderment. Together we can remember that Muhammad himself was called the Seal of the Prophets (khatam al naibya) not in the sense of the final closure of prophetic illumination, but rather in the sense of their authentification and endorsement, both in the past AND the future.

The point to make here is simply this: that within Islam there are innumerable, vast and largely unexplored resources for the advancement of the reunification of science and spirituality, in the form of innumerable manuscripts and writings, which are largely untranslated from both Arabic and Persian and Turkish, and which linger in libraries throughout the Middle East. The author was involved for a time with a project of the former oil Minister of Saudi Arabia, Sheikh Yamani, to classify and catalogue the innumerable number of such manuscripts (Al Furquan Institute). Tragically, the prevalence of wars in the region is putting this priceless heritage at risk. Secondly there are innumerable excellent universities throughout the Middle East which manage to combine a high level of Islam studies with scientific research and academic excellence, but there is always room for improvement and growth and further devotion to the need for each generation to discover Islam from within as well as from without. Further, with the radicalization and militarization of the Middle East, and the tragic polarization of forces, into both pro and anti Western camps, and into Shiia and Sunni factions within Islam itself, there is an urgent need for scholars from throughout the region to come together in peace and amity, above the battlefield so to speak, and to recreate the genuine conditions of lasting amity which genuine Islam ought to be about, irrespective of their own cultural or historical backgrounds, be they Shiia, Sunni, Sufi, Druse, or Ishmaeli, be they Christian, or Jewish or Bahai or secular or whatever. There must and will return a time when it is safe for anyone of any nationalist or creed to walk safely down the streets of Baghdad or Lahore or Kabul or Jerusalem or Bethlehem or Fez, so long as they themselves observe the laws of hospitality and good behavior, and be treated in turn as honored guests. The question is, what actions can our own governments in the non-Islamic world take to best hasten this advent? Philosophically speaking, however, the question is what can philosophers and intellectuals do to best hasten this advent? Here one must echo the words of Muhammad Iqbal who in his famous work *The Reconstruction of Islamic Thought* urged a reconciliation of traditional Islamic thought with modernity and modern philosophy, and who inspired the founders of Pakistan to establish their own Islamic state in the Indian subcontinent. Islamic and Islamophile philosophers and intellectuals need to say that what is needed for Islam is a reconstruction of its philosophical thought, a reformation if you like, and a renaissance and an enlightenment all at once; what is needed is courage on the part of its intellectuals to speak out against dogmaticians who would silence free speech and freedom of thought; what is needed is above all non-violence and the urgent inquiry into the lost or misplaced pacifist and non-violent tradition that hides at the very heart of Islam, and especially Sufism. For there is no other way than the way of enlightenment, by whatsoever name it is known, and whatsoever one's outer path.[12]

For Muhammad, although forced by outer circumstances to do battle, always said that the greatest jihad is the spiritual struggle. The jihad in which Muslims engage then, must be converted, through the advice and exhortation of their own sages and saints, to the way of genuine Islam, the path of persuasion, demonstration, reason and spiritual illumination, the

[12] Further details on the teachings of Islam on the nature of enlightenment can be found in the author's work Enlightenments: Towards A Comparative Epistemology Of Enlightenment In The World's Spiritual Traditions, which is available on www.lulu.com/iipsgp

way of irfan, of gnosis and the cultivation of true wisdom and enlightenment. And on this complex journey no one tradition, no one teaching, has the monopoly.

8. Exploring the Transpersonal – An Evolving Dimension of Research

This leads to the final field to consider in this Chapter, namely that of the evolving notion of the transpersonal. First explored by William James and Carl Jung and in archetypal psychology, the notion of a transpersonal field of knowledge has become the subject of serous scientific investigation in the decades of the late 19th and 20th centuries, since William James, published his work *The Varieties of Religious Experience*. Since then innumerable psychologists have argued that religious experiences are indeed a valid field of research for psychology, and that paranormal phenomena as reported in the lives of saints and highly evolved beings, such as telepathy, clairvoyance, prophetic ability, prophetic dreams, moral altruism and goodness beyond the range of normal human abilities, superhuman courage even in the face of persecution and indifference or ridicule, states of grace and bliss and ecstatic consciousness, the feeling of being in the presence of the divine source of existence etc. – That all these phenomena are reported across widely varying cultures and religious traditions, and that the implication is that they are indeed a universal factor of human nature. It begins to look as if the old teaching of the Buddhist, that we all have a Buddha nature seeded deep within us, some more easily accessible than others, but that it is there inside us all, might actually not only be true, but also empirically verifiably true. Figures such as Assagioli, Pitirim Sorokin, Maslow, Stansislav Groff, Carl Rogers, Ken Wilber, John Rowan, John Heron and many others, have, over several decades explored the many similarities and differences between transpersonal states of grace attained by practitioners, saints and sadhus in the worlds various living spiritual traditions and lineages, and have, on the whole., come to the conclusion that this mysterious faculty of illumination is indeed within us all, and furthermore, within each religious culture, there is an affirmed place for its exploration within certain mystical schools which are normally within reach of the careful seeker.

What is emerging from this field of study is that regardless of the religious labels the seeker attaches to his or her evolving journey, we humans have indeed a deep spiritual hunger which can be satisfied in different ways at different times, but which pretty much follows universal laws, and that the core golden ethics preached by all religions, not to do unto others what you would not have done unto yourself, stands good as a universal moral code of behavior. Furthermore different thinkers have succeeded over the past 150 years or more in developing maps and a cartography of altered states of consciousness, including out of body experiences and near death experiences, in which it is becoming possible to map the terrain of the transpersonal dimensions which surround and ensoul our human mundane realm. The field of transpersonal psychology is interested, inter alia, in : Past life regressions, therapeutic hypnotherapy, spiritual healing, so called miraculous cures and their phenomenology; the experience of illumination or enlightenment across different cultural and religious boundaries; the role of gender and sexuality on the spiritual path; the nature of love, altruism and compassion, and how some human beings seem naturally to gravitate towards an ahimsic personality and others do not; the psychic life of animals and indeed of plants and the study of whether sentience extends to all parts of creation; the nature of reincarnation and whether, when, how and where souls achieve liberation from reincarnation, or fulfillment from the journeys they are required to take to learn their lessons; the innermost workings of the laws of karma – how the laws actually operate, how where and why; why some peoples or souls develop malevolent qualities and seem to take delight in inflicting pain and harm on others, and what can and is being done about this, both in the mundane world and also in the spiritual worlds; the nature of education and whether transpersonal knowledge can be encouraged or taught or learned, and if so how; the nature of memory in relation to education – how some people, remember their learning and others forget, and why; the nature of gifted and talented children (and adults) and whether our schools and universities are making room for their education, or whether we tend to educate mainly for the middle and lower bands of achievement. The extent to which transpersonal awareness, or inspiration, plays a role in creativity and the arts, and to which our great poets, painters and musicians can be said to be divinely inspired and intoxicated with the flavors of celestial knowing, and if so, how can this faculty be shared more widely among the people; similarly, the extent to which the work of our great scientists in human history have been inspired and guided by transpersonal inspiration, bringing new ideas to minds seemingly in synchronicity, and if so, how this work of scientific inspiration and enlightenment can be encouraged, supported and educated for; finally, to what extent does transpersonal knowing encourage and support a peaceful life outlook, and bring about an attitude of non-violence and harmfulness on the part of those who adopt this viewpoint?

All the above are areas of current research by transpersonal thinkers in general, and by the author in particular. These questions are universal in nature, and not to be considered problems to be solved, but rather avenues of thought and research to explore. Furthermore, the combined spiritual and religious wisdom of humanity will be needed to solve them, as will the evolution of higher altruism and moral states by us both individually and collectively.[13]

Conclusion: Notes Towards a Map for the Future of Scientific Thought

Having surveyed briefly something of the way in which the philosophical systems of mankind have hitherto drawn the boundaries and relationships between the scientific and spiritual aspects of knowledge, it remains to wonder somewhat speculatively about the future map of this relationship, and to chart a possible cartography of the relationship between empirical knowing and spirituality intuition, and ethics, metaphysics and philosophy. Perhaps the one single conclusion that is important to make is that the work of science has always, all along, had a deep ethical component to it, and certainly as conceived by Sir Francis Bacon, it had a redemptive dimension, not as a rival to divine learning, but as a co-worker. It is to this cooperative dimension between knowledge workers that if anything, this essay has pointed us

[13] For further details of the history and scope of transpersonal theory and psychology see the Biographical Encyclopedia A-Z of Transpersonal Theorists, Historians, Psycholgists and Philosophers 1945-2001 which outlines something of the huge range and breadth of this evolving field of knowledge, available via www.lulu.com/iipsgp

– to an ending of the rivalries and contentions which too often affect the scientific community. Above all, with war, violence, inter-religious and intra-religious disputes, still threatening world peace, it is perhaps imperative to take up again the unfinished question posed by Bacon in his incomplete 9th book of *De Dignitate et Augmentis Scientiarum,* and to oppose with reasoned evidence and sound argument those who would say that war and violence are good for the advancement of scientific knowledge, the counter-evidence, that peace (achievable only through love and concord) is the final goal and end of all knowledge in and for itself, as the only Ararat on which the restless ark of human intellectual striving can at last come to rest. Without the achievement of peace for all, our human knowledge is incomplete, a misplaced adventure in the night; with it achieved, our true wisdom can at last unfold to its proper maturity. Bacon foresaw this problem, in saying, in that same incomplete Book 9 of De Dignitate, "But it is highly probable that whoever speaks of peace will meet with that answer of Jehu to the messenger "*What has peace to do with Jehu ? What has thou to do with peace ? Turn and follow me" (2 Kings, ix 19) For the hearts of most men are not set upon peace, but party. And yet we think proper to place among the things wanting, a discourse upon the degrees of unity in the City of God, as a wholesome and useful undertaking."*

So too this author would like to conclude this short essay with some similar observations, that of all the sciences most wanted at the present time, is the science of peace, which would be both a social and a natural science, and whose implications and reverberations would reach into all other domains of knowing (a kind of interdisciplinary endeavour). In terms of natural sciences, the science of peace would seek to solve the energy crisis of mankind and to ensure the ecological well being of mankind in the constraints of the global ecology which we are set to inhabit together, ourselves and our descendents, for some millions of future years, if we are fortunate and able to surmount our current crises. In terms of social sciences, the science of peace would seek to solve the many deadly quarrels[14] conflicts and violent disturbances that effect the collective political body at the present time. In terms of personal well being and bodily peace, the advancement of the sciences of medicine, psychology and psychotherapy likewise are needed, in order to establish the inner peace of individuals, which reflects in peace of the whole society. But these few sentences are but a crude approximation and a preliminary report on, the conversation of particular angels or intelligences, whose thoughts on the nature of peace it has long been both the privilege and the responsibility of those few with hearing finely tuned enough to hear, and whose minds are set not on party, as Bacon explained, but on peace.[15]

[14] One of the founders of the scientific approach to peace research was the eminent Scottish mathematician, Lewis Fry Richardson, who authored a paper on the Statistics of Deadly Quarrels – he maintained they could be studied mathematically and analyzed with the same precision as the complex weather patterns effecting the globe. Richardson also provided the basic equations which still underlay the work of meteorologists in predicting future weather events.

[15] The current author has written a long and detailed academic thesis exploring the nature of the history of thinking on peace from 1945-2001, in several selected intellectual fields of research and theory, including philosophy, historiography, psychology, theology and education. It is his intention of continuing this work into other fields of knowledge in due course, including the natural sciences, anthropology, geography, sociology, economics, law, political thought etc. Let us hope the author's admiration of Baconian thought does not eventuate in a similar catalogue of unfinished literary projects ! Whilst some readers might perplex at the invocation of angelic intelligences here, it is only done partly tongue in cheek – angelology is merely one of the theological sciences that a neo-Baconian survey of knowledge for the third millennium would have to consider. Certainly

But are we saying therefore that the future of science must somehow, in order to advance mankind in the direction of a peaceful planetary civilisation, attempt the reconciliation of the natural and social sciences with the spiritual sciences, so to speak? If so, and this indeed is part of the argument of this paper, then this is surely so important a point that it is worth consulting Bacon again as to exactly what his original conception of the nature of the scientific enterprise did with religious and theological and spiritual knowledge? In book 9 of *De Dignitate et Augmentatis de Scientarum*, he states that he has purposely left out theology from this survey of knowledge, not because such knowledge is invalid, but rather because it is so important, that he did not feel fully qualified to include it in his compass. Bacon does however discus the nature of divine knowledge, and theological insight, and argues that reason can and should be part of Christian theology, specifically, and he states that "*as God makes use of our reason in his illuminations, so ought we likewise to exercise it every way, in order to become more capable of receiving and imbibing mysteries; provided the mind be enlarged, according to its capacity, to the greatness of the mysteries, and not the mysteries contracted to the narrowness of the mind.*" Now this is a crucial passage towards the very conclusion of his great work, and sadly, one which many so-called *scientists* never seem to have persevered in the study thereof. Too much of what passes for science, seems to consist in a most un-Baconian contraction of the capacity of the human mind, and as Bacon puts it so eloquently, consists in the contraction of the mysteries to fit in with the narrowness of the human mind. It is precisely to restore Bacon's original vision of the genuine role of reason in inter-religious discourse and theology, that this current essay is arguing instead for a recommitment to the original scope and vision of the nature of science as conceived by the founding philosophers such as Bacon, who had a grander understanding of the mission of science, and its relation to spiritual knowledge.

We can safely conclude therefore that if science is to have a future, and indeed if we humans, Homo Sapiens Sapiens, are to have a future at all, at least one worth having, we are going to have to rediscover the capacity for wonder and childlike delight in knowing, in inquiring, finding out, researching, asking questions, exploring, connecting up ideas, and sifting false from true theories and modes of understanding. In short, we are going to have to rediscover the wonder that underpins all genuine science, the leaps of the imagination. We are going to have to remember that science is never far from art, and both are never far from spirituality, and to unlearn the notion, the false notion, arising from a misinterpretation of the whole Western scientific enterprise, occasioned by an ignorance of the actual aims and intentions of its founding thinkers, such as Bacon, that science is somehow a complete and entire way of thinking in and of itself, divorced from wonder, from imagination and from broader possibilities of human spiritual experience. It is this soulless misinterpretation of science which gave birth to the false materialisms of Marxism, from which the world has spent too long liberating itself, with great pains and bloodshed and grief. For scientific human progress cannot be engineered from central statist bureaucratic empires of control, but rather by the work of individual geniuses in their own knowledge fields, in which the state can play the enabling role as empowerer, enabler and connector (and sponsor). As Machiavelli observed, *Lo Stato* – The State, is merely a fixed concept, a term denoting stuckness, fixity, in the complex mechanisms for distributing power and knowledge, which all human societies

angelology is involved in the religious fragmentation, conflicts and spiritual chaos of mankind – so why not in its healing ?

require. Even Marxism, in its deeper theoretic insights, can be of service to the reconstruction of wonder and liberation which a recommitment to the original scientific revolution can initiate – only rather than the redistribution of material wealth and ownership, mediated through the heavy handed apparatus of the state, an authentically scientific neo-Marxism would rather be concerned with the redistribution of enlightenment, through the facilitation of the genuine Republic of Letters, whose diplomats all genuine scientists in some ways are. Such cooperation is surely the hallmark of genuine science. Rather than the violent competition and clash of world views and world systems that we have seen in the past several centuries, escalating now into the post 9/11 era of cultural wars-to-the-death between centre and periphery, legitimacy and illegitimate - the new scientific community of the future should be characterized by peaceful cooperation, and the surrendering of disputes to the courts of reason and wisdom alone, or not at all.

In final conclusion, and in the spirit of the advancement of scientific cooperation into the future, the author would like to invite all scientists and scholars interested in the future developments of scientific inquiry, and who are also interested in the meta-theoretical issues which arise from these questions, to participate in a project which it is the responsibility of the Institute of Peace Studies and Global Philosophy to propose, namely to undertake a complete revision and re-edition of Bacon's original *De Dignitate et Augmentis de Scientiarum,* consisting an up to date total overview and survey of all the various fields of knowledge, in terms of both natural, social and spiritual sciences, in the run up to a major international conference in 2023, celebrating the 4th centenary of the publication of Bacon's epochal work. To this end the author would welcome communications from all scientists and scholars who would feel able to undertake an overview of their own fields of knowledge, however small or great in scope, as a contribution to this wider common work.

BIOGRAPHICAL SUMMARY

Thomas Clough Daffern, is a British-Canadian philosopher, historian, educator, writer and poet. In the 1980's he launched an organization called Philosophers and Historians for Peace, aimed at defusing the crisis of the then cold war, and in 1990 was elected international coordinator of Philosophers for Peace, a network of scholars and philosophers from around the world, including Russia, USA, Europe, India. In 1991, he founded the International Institute of Peace Studies and Global Philosophy (IIPSGP) as a British based centre for excellence in peace research and peace theory and conflict prevention, and comparative philosophy. The IIPSGP continues its work worldwide to advance intercultural understanding and peace, through the synthesis of higher learning, philosophy, spirituality and science, and is always keen to support and encourage its members to undertake research and education in these fields. His various publications cover all aspects of his interdisciplinary theoretical research to date concerned with peace studies and intellectual history. In 2006, the author founded the Truth and Reconciliation Commission for Britain and Ireland, in Anglesey, Wales, which meets regularly in Britain or Ireland. In 2008 he travelled in Israel and Palestine to launch the Truth and Reconciliation Commission for the Middle East – an ongoing initiative.

INDEX

A

C

D

F

G

H

I

J

K

L

M

N

O

P

S

T

U

V

Z